Thomas G. Muth

Functional Structures in Networks

Thomas G. Muth

Functional Structures in Networks

AMLn - A Language for Model Driven Development of Telecom Systems

With 191 Figures

Springer

Dipl.-Ing. Thomas G. Muth
Frimans vägen 13
141 60 Huddinge
Sweden
thomas.muth@telia.com
www.amln.se

ISBN 978-3-642-06141-7 e-ISBN 978-3-540-27126-0

Springer is a part of Springer Science + Business Media

springeronline.com

Softcover reprint of the hardcover 1st edition 2005

Cover-Design: design & production GmbH, Heidelberg

Printed on acid-free paper 62/3141 Rw 5 4 3 2 1 0

Foreword

The architecture of "The Worlds Biggest Machine[1]" is of course expressed in the many underlying communications standards; however it is far from explicit nor readily accessible. While on one hand, marketing people are busy looking for their so called "Killer Applications" that will maintain the economic growth of this machine, the engineers are struggling to keep up with the myriad of networks, protocols and standards that interconnect an ever growing number of network services across a rapidly increasing variety of platforms and protocols.

Within the industry, it is commonly accepted that fewer than 10% of engineers working in the field have sufficient knowledge and experience to tackle the pre-study and feasibility phases; that is to say, only this group can process the knowledge and overview of the elusive architecture that allow them to identify the network nodes, network services, protocols and messages that will be affected by adding new network functionalities.

It follows that 90% of engineers are capable of performing the execution phase. This generalization has served as the motivation behind attempts to automate the generation of code for proprietary programming language, a major investment that ultimately failed totally. This long standing objective of moving away from implementing services, one line of code at the time, is however gradually becoming a reality for standardized languages. The software provider I-Logix, for example, has been successful in developing tools that not only generate fully executable Java and C++ code but also perform the critical round tripping.

Mr. Leif Edvinsson, winner of the prestigious Brain of the Year award in 1998, has developed a model to describe the hidden brain power of a company, the *Intellectual Capital* (IC) management model. In the context of the 10%–90% discussed above, these hidden tangible and intangible values describe a company's IC as the sum of the following five elements: 1) intellectual property (IP), 2) knowledge management (KM) systems, 3) the employees' knowledge, 4) the employees' experience, and finally 5) the employees' skills. Note that while knowledge is tangible and transferable, experience and skills are neither. In the case of proprietary systems and languages, tools support and training will be high. However, it does afford some measure of protecting (retaining) the skills and experience established.

---

[1] "The Worlds Biggest Machine" is the title of a white paper, "3G: Building the World's Biggest Machine" by Steve Jones, originally published on the 3G portal web page, http://www.mobiletech-news.com/info/2003/10/04/000100.html.

In many instances, knowledge management systems are at a bare minimum or non-existing, thus increasing the dependency on specialized skills and experience as the company's IP knowledge is held and managed by a very few individuals, that is, the "less than 10%" group. Do you know what your IC is doing?

The reality of today's market place is that being first on the market with a product can mean the difference between success and failure. On one hand the protection of a proprietary programming language quickly becomes costly to maintain and an overall burden. On the other hand, non proprietary languages, third party development tools and "free ware" are slowly but surely taking us away from the paradigm of developing applications one line of code at the time, reducing costs and shortening the time to market. All companies do of course have equal access to these same resources, including the pool of engineers that represents the remaining 90%.

In the context of the 10%–90% discussion above, the question remains "just where is a company's IC vested?" My personal view is that the majority (well over 50%) of a company's intellectual knowledge-based property is vested in the knowledge and experience of the less than 10%.

In response to this conclusion, this book by Thomas Muth, is the key to opening the door to that 10% critical mass of network knowledge that is crucial for the understanding of present and future architecture of the so called "Worlds Biggest Machine". Reduce your risk exposure by anchoring your IC with AMLn.

Montréal
October 2004

Jens J. Larsen, P.Eng
Senior Engineer
Ericsson Research, Canada

Preface

This book is about creating and using models as a means to describe and communicate the purpose and architecture of network systems faster and more accurate.

Today, the architecture of standardized systems is rarely described explicitly. An average protocol or interface specification may comprise several hundreds of pages. For a network system, such as ATM, GSM, ISDN, and others, the number of standard pages produced for its specification easily exceed 3000 pages each. For example, to become an ISDN expert, you have to read (and understand!) over 7000 pages. A complete specification of a network system also comprises a large number of system documents that specify how an implementor intends to realize the standard. Basically, all this information is presented textually, possibly accompanied by some ad-hoc figures.

A development process that is based on this way of describing a system is *document-driven*. It does not take much imagination to see the problems this creates for a network system designer: the facts the designer needs reside in hundreds of large documents that are far from easy to survey. In reality, the designer relies heavily on the few experts who have been engaged in standardization and early system-developing phases, and must repeatingly attend seminars and courses given by them, in order to understand a network system and its evolution. Moreover, it is also practically impossible to keep system documents updated since accurate tracing of requirements is very difficult. The architecture of the system therefore tends to deteriorate, its purposes get veiled in obscurity as time goes by, its evolution slows down, and the introduction of even minor improvements and service enhancements become major design projects.

After being exposed for several decades to the rather intuitive and ad hoc way we describe networks and systems, I decided to write this book in an attempt to make life easier for most network designers, whether you are a manager with technical background, whether you work with standardization, the architecture of implementors' solutions to standards, product design (constructing the software and hardware parts of operator networks), system testing, or whether you are an operator responsible for the configuration and maintenance of a network.

Modeling as a basic principle for producing specifications turns design from being document driven to *model-driven*. As humans, our creative thinking is based on modeling of almost everything we can think of (e.g., an artifact or a sociological system). In the context of this book, however, the only model type that is discussed is the *information model* that describes the purpose, structure, and behavior of a network.

Today we must deal with hundreds of documents and many thousands of pages in order to understand a system. My firm belief is that we would all vastly benefit if standards and implementors' solutions to standards were expressed as models instead. The potential benefits of a model-driven development process are enormous, in particular if models are created in the early phases, preferably already in the standardization phase: instead of interpreting documents, an implementor can build a model of his solution to standards by refining the standard model (using the same modeling technique). The potential for requirement tracing, shorter time-to-market, and many other benefits should be obvious to everybody.

I guess that you already know a lot of today's networks (or at least some aspects of them), so I am not trying to teach you anything about them Once you have read this book, you will find that now you are also the expert on modeling, at least as network modeling is concerned. You will then be the one who defines the requirements on the languages and tools you badly need, instead of being told what to use (by software system modeling experts). When modeling has become the general method in your work, you will also experience the satisfaction of being a system expert yourself.

The basic prerequisite for being able to enter the model driven path for network system development is a suitable *modeling language*, which is the subject of this book and my contribution in this respect. There is only one thing I must ask from you: you have to understand what modeling in general, and network modeling in specific, is about. Until now you have trusted and relied on (software) modeling specialists to provide the design languages and tools you use in your daily design work. This is just too much to ask. These specialist know how to design languages in general and they can design the most sophisticated tools for you. But you cannot expect from them specialist knowledge in your particular knowledge domain (i.e., networking). This is the reason that all you have been receiving are languages and tools for hardware and software system modeling (e.g., UML, the Unified Modelling Language[2]), but not tools for modeling network systems.

Systems modeling has been a rather hot topic for quite a few years. The wide spread use of UML is one result. UML is a standard and therefore perceived by many designers as a modeling language for everything. So, why did I have to develop another? The answer is simply that there is no such thing as "a language for everything" since formal languages are context-dependent. Different languages are needed in different application domains if they are to be useful. Every domain needs to describe its systems in terms of concepts and design rules that are characteristic for that domain. For example, you might be able to describe an aircraft system using the terminology and symbols of the telecom domain, but (obviously) nobody would understand anything. A language such as UML can however play an important role for modeling *system types (component systems) that exist in many*

[2] UML has been standardized by OMG in an effort to unify diverse object oriented languages for modeling software systems

other system types, such as software systems and hardware systems (electronic devices). Viewing a network system as consisting of such component systems is something you do first when the standardization phase and the initial phases of an implementor's system development is passed. We may therefore regard UML as a *common-value* language in relation to *added-value* languages, such as the one presented in this book.

To be useful, a common-value language must be completely transparent to the concepts and semantics of any other system domain, i.e., the model cannot tell you anything of the purposes with the added-value system. It only tells you what its software and hardware components are and how to connect them when building an added-value system (in our case, an operator network). An added-value language, on the other hand, is built on the concepts and architectural principles of a specific system domain (e.g., network systems), i.e., on the knowledge of what this type of systems do, what their functions are, and how to combine them into an architecture that is open for its evolution over a long time. In other words, an added-value language is based on, and preserves the *intellectual capital* of a system domain. Since mapping an added-value model on a common-value model means that the semantics and architecture of the added-value system gets completely (and deliberately) lost, creating and maintaining added-value models is extremely important to any system-developing enterprise.

The added-value modeling language that is presented in this book is called AMLn (*Abstract Modeling Language, network view*). To my knowledge, it is the first and, so far, the only attempt to create a modeling language for network systems. The background to AMLn is the enormous complexity and rapid evolution of today's network systems, caused by the introduction of new technologies in electronics, software, and transmission media over the last 25 years. As a result, new network systems appear in an ever increasing pace. One would expect that, over these years, well-proven and efficient architectural principles have been identified and commonly applied. On the contrary, only a single attempt in that direction has been made: the OSI RM (Open Systems Interconnection Reference Model), published in 1984. This generic model exhibits such a set of general architectural principles. In spite of the fact that the OSI RM was only a first attempt in the right direction and was constrained to experiences from a small class of pre-standard datacom networks (e.g., IBM's SNA, DEC's DNA, DARPA's Arpanet), the industry took it to its heart with 100% consensus, and with great enthusiasm. In retrospect, we can conclude that this was more an effect of a fear of losing control of the evolution than of a serious evaluation of the quality and applicability of the OSI RM.

Unfortunately, the industry still has not pursued such an evaluation. Even though the idea of "OSI networks" failed completely when confronted with actual needs in public networks and more sound architectures (such as the Internet), the industry continues to align the description of today's network systems to the OSI RM, particularly in standards. Since a competent network architect does design architectures for optimal maintaineability and openness to new technologies and requirements in mind, the strange situation exists that many standards continue to describe sound architectures in terms of a generic architecture (i.e., the OSI RM) that is not reflected by any network system in use today. Those standards are next to unintelligible.

I was as enthusiastic as anybody else over the OSI RM at the time. However, when entering the field of network systems design later, I gained a deeper knowledge of how networks systems were actually built and what the real architectural needs were. This knowledge and experience has been transformed into AMLn. In this endeavor, I had to deal with two problems:

1. Finding the minimum set of concepts that could be the corner stones for the language. For a system domain that has been developing for over 100 years, those concepts should have been available for a long time. Except for concepts defined by OSI (which to a large extent are insufficient for modelling purposes), this is not so, however. The ITU–T (a leading standardization body for telecommunication standards) tried for many years by publishing a steadily expanding list of thousands of terms. It gave it up some years ago when it found that, in reality, every standard defined its own concept definitions. Thus, this issue became a major research topic in developing AMLn.
2. Identifying the architectural principles that result in a stable system structure over a long time, and a model that is easy to understand, manage and relate to common-value models.

The specification of a system type must cover two aspects: its *behavior* and *structure*. Behavior specification is a topic that can be covered quite acceptably by most common-value languages. Structure cannot, however, since it requires the identification of efficient architectural principles of the actual domain. As network systems are concerned, a few principles are common knowledge, e.g., separating access network problems from core network problems. Some new principles (e.g. the "layer" concept) were defined by the OSI model. Over the last two decades a number of new principles have emerged (e.g., the concepts of "horizontal networks" and "logical networks"), which cannot be defined in terms of common knowledge and the OSI model.

In AMLn, all essential architectural principles (old, useful ones and new ones) are built-in. The general principle applied for managing complex system is "separation of concerns." AMLn is based on two main concerns:

- Separating *functionality* from *connectivity*. These subjects are covered in Chaps. 2 and 3 respectively.
- Separate the system that offers services to system users (the " managed or *traffic system*") from the system that manages that system (the "*management system*"). How these systems are related is described in Chap. 5, that also discusses existing management systems (TMN, SNMP, CORBA and OMAP) through AMLn models.

Whenever you separate concerns, you must also define how to *integrate* the parts, since a system consists of related parts. Chapter 4 discusses this subject as regards the separation of functionality from connectivity. The chapter also considers that the specification of a network system is always created by breaking down its structure step by step. The structure of an AMLn model that results from this is characterized as its *network-level* structure.

Chapter 6 presents some case studies where AMLn has been applied as an analysis tool on well known network systems.

An overview of AMLn is given in Chap. 1. A brief introduction to modeling in general is also included, as well as the definition of the *logical dimensions* in which an AMLn model may be viewed. Logical dimensions are conceptual tools for creating and viewing the information structure of an AMLn model. They also define the structure of an AMLn model data base.

This book is a successor to my previous book (see Muth (2001)), which describes two other views of AML: AMLs (*AML, service view*) and AMLp (*AML, protocol view*). Both these views deal with behavior specification and are applicable when one wants to refine an AMLn model for simulation, or for generating code directly from such a model. A brief overview of AMLs and AMLp is given in Appendix D at the end.

During the time I have been struggling with this book, many people have encouraged me and given me valuable feedback. I would especially like to thank Per Dahl, Anders Eriksson, Lennart Holm, Jens Larsen, and Anders Olsson.

Stockholm
October 2004

Thomas Muth

Contents

4 Modeling Vertical and Horizontal Partitions

5 Management and Traffic Systems

6 Applying AMLn

1 Introduction to Network System Modeling

1.1 Systems Modeling in General

This book presents AMLn (**Abstract Modeling Language, network view**). AMLn is a proposed language for modeling **network systems.** We regard as "network system" any system that exhibits a structure of nodes and provides connectivity services for humans, machines, and different types of application systems. Network systems are commonly defined by international standards (AMLn is, however, not restricted to standardized systems only). Examples of standardized network systems are the Internet, UMTS, PSTN/ISDN, SS7, GSM, GPRS, ATM, SDH, and diverse LAN concepts (see Appendix A for definitions of acronyms).

In this chapter we provide a general background to modeling. When we want to describe something, whether it is an artifact or something more abstract, we can do it verbally, using natural language text in documents (perhaps with some supporting pictures) or we can build a model of it. A model may be a miniature of an artifact, or it may be an information model (the only interpretation of modeling that concerns us in this book). In an information model, parts and relations between parts of the thing we model are represented by information elements.

An information model is built by using a modeling language. A modeling language is designed so that a model in that language can be understood not just by a human interpreter, but also by an interpreter that runs in a computer system. Thus, a modeling language is a formal language with special qualities for human interpretation of models that are constructed in that language. For example:

- It includes concepts and terms that are familiar to designers and users within a particular system domain (which may be networks, power plants, missiles, software systems, etc.).
- It includes graphical representations of important basic language concepts.
- It allows names and acronyms to be used instead of neutral identifiers for entities in the model.

Since specifying network systems is still document- and not model-driven, we need to emphasize the enormous power of modeling in enhancing creativity and improving communication between people, compared to providing text in documents. *Modeling* is the creative thinking that you do when trying to find a solution to a problem. *Documenting* is what you do when you describe your solution. Thus, if you create a model of the system instead of just documenting it, not only do you improve your own thinking, *the final model also becomes the documentation* (provided that you use a known modeling language and some discipline). Another important side effect of a modeling approach is the possibility to add information to the model that explains *which problem* a certain feature solves, and *why* the particular solution has been chosen in favor of already known solutions to the same problem (something you seldom find in traditional system specification and description documents, in particular not in network standards).

Every modeling language uses the same basic paradigm for describing the *structure* of a system, which is that "a system consists of related parts." Considering what we know about the complexity of most systems today, this view seems to be too simplified to be useful (especially since the *behavior* of systems is not included in the definition). It is, however, of value for something we may call a *universal abstraction*.

This abstraction has been the paradigm for modeling languages that produce things such as *semantic diagrams*, *conceptual models*, *entity-relation diagram* (ER) or *entity-relation-attribute diagrams* (ERA). Whatever name has been used, these languages all support modeling on the level of universal abstraction:

- Semantic diagrams were the first to be used, in particular for modeling concepts (e.g., the concept of "religion") and conceptual systems (e.g., the system of numbers). The purpose with this type of models is primarily to explain the meaning with concepts for human readers (the model may, for example, model the statement "a car has four wheels" as the *relation* "has" between the concepts "car" and "wheel", also adding the *attribute* "four" to the relation "has").
- With the advent of electronic computing, the things one needed to model was a part of the physical world (e.g., the business control system in an enterprise) that had to be supported by the computer. This part was modeled as a conceptual model (or equivalent: an entity-relation diagram). The goal with such a model was to be able to first understand the actual part of the reality, and then translate that model into a data model (i.e., a data structure) that could be stored in the computer. The things that appear in such models are now no longer any type of concept (as in a semantic diagram), but representations of *entities* of the physical world. Some type of relations seemed to recur in models of all system domains, such as is_part_of and belongs_to_the_class, and are therefore often defined as parts of the modeling language itself.
- The next step in the development of modeling languages took place when people started to use modeling not just to define data structures in computers, but for modeling the computer system itself. Some early modeling languages tried to cover all aspects of a computer system in a model. With the enormous increase of software in systems in general, however, successful modeling languages have focused on modeling the software part of it.

During the 1980s, the first step to include behavior in models was introduced. It became popular to regard the entities in models as *objects* and the relations as *object relations*. An object is an entity that hides its design behind an *interface*, and an entity which can execute *operations* upon requests from other objects. The structure of the system can now be described by interfaces between objects, and the behavior as interactions between objects, requesting operations and responding to requests (sometimes described as "protocols" between objects).

Object oriented (OO) modeling languages in use today have, in reality, only been developed and used for software system modelling. A well known language of this kind is the Unified Modeling Language (UML).[1] The fact that such languages are constrained to systems that have a single object type (which is a piece of software or data), makes it possible to pre-define a larger number of relations than just is_part_of and belongs_to_the_class. For example, software objects exist both as classes, types and as instances, where an instance is related to a type by an is_an_instance_of relation. Most operating systems also offer connectivity relations between functional elements, which can be modeled as an is_connected_to relation, etc.

OO modeling languages were preceded by OO software design languages, such as Smalltalk, Ada, Eiffel, C++, and Java. Since you can express almost the same thing in these languages as you do in UML and similar modeling languages, the latter tend to be merely graphical versions of OO design languages. The extra value they give, besides a graphical notation and concepts for the universal abstraction level, is when diverse packaging concepts are added that make the model useful in managing the overall complexity of a large software system.

When modeling languages have the ambition to be useful beyond the level of universal abstraction and/or include behavior modeling, the complexity of the models normally require the language to define some kind of *views*. One distinguishing aspect between languages is therefore their choice of views. Some languages that are called "modeling language" offer no views at all, and should therefore be called "drawing languages" instead. A view is the picture you get of a system model when looking at it from a particular "angle." For example, a mechanical design exists in 3-dimensional space and the model of it is normally presented in separate drawings for a number of cuts in that space. In the case of more logical (or abstract) types of systems, such as networks, computers, or software systems, the selection of views is not that obvious.

Basically, the selection of views depends on the type of system that the modeling language targets. For example, UML (a language designed for software-system modeling) defines a large number of views, called "diagrams" (use-case diagram, class diagram, state diagram, activity diagram, object diagram, etc.). These are (hopefully) relevant and of interest to software system designers. A language for computer system modeling may define additional views that show other properties of the system, such as its system bus and connected hardware, memory organization, power supply subsystem, and mechanical design, including cabling.

---

[1] See The UML Reference Manual. Object Management Group, Framingham, MA. http:// www.omg.org/. Cited 1998

Note however that there is no consensus between similar languages regarding which views should be supported for a particular system domain (the choice of views seems to be a means for competition between languages instead).

1.2 Added-Value versus Common-Value Languages

1.2.1 General

The previous chapter discussed how far you can get on the level of universal abstraction and/or by using *common-value* languages (e.g., languages for software and/or hardware system modeling). For productive modeling of systems within a particular system domain (e.g., networks) we must define more specialized languages, i.e., modeling languages that are based on the concepts and relations that are unique to a domain (such as the relation "connection" and the entity "layer" used in network modeling, or the entity "exhauster" in modeling cars). This is so because we neither want, should, nor can model the system in terms of its software and hardware components before it has to be implemented. We will refer to such modeling languages as *added-value* languages.

It has been advocated (e.g., by UML proponents, see Chap. 1.2.2) that UML (a common-value language) should also be used for modeling systems in domains other than software. The fact that this is theoretically possible does not, however, imply that it is realistic and meaningful. Let's take a look at the implications of such an approach. Figure 1.1 shows the general paradigm for all modeling languages.

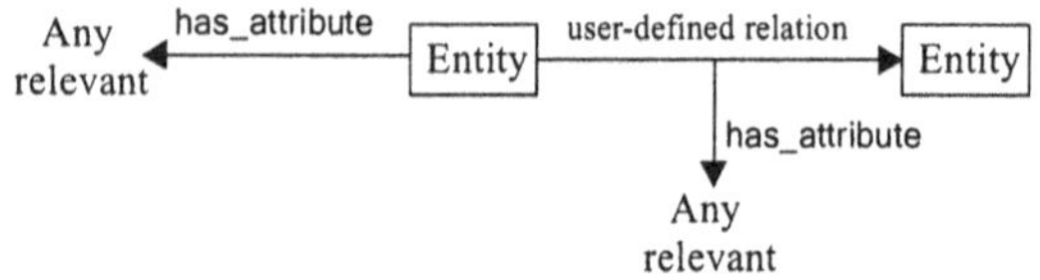

Fig. 1.1 The model for all modeling languages

The model describes the language as defining *entities* that may have any kind of user-defined *relations*, some of which may be *attributes* to entities (attributes are things that have a value range and no other relation to entities in the actual model). Some languages allow relations to have attributes as well. Common-value languages normally define only two symbols, one for relations and attributes and one for entities. Some languages (e.g., the language component in AMLs for modeling resources) do not even use any symbol for entities, i.e., entities are just text strings. Since a common-value language cannot define symbols for entities and relations of added-value systems, most common-value languages realize that they must define a relation that tells which kind of entity is meant. This is the belongs_to_the_class relation mentioned previously. Looking for similarities cross many system domains reveals that another common aspect is to describe that something is part of something else. Consequently the relation is_part_of is normally included in common-value languages. For example, UML includes both these types of general relations.

Let's now use UML notations (as an example of a common-value language) for creating a very small part of an added-value model in the network system domain. This model (see Fig. 1.2) tells:

> *Two nodes, N1 and N2, are interconnected by a route, R1. N1 includes a layer element L1. N2 includes a layer element L2. L1 and L2 communicate according to the protocol P1.*

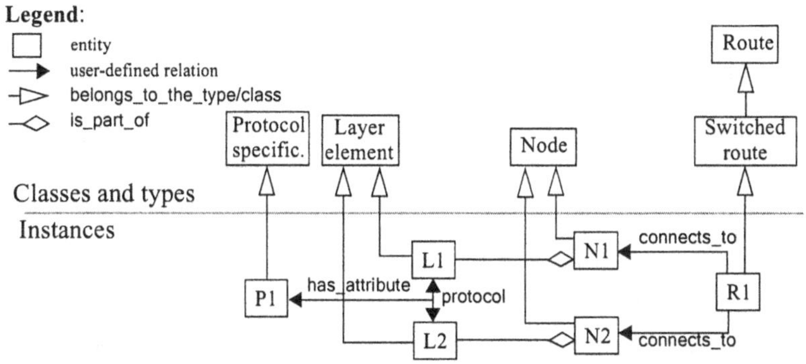

Fig. 1.2 An entity-relation diagram of a network system statement

Using a common-value language implies that added-value concepts (here **node, route, layer element, protocol, protocol specification** and **connects_to**) must be defined by the modeller by including specific type and class entities, and by textual annotations of user-defined relations. The model also demonstrates that the meaning of the statement we model is not given by concepts defined by the common-value language, but (at best) by the added-value model. In other words, *the meaning with the model is (to a part) to model meaning.*

The fact that common-value languages use very few symbols may make the impression that they are easy to use and give intelligible models. The model in Fig. 1.2 clearly shows that this is a very deceptive appearance. It does not take too much imagination to see that modeling a larger part of a network system, with all its domain-specific entities, relations, and attributes in a common-value language would create a model that would be extremely complex and hard to understand (and therefore useless). The model would be cluttered with class and type objects, class and part relations, and user-defined annotations that exist only to define what type of entity, relation and attribute is meant. This is definitely something we do not need when modeling a system that is complex in itself. Therefore let's look at how the same statement is modeled in AMLn (see Fig. 1.3).

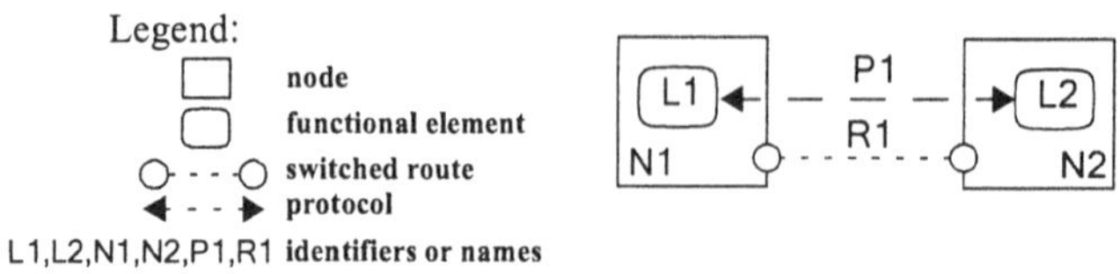

Fig. 1.3 The AMLn model that corresponds to Fig. 1.2

We leave to readers who work in the networking industry to judge which of the models in Figs. 1.2 and 1.3 they prefer. Here we will only summarize the main differences:

- All entity symbols are domain specific.
- All relation symbols are domain specific, i.e., no user-defined relations are allowed. Therefore, the modeller need not add any text or code to explain what is meant with a particular relation. The excess of textual annotations that characterizes models in common-value languages is thereby avoided.
- The need for the general relation is_part_of is eliminated by graphically including entities that are part of another entity in the latter.
- The need for the general relation belongs_to_the_type/class is eliminated by using domain-specific graphical symbols for entities and relations.

Obviously, an added-value modeling language will have to define more graphical symbols than a common-value language, which could be an argument against it. For example, in the very early proposal for AMLn, the author did use an excess of network-specific symbols, in order to allow for models with high semantic precision. This is counterproductive, however, since a well known fact is that humans have a short memory span that cannot cope with more than something in the range of 5–9 concepts (in this case: symbols) at the same time. For most system domains, this is a tough challenge to the language designer. For the case of AMLn, the author has solved this problem as follows:

1. Carefully analyse and select the minimum set of concepts and relations that are used in all network systems.
2. A specialization that cannot be included in the range of symbols can always be defined by attributes of an entity or a relation.
3. Define a set of model types ("views"), each set using only a small part of the complete set of symbols.
4. More important, most entity types are not defined by symbols of their own, but indirectly by the type of relation they have. Figure 1.4 gives some examples of this approach. Five entity types are used here, relying on only two entity symbols (AMLn defines only two entity symbols: a squared shape, denoting the general entity type **node**, and a round shape, denoting the general entity type **functional element**, which is any entity that shows a pattern of behavior).

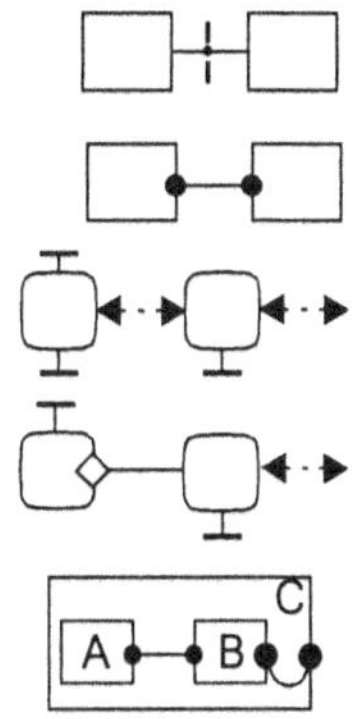

These entities are **derived nodes**
(The relation is a **network interface**)

These entities are **logical nodes**
(The relation is a **link route**)

These entities are **layer elements**
(The relation is a **message protocol**)

The left entity is a **relaying actor**, the right an **agent**
(The relations are two **layer interfaces** of type CSP, an **actor-agent interface** of type SAC, and a **message protocol**)

These entities are all **logical nodes**
(The relation between B and C is a **substructure** relation. The model tells that A and B are parts of C and that only B connects to external nodes)

Fig. 1.4 In AMLn, most entity types are defined indirectly by the relations they terminate

The actual type of entity is defined by the relation types that it terminates. The number of predefined relations increases of course. There are nine basic relation symbols in AMLn. Some symbol specialization possibilities exist, which increases the total number of specific symbols to around 20. Since an AMLn model consists of several model types (or views), however, only a few relation types are used in each model type.

1.2.2 AMLn versus UML

The author has been told that AMLn takes a controversial path to systems modeling because it does not rely on UML. Proponents of this opinion do not understand that there cannot be any controversy since UML and AMLn are used for modeling completely different things.

- AMLn is an added-value language, used in early development phases.
  UML, a common-value language, is used first when a supplier wants to realize an AMLn model.
- An AMLn model describes a network system in a way that is independent of how it will be realized (sometimes denoted as its "architecture"), i.e., the entities and interfaces of the model are undefined as regards their realization.
  A UML model describes a system of software elements and interfaces. Such a system is itself a component that realizes a part of an added-value model. Although an AMLn model and a set of corresponding UML models are strictly related, there is no simple one-to-one relation between software elements and entities of an AMLn model. An AMLn entity is often realized by an aggregate of software elements, or by a blend of software and hardware elements.

An added-value system belongs to a particular system domain (e.g., "network systems") with its own vocabulary, language, and to some extent also a de facto modeling notation. This is of immense value and importance to the actual system domain. Since using a common-value language for modeling added-value systems means that this vocabulary and notations cannot be used, it is not just inconceivable, but also counterproductive to use a common-value language for modeling added-value systems.

The impact UML has had on the industry is partly an effect of that it tries to unify several pre-UML object-oriented languages, but mostly because it became standardized. UML has been much criticized for its complex and imprecise semantics. However, since it is a standard, it is regarded as something good: it is assumed that communication is easier when everybody talks the same language, pointless notational differences are avoided, and it is easier for the industry to hire people and for engineers to reuse their language knowledge.

UML was designed for being a common-value language for modelling all kinds of software systems in an object-oriented way. Due to its wide-spread use today, some people believe that it can also become the modeling language for everything. We must therefore ask how UML deals with the problem of modeling different kinds of added-value systems. The UML solution to this problem is to extend the basic language with meta-language mechanisms (called "stereotypes" and "tags"), intended to allow users to include added-value semantics in their UML models. UML also assumes that such extensions are isolated in a UML "profile" for each system domain. Adding these mechanisms increase the complexity of an already complex language, of course. As a matter of fact, there is nothing but confusion to gain by trying to incorporate added-value semantics in a common-value language such as UML, not to mention the costs of maintaining the language. A recommendation increasingly advocated, therefore, is to scale UML down to a simpler and more general modeling level. The author's suggestion (since the birth of UML some seven years ago) is to divide UML into three general parts, and leave specific system domains to define their own added-value languages. These general parts are:

1. One part that covers the *universal abstraction* level we discussed previously (in UML, comprising concepts such as "class," "classification," "collaboration," and "association"). This part is useful for producing semantic diagrams, conceptual models or entity-relation diagrams of almost anything.
2. A *common-value* part for software system modelling, with and without object oriented features.
3. A good *meta-language* (replacing UML's stereotypes, tags and profiles). Such a tool can be used for creating interpreters that can translate a model in an added-value language to UML models.

Considering this perspective, the author has always regarded any effort to extend UML to also cover the modeling needs for network systems as a dead end.[2]

[2] This statement has to be made, since it explains why AMLn has been developed and why this book and its predecessor (see Muth (2001)), have been written.

1.2.3 The AMLn Process Context

To develop network systems, producing network solutions, operating and maintaining networks are the core tasks of the telecom industry. These tasks are performed by a variety of businesses, traditionally called (telecom-) supplier, operator and standardization body. The roles these players take are not so distinct as the mentioned tasks suggest, however. They also change over time:

- A supplier may restrict his role to deliver network components (switches, transmission equipment, etc.), together with instructions of how to interconnect them and how to configurate, operate and maintain the network. In that case, the supplier leaves to the operator to design the network, to connect it to other networks as well as to handle it in all respects.
- With the advent of an increasing number of operators not having the necessary skills in network design, traditional suppliers tend to also deliver "total solutions", i.e. they design the particular network, they build it, connect it to other networks and they train the operator's staff in network operation and maintenance.
- Suppliers of today may even take the network operation and maintenance role as a service to operators, who mainly deal with the business aspects of providing network services.

This insight tells that we must discuss the needs for modeling languages and tools in the context of the overall network system development process. Figure 1.5 shows the main features of this process.

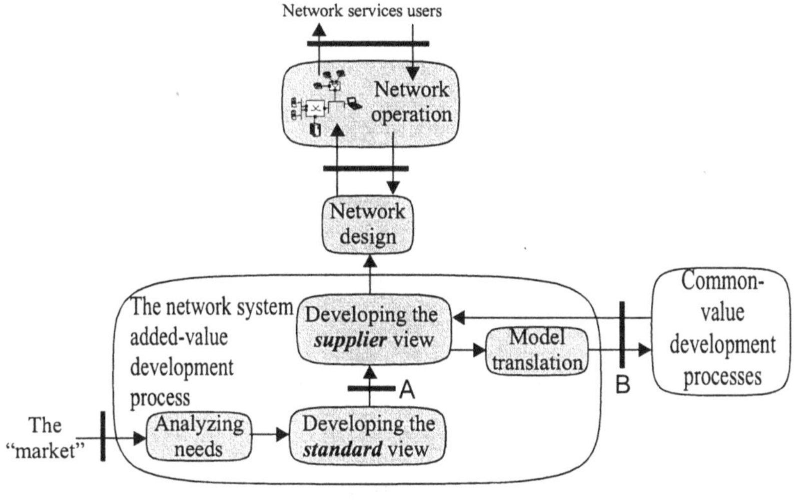

Fig. 1.5 A flow-chart model of the network system development process

Added-value development produces generic network solutions (such as PSTN/ISDN, GSM, the Internet architecture) from which particular operator networks are designed, produced and operated. The development process has two outstanding features, symbolized by interfaces A and B in the model:

- The cooperative nature of networks and network products implies that the added-value development process is divided in two distinct parts: one process that provides a *standard view* of a network system (in interface A), and other processes that, for each supplier, provide a *supplier view*, i.e., a solution to the standard view. A supplier view includes parts of other systems, as well as solutions to parts that are not considered for standardization.
- Networks and network products are constructed based on software (SW) and hardware (HW) component systems that realize supplier views. Provided that the part of the development process that requires networking competence (i.e., the added-value part) can create unambiguous specifications of SW and HW component systems in interface B, the whole development process can always be divided in an *added-value process* and a number of *common-value processes* (true for many other domains as well). As a result, added-value and common-value processes can also be run by different businesses. Telecom suppliers can then focus on their core business (network system development, network design and perhaps operation and maintenance) and leave a major part of common-value design to sub-contractors who are specialist in SW and HW engineering. This insight lies behind UML, a common-value method that is applicable to telecom as well as to other industries.

Considering the interfaces defined in Fig. 1.5, we can identify four business systems that cooperate in the telecom industry (see Fig. 1.6).

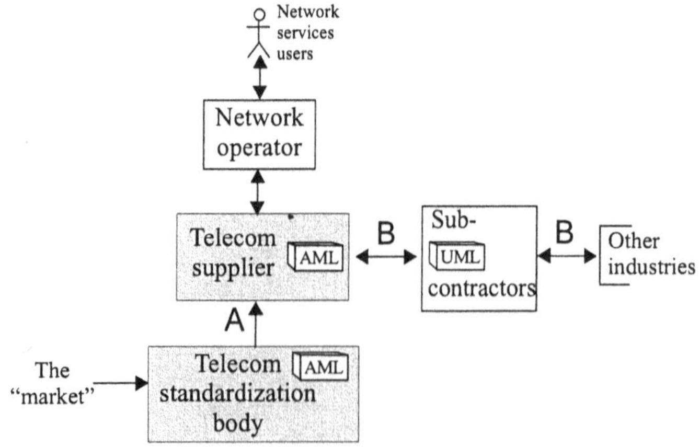

Fig. 1.6 The business systems in the telecom world; AML, UML=modelling tools

Presently a major part of common-value development is performed by telecom suppliers themselves. A normal estimate is that only 10% of a supplier's engineering staff work in the added-value process, the 90% with common-value development (designing SW and HW component systems). Thus, 90% work in processes that could be handled by sub-contractors.

Assuming that SW sub-contractors use UML, B-interface information must be expressed in UML. Since, in reality, UML cannot be used for providing added-value information, its role is only to act as a common language interface between added-value processes (telecom just being one of them) and processes that produce SW component systems. Consequently, *added-value development processes do not become model-driven by using UML. They only gain the ability to communicate with sub-contractor processes.*

The network system added-value process is divided between a standardization process and suppliers' development processes. Both processes deal with the added-value view of the system, the supplier selecting between standardized options and adding features and designs that are not defined by the standard model. Presently, both standard and supplier views are described in *documents*, using natural language text and intuitive pictures. Consequently, the translation from the standard view to a corresponding UML model is done in two steps by human interpretation of a large amount of documents.[3] In this scenario, it is an almost impossible task to guarantee that the UML model is an accurate realization of specified added values. However, by using AMLn both in the standardization process and suppliers development processes, a new scenario appears, where the added-value process becomes model-driven as well (see Fig. 1.7).

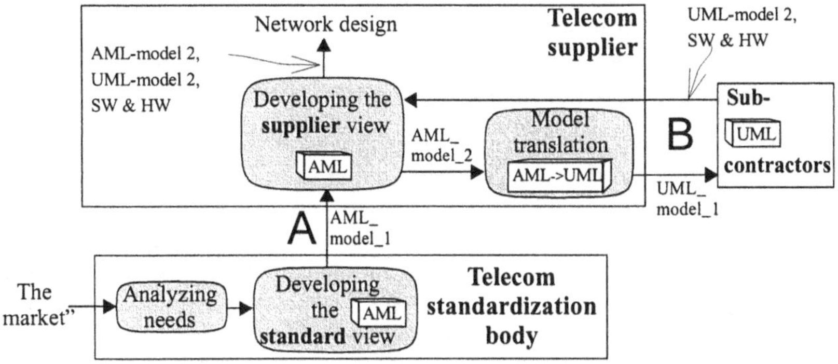

Fig. 1.7 Models and model translations in the telecom development process

[3] As an example, we can look at the standard view of the ISDN network system: only the SS7 subsystem that supports control signalling in ISDN comprises about 2500 pages. If we add all other standards that describe the ISDN, we land at about 7000 pages (quite a few books) to interpret in order to fully understand this view. Still, this is not the only knowledge to be mastered by suppliers' designers. They have to know SDH, IP, ATM, TMN, etc., as well.

The scenario tells the following:

- AML_model_1 (in interface A) is the standard view, created by a standardization body.
- This model is refined (in a supplier's development process) to a supplier view (AML_model_2). This model is used for validation and verification of the supplier view. If behaviour specification of interfaces and entities is included in the model, it may be used for simulating properties of the network system as well.
- When the supplier decides to realize the network system, the verified AML_model_2 is (automatically) translated to one or several UML_model_1 (in interface B). Objects and relations in an UML_model_1 will be (more or less) equivalent to entities and relations in the AML_model_2. A UML_model_1 is used as a requirement specification for sub-contractors.
- A sub-contractor refines a UML_model_1 to a UML_model_2 that on its lower structural levels refer and specify SW elements, and on its highest structural level is equivalent to the UML_model_1, i.e., refers entities and relations of AML_model_2.

This scenario guarantees that requirements can be traced over the whole development process (from both ends), thereby preserving network system integrity over a longer time.

Although it may take some time before standardization bodies will create standards as AMLn models (instead of conventional standard documents), suppliers can benefit at lot of manually translating standard documents to AMLn models:

- The interpretation quality is considerably improved by creating an AMLn model from standard documents.
- Analyzing the effects of new requirements on the supplier view is facilitated.
- A supplier who applies this approach will not risk erroneous and/or ambiguous interpretations later on in the system's lifetime.

Note that network developers do not need to work in both AMLn and UML, since the entities, relations, and the behavior described by an AMLn model can always be automatically translated to objects and relations, and the behavior described by an UML model. The translation is done by using an *AMLn-to-UML translation* tool. Such a tool can have an enormous impact on the development process as a whole, since it has the potential for eliminating many sources for misunderstanding and errors in the translation process, not to mention the potential for time and costs savings.

Even if automatic translation to UML models is not applied, an AMLn model is an important tool for facilitating communication between all designers in the added-value process, as well as with sub-contractors and network designers.

1.3 Contributions to AMLn

AMLn has been developed over the past eight years, based on the author's experiences of over 20 years of systems development in the telecommunication industry. AMLn is used for modeling structural relations in network systems. Other parts of AML (AMLs and AMLp) can be used for specifying behavioral aspects of an AMLn model. These language components have been described in details in Muth (2001). A brief presentation of AMLs and AMLp is given in Appendix D in the present book.

The development of AMLn has been influenced by improvements of the architecture of network systems over the last 20 years. We can summarize this evolution in seven major steps:

1. *OSI RM* (Open Systems Interconnection Reference Model) introduced a layered functional structure for network systems in the early 1980s. As a result the concepts "layer," "layer element," "layer interface," and "protocol" became widely used. OSI RM also introduced an operation-oriented technique for communication and behavior description (the Remote Operation Service Element concept (ROSE), supported by Abstract Syntax Notation No. 1 (ASN.1), used for typing data and other types of entities).
2. *ISDN* (Integrated Services Digital Network) introduced the separation of streams of user data from streams of control data in the mid-1980s. As a result layers were classified as either belonging to the "user plane" or the "control plane."
3. *TCAP* (Transaction Application Part) was the first layer outside the OSI RM context that used operation-oriented communication. More important, however, is that TCAP devised an architecture where a single protocol could support many different and independent network functions in parallel. AMLn has inherited this technique by defining the "common-agent layer" concept as part of the language.
4. *TMN* (Telecom Management Network) introduced a generalized management network solution (based on a framework for management that was defined within OSI) in the late 1980s. This solution was based on the notion of a "managing system" separated from a "managed system," (or "traffic system,") and on "managed objects." The managed object approach was of course very much influenced by the development of object-oriented systems during the 1980s. The same can be said about AML, which defines object oriented communication in AMLs.
5. *SDH* (Synchronous Digital Hierarchy) introduced the notion of a "stratum" in the early 1990s. A stratum was a functional element in a network that could be operated as a network of its own (i.e., it had its own internal layer structure and management plane). To the environment, however, an SDH stratum appeared as just an OSI physical layer.

6. The *Internet* and the *UMTS* (Universal Mobile Telecom System) introduced a new paradigm for network architectures, which we may call "service-resource separation." While in the OSI RM and ISDN, a layer must include the resources it needs (e.g., multi-party or multi-media bridges) for the services it provides, the new paradigm allows service control layers to be spatially separated from resources, which are pooled in special "media gateway" nodes.
7. Both the *Internet* and the *UMTS* also introduce a clear separation (in different strata) between functions that handle end-user services and functions that provide raw connectivity (sometimes denoted the "call-connection separation"). This evolution developed from new transport technologies, such as IP and ATM, that could support all kinds of services (including transport of control signals) and could be used not only inside operator networks but for terminal access as well.

OSI RM, ISDN, the Internet and UMTS show possible, but different, ways to build layer structures for network systems. What distinguishes them are the fundamental principles that are applied on the highest level of functional structure:

1. The OSI RM emphasizes *technology separation*. It presents an abstraction hierarchy where transmission services are refined to more useful communication services step by step by adding specific layers. OSI RM does not define any other separation principle for the structure of network systems.
2. The ISDN emphasizes separation of *control-data streams* from *user-data streams*. It supports neither call-connection separation nor service-resource separation however. On the contrary, all services are controlled from a common "control plane."
3. The Internet network system (as well as the UMTS) supports both call-connection separation and service-resource separation.

Note that, except for the OSI RM, none of the network systems mentioned are based on a generic set of layers (as the seven layers in OSI). This partly explains why OSI networks did not survive. Both the ISDN and Internet define important "higher level" separation principles that do not exist in OSI RM, which is another reason for the disappearance of OSI networks. The success of the Internet indicates that the call-connection and service-resource separations should be the primary principles for functional structures. The same principles now controls the evolution of public telecom networks into network systems such as the UMTS. AMLn supports all principles mentioned, each on a specific level in the model of a network system.

Many of the new concepts that were introduced by the mentioned network systems were commonly used without first being unified, i.e., formally defined and agreed upon in the telecommunications domain in general. As a consequence, the usage of a concept in descriptions of network systems other than the one for which it was initially defined is often ambiguous, sometimes misunderstood, and even misleading. In the software system domain, the industry has managed to achieve consensus on a reasonably small common set of concepts just over a couple of decades. It is therefore astonishing that in the network system domain, which has been developing over 100 years now, there still exists no consensus on concepts, to the extent that they can just be chosen for a modeling language. One of the main standardization bodies for network systems (ITU–T) tried for many years to find a common terminology by publishing a steadily increasing set of the thousands of terms that were used in standards. However, ITU–T gave up some years ago when it found that almost every term had two or more contradicting and/or overlapping definitions. This fact has been a major challenge when developing AMLn.

1.4 Modeling Network Systems in AMLn

In this overview we describe briefly how AMLn deals with modeling the *structure* of network systems, how an AMLn model may be *viewed*, the AMLn *concepts* and *notations,* and how *behavior* of a model is specified.

As an added-value language to be used for modeling systems that are often extremely complex, AMLn must deal with the subject of how to structure a network system model very seriously. This issue is not a matter of defining a structure of documents (as we have commented on previously), but to define a structure that is in line with a sound *architecture* of network systems.

1.4.1 Structures in Network System Models

1.4.1.1 Layer Structure and Node Structure

The term "network system" indicates that such a system is something more than just a number of related functional elements or objects. It is also a system that has a structure of interconnected elements that we call logical nodes. These two aspects of a network system, which we call **layer structure** and **node structure**, are completely separated from a specification point of view.

- A layer structure (a refinement of the OSI RM view on this subject) consists of a structure of **layers** that communicate over **layer interfaces**. This structure may be simplified to a structure of **strata** by aggregating layers into strata. A layer consist of layer elements that communicate according to **protocols**. Communication over layer interfaces and according to protocols represent the behavioral aspect of a network system.
- A node structure consists of a structure of **nodes**, interconnected over **routes**. "Route" is a general concept used for all types of connections (single or multiple channels, synchronous or asynchronous, connection-oriented or connectionless).
- When nodes of an AMLn model are mapped on elements of a common-value model (e.g., a UML model) they also exist in a *processing view*, which means that they have access to processing platform functions. Such functions are

defined first in common-value models, and are therefore of no concern to an AMLn model.

- Layer structure and node structure are two views of the same thing. By defining how these views are related, we create a complete model of a network. This relation is simply that a functional element must be *allocated* to a node since otherwise it has no access to connectivity (i.e., it cannot communicate with other elements), nor can it have access to processing functions (i.e., it cannot execute).

We can depict the relation between the two views as that a network system exists in a plane defined by two logical dimensions, called the **layer dimension** (L) and the **node dimension** (N), se Fig. 1.8.

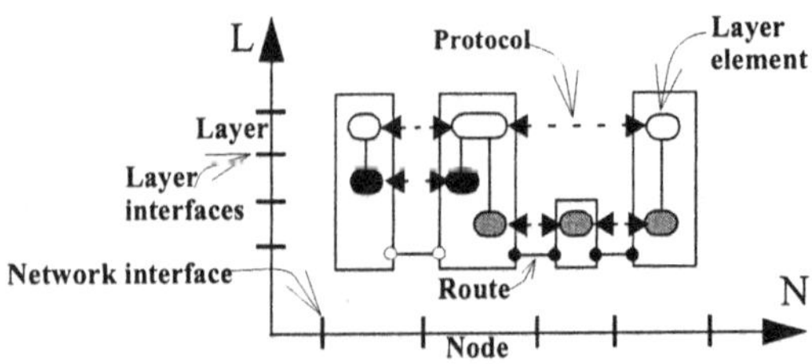

Fig. 1.8 A network system in a plane defined by the layer and node dimensions

Each view is maintained separately in a network system model. For example, the network system in the L dimension is depicted by eliminating the N dimension (see right side of Fig. 1.9).

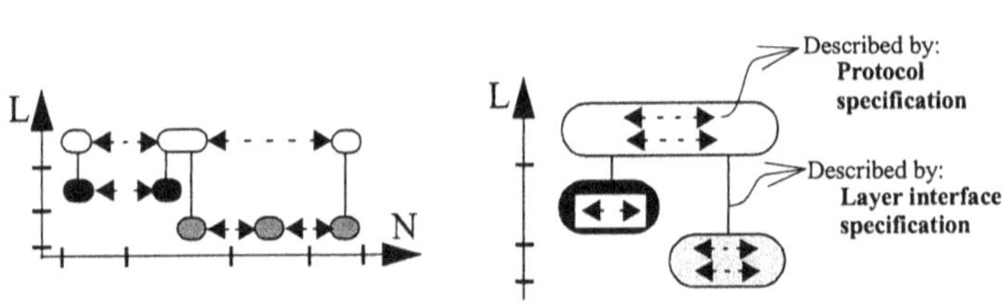

Fig. 1.9 The network system in the L dimension

Associated to this model are all **layer-interface specifications** and **protocol specifications**, since these are independent of the actual structure of interconnected nodes in a network of the kind. Figure 1.10 shows the network system of Fig. 1.9 in the N dimension.

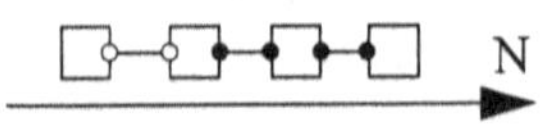

Fig. 1.10 The network system in the N dimension

This is obviously a model of one of many specific networks that can be built from the definition of the network system in the L dimension. This indicates the primary reason for keeping layer structures and node structures apart. Defining and maintaining the L- and N-dimensional views of a network system are two very different activities, regarding the incitements for modifying these structures, when in time modifications are made, and who manages the changes.

- Layer structures are modified as a result of functional requirements, while node structures are modified as a result of nodes being introduced or removed in a network and/or routes being changed, which is not necessarily related to any modifications that affect the L-dimensional view.
- The L-dimensional view is modified as a result of functional requirements defined by implementors or their customers. These are relatively rare events, which implies that the L-dimensional model is rather stable. Furthermore, modifications to the L-dimensional view may in due time result in modifications of some networks, but not necessarily in all existing networks.
- The N-dimensional view is modified by a network operator, mainly to respond to needs of increasing capacity and for dealing with reliability problems in his network. The N-dimensional view therefore changes more often than the L-dimensional view.

Figure 1.8 showed that a **network interface** includes both routes and (one or several) protocols. When we show the system in the L dimension only, we do not know or say anything about routes. When we show the system in the N dimension only we do not know or say anything about the contents of layer elements in nodes. The N dimension only defines which logical nodes exist and how they are connected. The nodes that exist in a network, however, are both connected to other nodes and run a number of layer elements that are interconnected over layer interfaces. To support mapping an AMLn model on a model that defines node objects in real networks, AMLn also defines how to model network structures of **derived nodes**, as shown in Fig. 1.11.

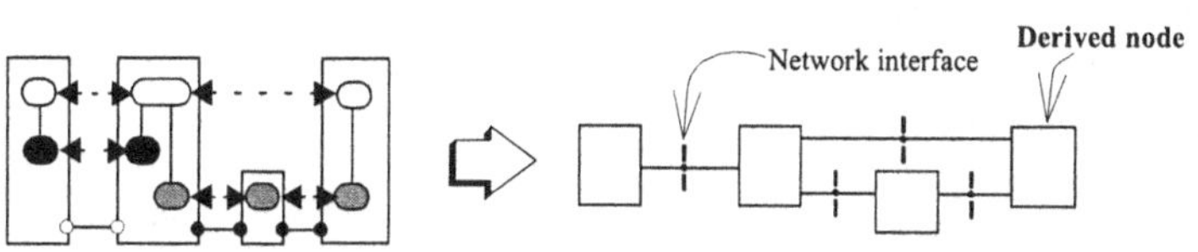

Fig. 1.11 Derived nodes and network structures

Derived nodes are entities that exist both in the L and N dimensions. The relation between derived nodes are network interfaces that comprise a route as well as one or several protocols. The term "derived" indicates that this kind of model is not specified separately. It can be extracted by a modeling tool out of the layer and node structures upon a user's request, e.g., when an AMLn model is to be translated to a UML model.

1.4.1.2
Management Plane and Managed Plane

A network system concept such as PSTN/ISDN or ATM, actually comprises two separate planes, a managed plane (we prefer "traffic plane") and a management plane, since no network can survive without some degree of control from management functions. Each of these planes describe its own L and N dimensions, which implies that the two planes define different layers and layer interfaces as well as nodes and routes. The relations between these planes is therefore of a special kind that we define in a separate dimension called the **management dimension** (M), conceptually depicted in Fig. 1.12.

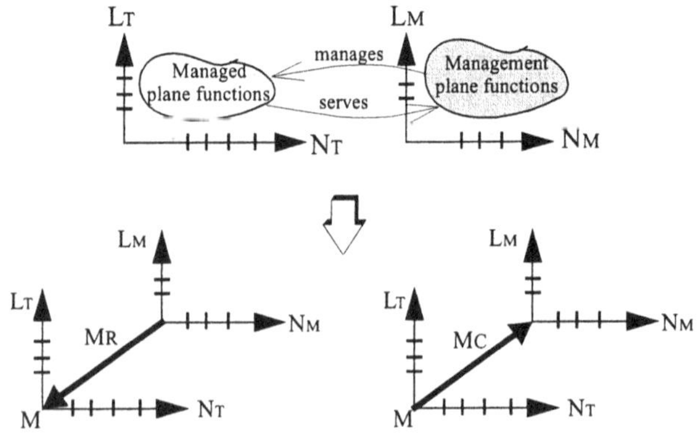

Fig. 1.12 The management dimension (M). LT-NT=managed plane; LM-NM=management plane; T=traffic (plane), another term for managed (plane)

The management dimension consists of two separate relations, called MR and MC in the figure:

- The management plane consists of functions that manage resources existing in the traffic plane. These resources are commonly accessed via objects in the **traffic system** that are called **managed objects** (MO). Communication between the **management system** and MOs may also be object oriented. Interfaces that define how management plane functions access resources in the traffic plane are interfaces between layers in different planes. We therefore denote them MR relations in Fig. 1.12.
- Management functions are frequently integrated in the traffic system in the sense that they rely on connectivity services defined in the traffic plane. Such functions operate more or less autonomously from other functions of the management plane. The relation between these functions and the traffic plane are normal connectivity layer interfaces, used for management purposes however. We therefore denote them MC relations in Fig. 1.12.

1.4.1.3
Network Levels

By modeling a network system separately in the L and N dimensions, and by defining the distinction between managed and management planes, we achieve a considerable reduction in model complexity. Further reduction of model complexity is needed, however.

Any of these views can be fragmented in smaller pieces, primarily applied to node structures (i.e., the N dimension). Fragmentation is also needed in order to allow for piecewise development of a model, and by different modelers. Such fragmentations must be done so that the internal structure of a fragment is not revealed outside it, which implies that interfaces between fragments must be defined by architectural relations such as layer interfaces, routes, and network interfaces.

Since different specification bodies may define structures on different levels, it is also very important that a fragmented view shows to which level of fragmentation a particular functional element or node belongs. We call such levels **network levels**. Typically:

1. A standardization body will define the top network level, where a whole network (including terminals) is regarded as a single node, and only the services to network users are defined. This network level is obviously valid for any kind of internal structure of the node. We call this level "network level 0" since it is a model with a collapsed N dimension. The specification of services are associated to layer interfaces (i.e., the L dimension) that are accessible in terminals. Realization of such interfaces is part of the network level 0 model, but normally not specified by the standardization body.
2. On next network level (level 1), the standardization body describes the structure of different types of terminals connected to the network, where the latter is defined as a single black box. All access network interfaces are specified on this level.
3. The standardization body may stop there as far as terminals are concerned, and leave to implementors to define the internal structure of terminals (on network level 2), including how to realize network level 0 layer interfaces.
4. On network level 2, however, the standardization body will most likely define an internal structure of the network defined on network level 1 in terms of operator networks, each having its own terminals. These networks operate autonomously but need to interwork with each other. The network level 2 model shows interworking networks as black boxes, defining everything that is included in network interfaces between such black boxes.
5. At least one of these networks will be of the actual network system type. For this network the standardization body will most likely define an internal structure of types of derived nodes, such as access nodes, switches, service controllers, media gateways, etc. This model is then a network level 3 model.
6. Standardization bodies normally stop there. However, implementors will no doubt develop at least one lower network level (level 4), where they describe an

internal structure of the derived nodes defined on network level 3. This structure may be a node structure as well, since nodes that are defined on network level 3 are commonly constructed as (local) networks of their own.

Thus, documenting network levels in a network system model is necessary since it defines which structures are independent from each other, where interworking is an issue and who is responsible for which network level fragment. AMLn provides the general relation **sub(-structure)** for identifying network levels (see Fig. 1.13). A sub relation has no attributes or properties. Its only purpose is to define how a relation on a higher network level is terminated on a lower one.[4]

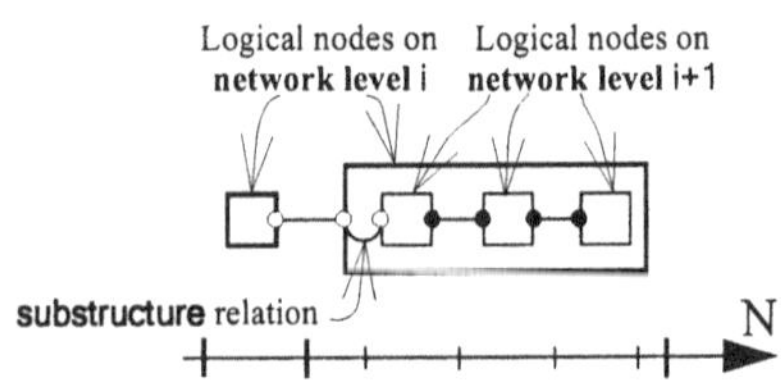

Fig. 1.13 Network levels and the substructure relation

This model uses a sub relation for defining model fragments on different network levels. Note that both fragments will be described in the same way, i.e., by models in the L and N dimensions. The sub relation can also be used for fragmenting strata into layer structures and derived network models into network partitions (i.e., terminals and different operator networks).

1.4.2 Modeling Layer Structures

The L dimension defines the functional structure of a network system, including its behavior. Since we assume that network systems are layered, we say that the L dimension shows the **layer structure** of the system. Layer structures are of course based on the "layer" concept, originally defined in the OSI RM (see Figs. 1.8 and 1.9). However, the OSI RM layer concept is too primitive to be useful for modeling as it is (which partly explains why it is often misused).

AMLn introduces four basic improvements of the OSI layer concept. They make it possible to model important structural distinctions in layer structures that

[4] The author has seen other modeling languages where the need for model fragmentation is satisfied by references between pages that say "the description of this system continues at point X on page Y." Such a relation serves no purpose other than fragmenting a large description into multiple pages, and has nothing to do with the AMLn substructure relation. In fact, such "page references" are symptoms of an insufficient structuring approach and should never be allowed in a modeling language.

cannot be defined in models based on the OSI RM layer concept. These principles are limited in numbers and easy to understand (see Sects. 1.4.2.1–2.4):

1. The *control–connectivity* separation separates **control layers** from **connectivity layers**.
2. The *actor–agent* separation separates message handling elements (**agents**) from acting elements (**actors**).
3. The *actor–resource* separation separates **resources** from actors that use or control them.
4. The *LSM–LPM* separation separates the stable states of a layer from the volatile states that are associated to message handling.

Three additional principles are derived from the above:

1. Substituting an OSI RM layer with an **agent layer** and supported actors allows a layer structure to be divided into a **control structure** and one or several structures of agent layers and connectivity layers. We call the latter **connectivity structures**.
2. Defining the **common-agent layer**.
3. Defining **common actors**.

These principles are described in Sects. 1.4.2.5–2.7.

1.4.2.1 The Control–Connectivity Separation

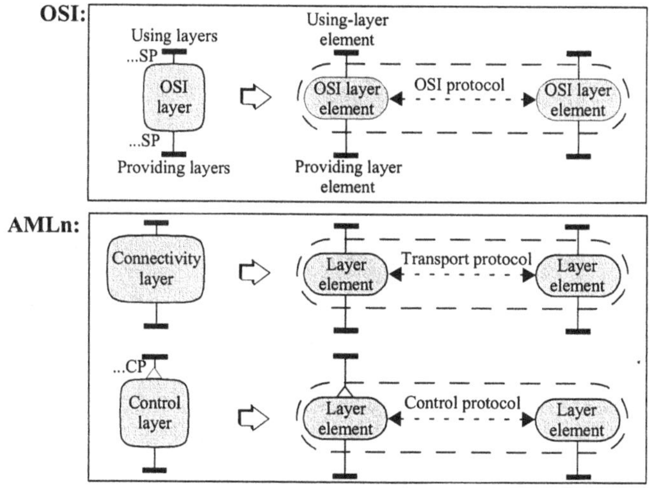

Fig. 1.14 Layer definitions in the OSI RM and in AMLn

With the exception of the OSI RM "application layer" (that does not offer any layer interface at all), all OSI layers offer some type of connectivity service. We know, e.g., through the ISDN and ATM architectures, that most layers in real networks do not offer connectivity, since they deal with different forms of control (e.g., controlling service access, controlling media resources, controlling managed resources). AMLn therefore defines the difference between control layers and connectivity layers by providing different layer interface symbols for these layer types, as depicted in Fig. 1.14.

The layer interfaces represented in these models are of two classes: **connectivity service points** (...SP) and **control points** (...CP). A reasonably large set of symbols for denoting different types of layer interfaces within these classes are suggested in Sect. 2.1.1. A notation for denoting layers that combine control and transport functions is also included.

1.4.2.2 The Actor–Agent Separation

A protocol in the OSI RM is defined as an unstructured description of what a layer element does when receiving a message. This may be a reasonable approach for protocols of non-switching connectivity layers (transmission and link layers), but not for switching and control layers. AMLn therefore separates layer elements in one part that handles message reception and sending (the **agent** part), and another part (the **actor** part) that takes actions as a result of users requests and reception of a correct message, as depicted in Fig. 1.15. Through this separation, agents handle interfaces to connectivity layers and actors handle interfaces to using layer elements.

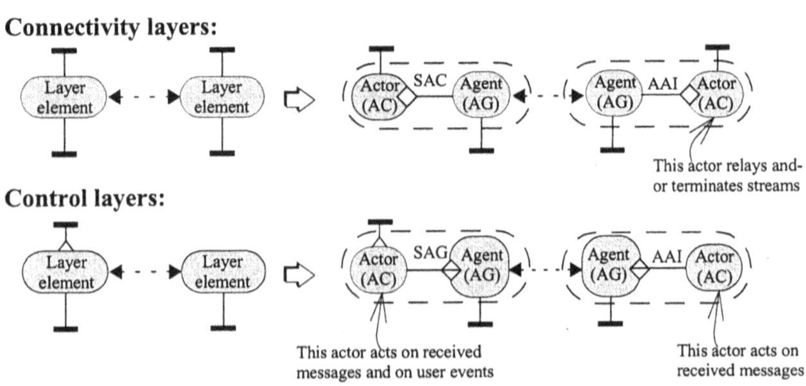

Fig. 1.15 The actor–agent separation is equally applied on connectivity and control layers. AAI=actor–agent interface; SAC=specified actor interface; SAG=specified agent interface

Messages carry data that are partly handled by actors, but actors are completely independent of how agents format and encode messages. The coupling between agents and actors is therefore very weak. The same actor could be used over different protocols, provided that the **actor-agent interface** (AAI) is kept intact, which is why agents and actors should always be modeled separately. AAIs come in two flavours: **specified agent interface** (SAG) in control layers, and **specified actor interface** (SAC) in connectivity layers (see Fig. 1.15). The difference between SAGs and SACs is how interactions between agents and actors are defined (see Appendix B).

If you look at implementors' network system models, you will find that actors are commonly separated from agents. Standards do not apply the actor–agent separation, however. It should be specified already in the standard view of a network system, which is why it must be part of AMLn.

Figure 1.15 shows that this principle is applicable on any type of layer. There is no real differences when it comes to what the agents do and how message exchange works. This is not so where actors are concerned, however:

- When an actor of a connectivity layer receives some data from its agent, it can only choose between two activities: either it decides to relay the data to another agent, or it decides to terminate it in a user layer element. In no case, however, does the actor *interpret* the data.
- When an actor of a control layer receives some data from its agent, it *interprets* the data and chooses an *action* that is relevant to the received information. This may or may not result in a communication with a using layer element.

A consequence of this approach is that while actors of connectivity layers have no states (at least not lasting states), actors of connectivity layers have. As a result of received information, a control-layer actor may affect resources in the layer, which changes the state of the layer for a shorter or longer period.

It is important to model the distinction between these two types of actors, which is why we use slightly different symbols for actor–agent interfaces. We also use different names (SAC and SAG) that associate certain (generic) AMLn primitives to those interfaces, as was mentioned before.

1.4.2.3
The Actor–Resource Separation

The OSI RM has no notion of **resources** that a layer needs for fulfilling its task, since it assumes that such resources exist locally in nodes. This view has never reflected the architecture of real network systems. Resources in terms of data are often pooled in separate data base nodes in a network. With the event of both the intelligent network concept (IN) and the Internet and UMTS, other resources (such as switches and diverse media devices) are pooled separately as well. Furthermore, resources that exist locally in layer elements (diverse tables and data) must be manageable from remote management nodes. In AMLn, therefore, all resources are made visible in a layer-structure model, with some as local resources, others as resource layers (see Fig. 1.16).

When you look at a conventional message-protocol specification, the behavior of an actor is normally described in terms of states only. For example, one state may be "service request is accepted." However, this "state" represents that certain resources on the performing actor's site are used (e.g., a subscriber data record) and other resources are dedicated to the service request (e.g., a path through a switch). Which these resources are is of no interest to the requesting actor. Therefore, these resources must be defined separately in an extra refinement step, according to Fig. 1.16. Such resources may be "local," i.e., they exist in the actor's node, or "remote," i.e., they exist in a remote node, or are distributed to the actor's node and remote nodes. Remote and distributed resources are parts of a **resource layer**.

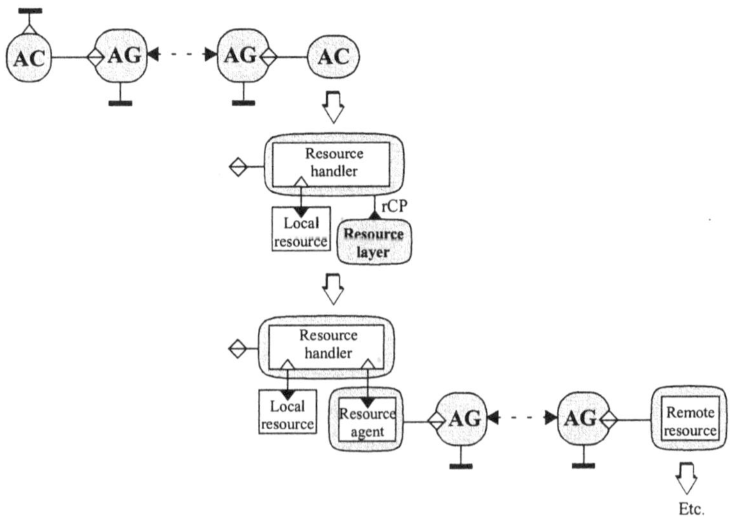

Fig. 1.16 Resource identification and modelling. rCP=resource control point

A resource layer may further depend on other resource layers, i.e., part of the layer structure of a network system may be a layer structure of resource layers. Thus, resource layers appear as any other layers in a layer structure, which is why we define a special control point symbol, the **resource control point** (rCP) for layer interfaces to such layers. The actual resources, resource agents, and the resource controller in an actor are modeled as state machines however, using the AMLs notation (see Appendix D).

Resources may be used by actors both in connectivity layers and control layers. Most resources of interest to model are controlled by the latter, however.

1.4.2.4
The LSM–LPM Separation

When we distinguish what agents do and what actors do in control layers, we actually separate the behavior of the layer in two independent parts, since actors do not know how agents handle messages, and agents do not know what actors do with the information they deliver to them. Since the agents and the message protocol between agents make distribution transparent to actors, actors communicate with each other only through the information they get from their local agents. This information may specify an **operation** that one actor wants a remote actor to perform. Each interaction of this kind is therefore an **event** that is an instance of an **abstract service primitive** (ASP). When we describe the behavior of actors in terms of ASPs, we are describing an **abstract protocol** (see Fig. 1.17).

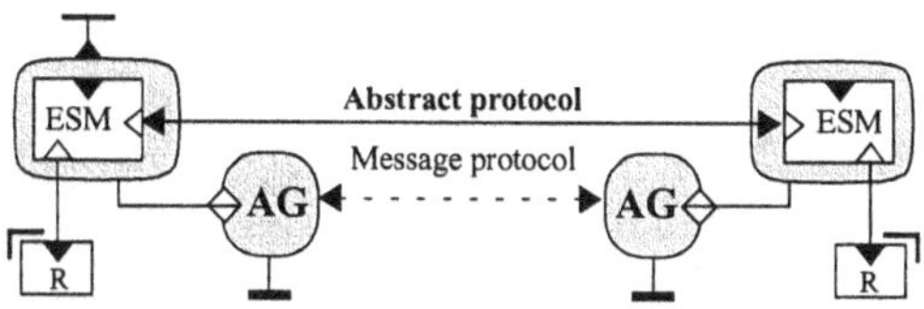

Fig. 1.17 Actors of a control layer handle element state machines (ESM) that interact within the layer according to an abstract protocol. R=local resources and/or resource agent

Since the communication between actors is an abstraction of events in actor–agent interfaces and some message exchange between agents, we use the AMLs specification technique to define a state machine that is handled by an actor, and that represents the behavior between actors, expressed in terms of events. This state machine is a (layer-) **element state machine** (ESM). As a result, a control layer becomes separated into one or several **layer state machines** (LSM) and a **layer protocol machine** (LPM), as depicted in Fig. 1.18.

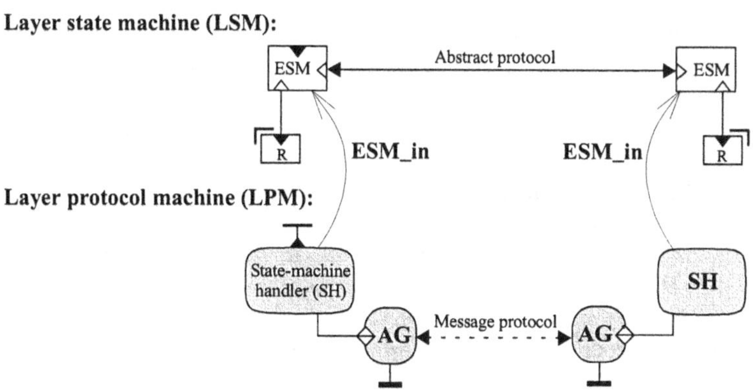

Fig. 1.18 Layer state machines and layer protocol machines are parts of the same control layer.

An ESM also controls all resources on which the layer relies, in the node in which the ESM resides. The separation of a control layer into one or several LSMs and an LPM implies that we must define **ESM_in** relations in the model. We also get two views on actors: when we talk about actors in the LSM context, we refer an ESM, while in the LPM context it is a **state-machine handler** (SH).

An SH will communicate with its agent by events as well. These events carry the ESM events. AMLn defines them as **meta-primitives** (see Fig. 1.19).

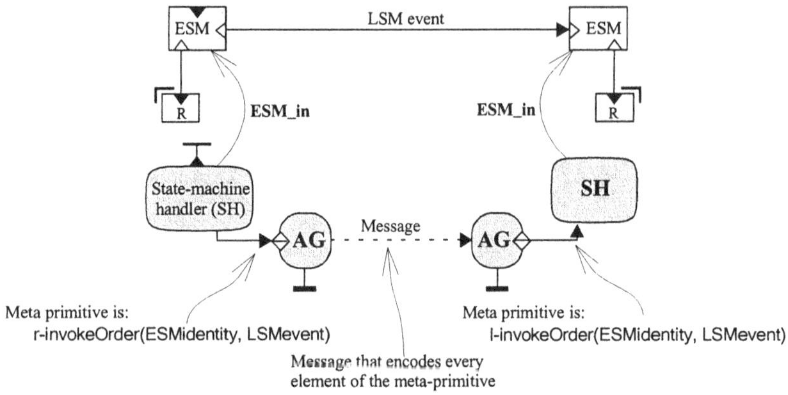

Fig. 1.19 Primitives in actor-agent interfaces of control layers are meta-primitives

One meta-primitive is the **r-invokeOrder(...)**, the other is the **l-invokeOrder(...)**. Every actor–agent interface of a control layer can be specified by using these two primitives (or some user-defined specialization of them). Figure 1.19 shows the simplest form of meta-primitives (AMLn defines a few more parameters, see Appendix B).

The introduction of meta-primitives explains the role of the LPM as a "carrier" of LSM events, i.e., to encode every information element of a meta-primitive into some message format. As a result, an LSM event that is defined by an r-invoke in one actor–agent interface will be reproduced by an l-invoke in another, which implies that an LPM maps an abstract protocol on a message protocol.

The LSM–LPM separation principle has a very simplifying effect on the complexity of some protocol specifications. Since an OSI RM protocol for a control layer does not apply the actor–agent distinction, its state machine can become very complex, if its states are defined by multiplying the stable state space of an LSM with the volatile states of the LPM that supports the LSM.

The LSM–LPM separation principle is *not* applicable on relaying and switching layers, since actors of such layers do not interpret the data they relay or terminate, and therefore control no layer states. Thus the behavior of such a layer is described by a single machine, which is an LPM. Figure 1.20 shows the generic operations (**send, terminate, relay** and **receive**) that can be associated to layer interfaces and actor–agent interfaces of all connectivity layers.

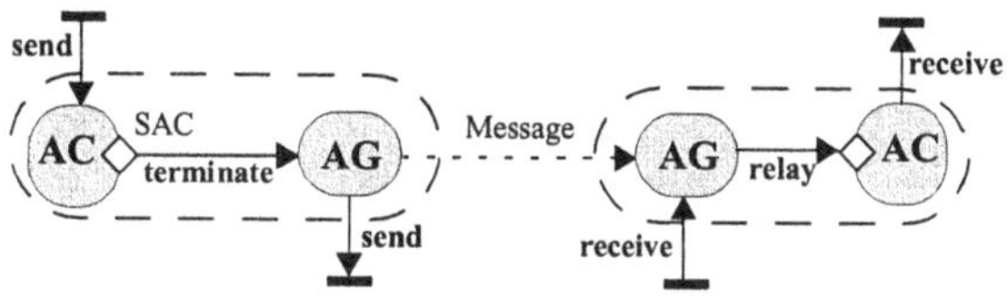

Fig. 1.20 The LPM of a connectivity layer

The semantics of these operations are:

- **send**: an explicit or implicit request from a user to send some data to another user.
- **receive**: an explicit or implicit report to a user that some data has been received from another user.
- **terminate**: an explicit or implicit order from an actor to its agent to deliver some data to a defined user.
- **relay**: an explicit or implicit order from an agent to an actor to relay some data to a defined user.

The reason for making a distinction between relay and terminate is that an agent has no business of knowing if the actor is a relaying layer element or if it is just a terminating layer element. This implies that an actor that, at one point in time, is just terminating all incoming data in layer interfaces to local users, may later be replaced by a relaying or switching element, that can relay data to remote users as well.

Note that the realization of these operations depends on the type of connectivity service: when the service is asynchronous, it depends on *explicit* primitives in interfaces; when the service is synchronous, the primitives are *implicit*.

1.4.2.5 Agent Layers

From these basic separation principles we can derive another, very profitable principle for modeling: to redefine an OSI RM layer, consisting of layer elements, to a structure of an **agent layer** that support one or several actors. Figure 1.21 shows this principle applied to simple layers of only two actors.

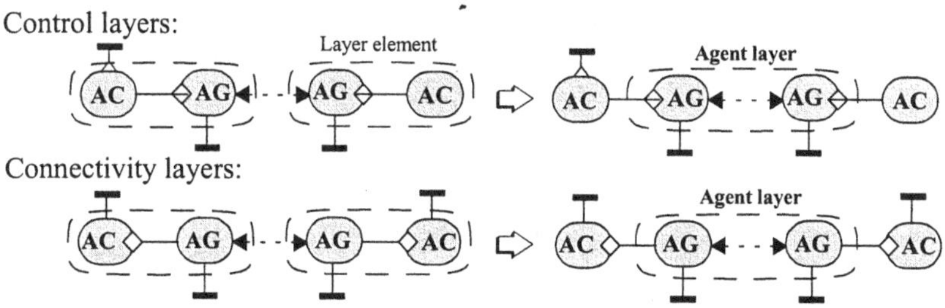

Fig. 1.21 The agent layer

As can be expected, this principle is equally applicable on control and connectivity layers. What differs are the primitives that are used in actor–agent interfaces, and the functions of actors:

- Actors in control layers handle state machines and may serve users with control points.
- Actors in connectivity layers are switching, relaying, or just terminating elements that may serve users with service points.

When modeling, we can use this principle to separate control structures from connectivity structures. In Fig. 1.22 we have applied it on the ISDN layer structure (L dimension only). The boundary between control and connectivity structures are defined by actor–agent interfaces only, which allows us to store and handle separate model fragments for these structures. As long as actor–agent interfaces are not modified, a modification of a connectivity structure, for example, cannot affect a control structure, and vice versa.

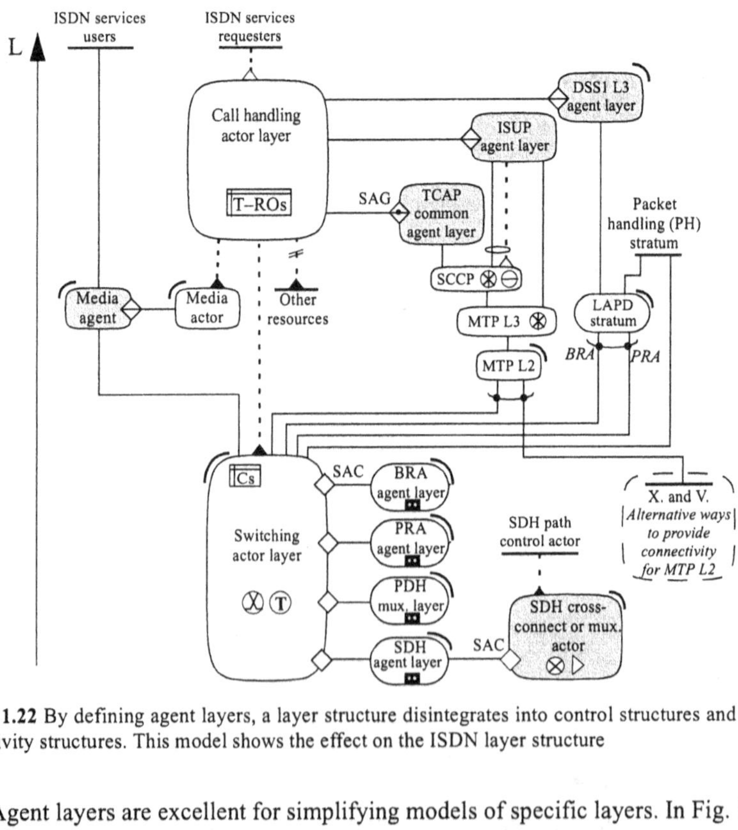

Fig. 1.22 By defining agent layers, a layer structure disintegrates into control structures and connectivity structures. This model shows the effect on the ISDN layer structure

Agent layers are excellent for simplifying models of specific layers. In Fig. 1.23 we have used it to show the LPM of an ISDN call-handling layer.

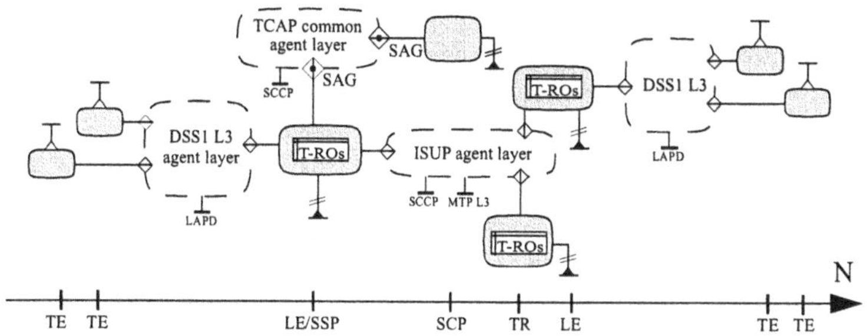

Fig. 1.23 The LPM of an ISDN call-handling layer (state-machine handlers and agent layers)

1.4.2.6 Common Agent Layers

Agent layers are just transport mechanisms. An agent layer can therefore be designed to support many actors concurrently. If it does we call it **common agent layer** (a well known example is TCAP). This principle is also valid for both control layers and connectivity layers, but normally only used for control layers.

Figure 1.24 shows an example of a common agent layer that supports two **actor layers,** {11..13} and {21,22}. The messages of a common agent layer must include parameters for identifying actor layers (ALi) and actors (As). A common agent layer may also support **associations** between actors (read more about this in Sect. 2.3.2.3). Note that a common agent layer must provide identical actor–agent interfaces (SAGs) over the whole network in order for actors to be allocated freely in the network. Thus, the SAG for a common agent layer is defined by the agents.

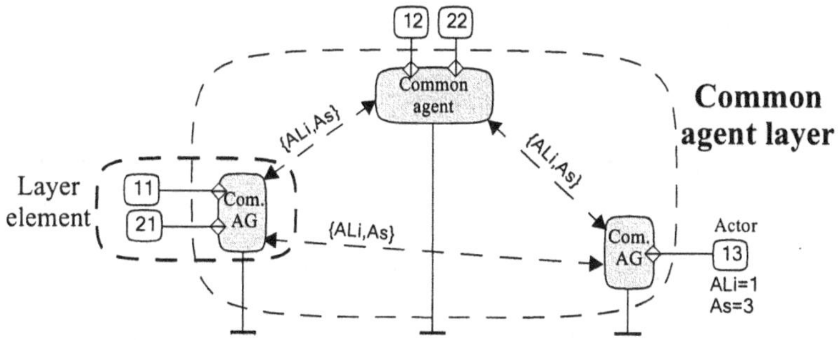

Fig. 1.24 A common agent layer

When the common agent layer principle is applied, the OSI RM layer concept breaks down completely. The actor layers perform separate and independent functions in the network. If the actors perform control functions, each actor layer is a control layer of its own, defined by one or several abstract protocols.

It is also conceivable that a common agent layer serves a number of independent actor layers that are separate relaying or switching layers. This principle is applied by the PPP layer in the Internet, for example, where PPP supports the IP switching network, as well as a number of other packet-switching networks.

1.4.2.7
Common Actors

A **common actor** is a switching or relaying element. While a common agent is used by several actors, a common actor relies on more than one agent. These agent layers may be of the same or different kinds, depending on the underlying connectivity layers that are used. Figure 1.25 shows the LPM of a typical PSTN/ISDN switching layer, which relies on many different types of agent layers.

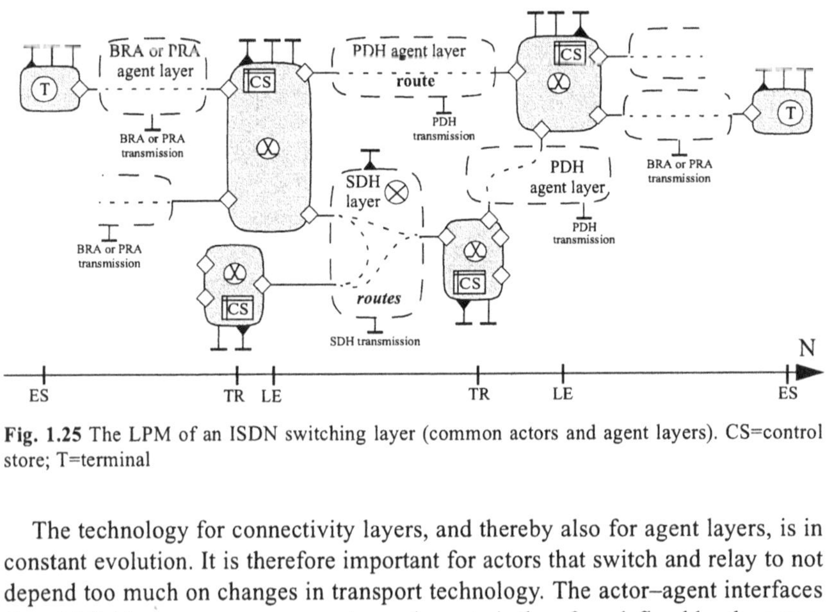

Fig. 1.25 The LPM of an ISDN switching layer (common actors and agent layers). CS=control store; T=terminal

The technology for connectivity layers, and thereby also for agent layers, is in constant evolution. It is therefore important for actors that switch and relay to not depend too much on changes in transport technology. The actor–agent interfaces (i.e., SACs) between a common actor and agents is therefore defined by the actors and is uniform over the network. In most switching networks, a small set of different SACs are defined in order to connect to different types of agent layers (e.g., one or a few SACs for access and one or a few for network internal transport).

1.4.3
Modeling Node Structures

The rationale for separating layer structures from node structures was discussed already in Sect. 1.4.1.1. Basically, for all network systems and networks, the things that are depicted in these structures are handled by different activities, at different points in time and with different frequencies of occurrence. While layer structures describe how layers are related over functional interfaces called "layer interfaces," a node structure shows some kind of nodes, interconnected by some kind of "routes."

Packaging layer elements of a layer structure (thereby producing node definitions), can be made in many ways however. Furthermore, there are not only many types of connections around, we also know that one type of connection can be a bearer of other types of connections, such as 64 kbps bitstream connection that carries a voice connection, or a route in a switching network that defines a number of channels.

We discussed in Sect. 1.4.1 the node concept that is defined in the L–N plane. It can be used to define node structures where nodes are derived nodes, related over network interfaces. We also discussed the effect of using sub relations to define network partitions as nodes and nodes as node structures, only on different network levels. Both of these modeling techniques are used to define node structures.

A more basic node concept, however, is the **logical node**, used to describe **logical networks**. Logical networks are the result of a **horizontal partitioning** approach to network systems. This implies that a network can be sliced into a number of logical networks that rely on other logical networks for services, primarily for connectivity. Since this is more less the idea with the OSI RM layer concept, one may ask what the difference is between a layer and a logical network. An important aspect of logical networks is that a particular type of layer interface is specified as a standardized and configurable interface (see Fig. 1.26).

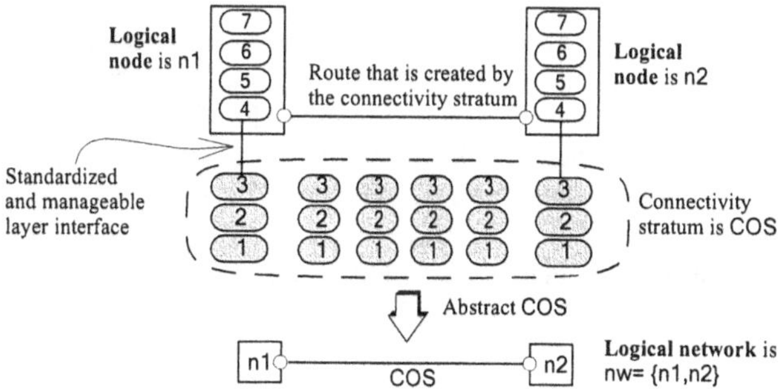

Fig. 1.26 The definition of a logical network

Here we have created a logical network out of logical nodes that comprise layer elements 4 through 7 of an OSI network. By defining these nodes we say that they can be replaced in the OSI network by other nodes (through a management activity). Note that horizontal partitioning is applied on a given layer structure without changing anything of that structure. Since the only thing the logical nodes in the model see of the connectivity stratum (in this case OSI layers 1 to 3) is their interconnection, we can produce a separate logical network definition that consists of the two nodes (n1 and n2), interconnected by a route symbol.

The benefit with horizontal partitioning is that we can describe and implement a complex network as a structure of less complex, configurable logical networks, i.e., as a **logical-network structure**. Let's make an example out of the OSI RM model, by defining another logical network that comprises OSI layers 5 through 7 only (see Fig. 1.27). The two logical networks defined in this way are nested like a Chinese box, since logical nodes of network nw2 must exist in logical nodes of network nw1. This fact is modelled by **hosts** relations between nodes of different logical networks, that relates a **hosting node** to a **hosted node**.

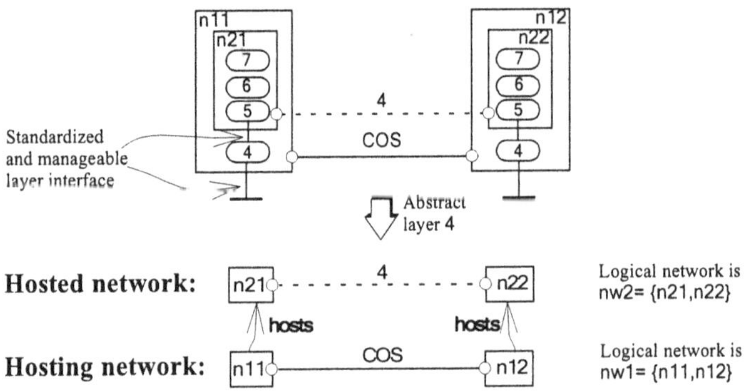

Fig. 1.27 The definition of a logical-network structure

Obviously, this nesting of logical networks can be repeated until every layer defines a logical network that is either a hosted network only, a hosted network that hosts other networks, or just a hosting network. A logical network that does not host any other network is therefore the counterpart to an OSI application layer. The only logical network that cannot be a hosted network is a network that connects to physical media.

Logical networks are models in the N dimension only. By describing a network as a logical-network structure, we actually separate the N dimension into a plane defined by a **logical node dimension** (NO) and a **logical network dimension** (NW) (see Fig. 1.28).

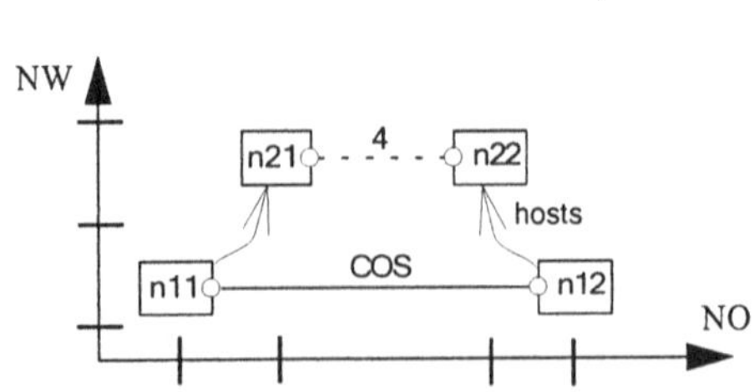

Fig. 1.28 A logical-network structure is defined in the NO–NW plane. NO=logical node dimension; NW=logical network dimension

The model of a logical network defines logical nodes and connectivity relations only (i.e., routes). However, it says nothing about which layer elements are included in logical nodes. A logical-network structure adds **hosts** relations, which do not reveal anything about the allocation of layer elements to logical nodes. We can therefore view a logical node as some kind of "package" for layer elements, and a logical-network structure as a specification of how logical networks are configured in relation to each other. To be able to operate logical networks independently, the following requirements must be fulfilled:

- Hosted nodes must be defined as packages that do not reveal their layer element contents.
- Layer interfaces to be used by hosted nodes must have a standardized realization.
- A hosting logical node must support multiple layer interfaces for hosted nodes.
- It must be possible to bind a hosted node to a layer interface, offered by a hosting node, through a configuration activity from the management system.

Let's therefore look at how a hosting logical node must be designed. The generic model in Fig. 1.29 defines all concepts that are associated to logical network structures.

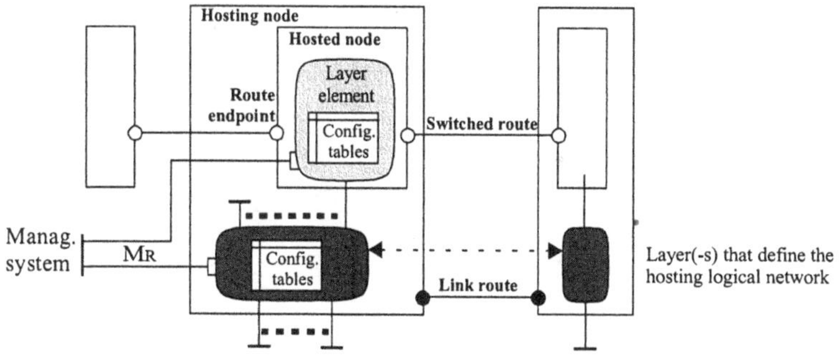

Fig. 1.29 Concepts that define logical network structures

- A hosting node contains a functional element that must be able to support changeable contents of multiple hosted nodes. It may also be a hosted node in one or several hosting nodes. The functional element may be a single layer element (as in Fig. 1.29) or a layer structure.
- Since the contents of hosted nodes is changeable, it must be defined by data in configuration tables that are maintained from a management system. This configuration function relies on the **hosts** relations, defined in the logical-network structure model.
- A logical node communicates with peer nodes over routes. The node knows the **route endpoint** to each peer. A route endpoint can be the name of a remote node

and such names are only locally known within the very logical network (sometimes called "global titles").

- Some hosting networks support **global titles** (GT). Most hosting networks do not, however. A hosted node must therefore also include configuration tables that relate route endpoints to address parameters to be used in the layer interface to the hosting node. These tables are also maintained from the management system.
- The routes used by a hosted network rely on routes used by the hosting network(s). A hosted logical network may therefore use more refined routes than the hosting network does. For example, the hosting network may rely on **link routes**, but offer **switched routes** to hosted nodes. AMLn provides route symbols for a small set of route types.

Figure 1.30 gives an example of a logical-network structure which shows a small part of the Internet (or some other IP-based network).

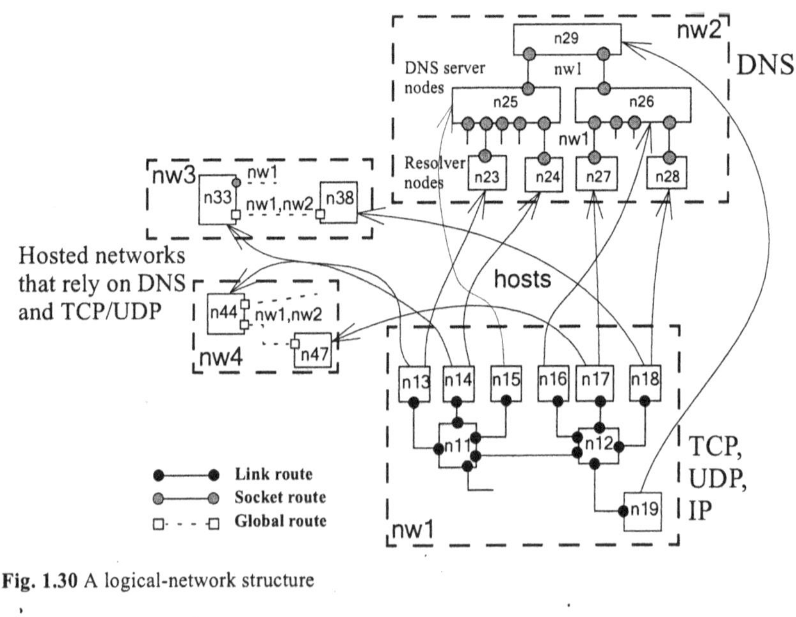

Fig. 1.30 A logical-network structure

This structure shows four logical networks, nw1 through nw4. The network nw1 is a switching network, while nw3 and nw4 are logical networks that run as applications on TCP/UDP/IP. The network nw3 relies on **socket routes**, offered by TCP/UDP. Both nw3 and nw4 rely on **global routes** as well. To that regard they need to use the services of nw2, which is a DNS network ("domain name system;" nw2 is a network that can translate global titles to socket addresses). The DNS network is accessed over "resolver" nodes that exist in the same nw1 nodes as nodes of nw3 and nw4. The DNS network uses socket routes as well.

This logical-network structure contains all information that is needed for configuration, provided that its relation to layer elements in the L dimension are known. Such relations are stored as **allo(-cates)** relations in the model, as exemplified in Fig. 1.31.

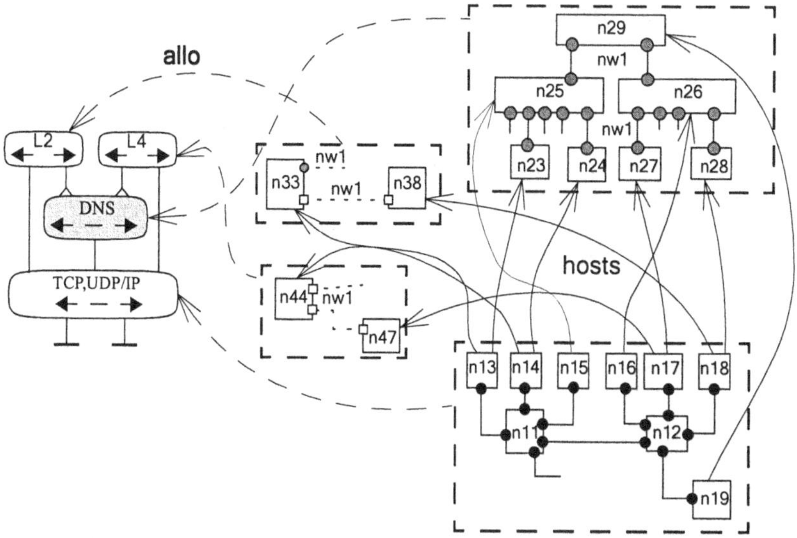

Fig. 1.31 A part of an IP-based network in both L and N dimensions. L..., DNS, TCP, UDP, IP =layers or strata that are implemented in the logical networks

A logical-network structure separates configuration activities in tasks that concern only single logical networks and tasks that concern **hosts** relations between networks. Relations of type **allocates** show which layer functions are included in different networks and nodes, and thereby if and how functional modifications concern a particular logical network. These are all properties of the modelling technique that makes it possible to maintain correct models over time.

Note that the logical-network structure in the previous figures is developed within a single **network level**. When layer interfaces between logical networks are realized as network interfaces, some logical nodes will have an existence in more than one hosting logical network, defined on *different* network levels. This requires that an additional relation, the **term(-inates)** relation, has to be maintained. This relation relates the allocation of a layer element in a node on a higher network level to one or several nodes on a lower. Read more about this in Chap 4.

1.4.4
The Boundary Between Traffic and Management Systems

All functions that are not executed on-demand by requests from network users are, by definition in AMLn, management functions. As discussed in Sect. 1.4.1.2, management functions belong to a management plane defined in an L–N plane, similar to but separate from the plane where we model functions of the traffic system. Interfaces between these two planes belong to the **management dimension** (M).

Through these definitions, interfaces in the M dimension exist in nodes of the traffic system, where the resources to manage exist. Therefore, the boundary between the management plane and traffic plane on the one hand, and between the management system and traffic system on the other, do not coincide, as depicted in Fig. 1.32 for an imaginary node of a traffic system (note that we call managed system "traffic system", to avoid confusing the two systems).

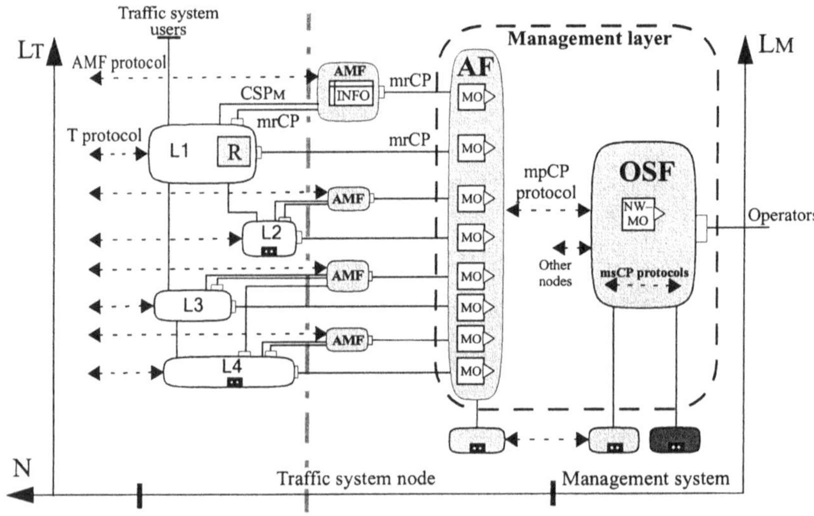

Fig. 1.32 The boundary between traffic plane and system, and management plane and system

Figure 1.32 shows that a management plane consists of two parts: a (probably) rather large set of **autonomous management functions** (AMF) and a **management layer** that runs on some suitable connectivity strata, one of which must be common for the two systems in order to support communication between **adaptation functions** (AF) and the **operations system functions** (OSF).

- The boundary between planes consist of two relations denoted M_R and M_C in Sect. 1.4.1.2. In a model, these relations appear as control points and connectivity service points, that we denote **mrCP** and **CSP$_M$** respectively.

- A traffic system node always includes an interface function to the management system (called AF in the model). These functions define the interface between the two systems, normally in terms of **managed objects** (MO).
- Managed objects create a view of resources ("R" in the model) in a traffic system node. This view is constrained to properties of resources that are accessible. A managed object hides the actual implementation of resources, and thereby of their interfaces (mrCP), since this interface is of concern only to implementors of traffic system nodes.
- The management plane may also include many autonomous management functions (AMF) that perform their tasks without control from the management system (they may, however, deliver data for statistics and other information to the management system over AFs). AMFs use special connectivity service points (CSP_M) in the traffic system for communication between AMFs.
- Interfaces between nodes of the traffic system and the management system are control points called **mpCP**. These are defined within the management layer and normally standardized protocols. SNMP defines the mpCP in the Internet management system, while the OSI RM solution to mpCP is called CMIS/CMIP (part of TMN).
- OSF is the part of the management layer that exists in the management system. This part is normally a network of its own where different management functions are allocated to different kinds of management network nodes. Interfaces between such nodes are also control points, called **msCP** in AMLn. Interworking management systems (not in Fig. 1.32) are also interfaced over special msCPs.
- An AMLn model, e.g. the logical-network structure model of a traffic system, may be used to create **network managed objects** (NW–MO), accessible by network operators for configuration activities.

The boundary between the two planes becomes more visible if we collapse the N dimension for both systems (see Fig. 1.33).

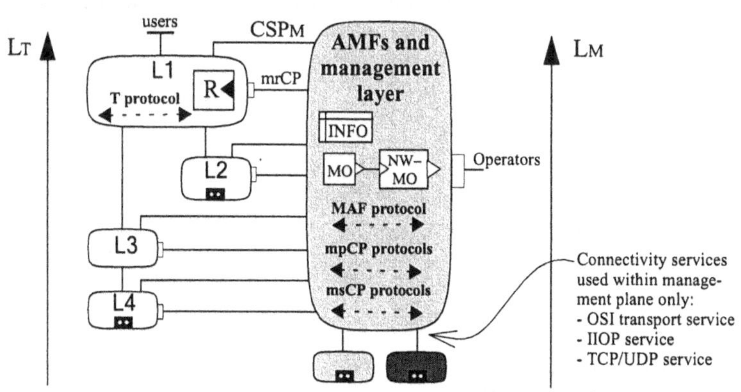

Fig. 1.33 A management plane defines all functions needed for a particular traffic system

This model is used for defining which functions and MOs must be specified for managing a particular traffic system. This is especially valuable since a management system may be used to manage several traffic systems in parallel.

A management layer is a single control layer. We can therefore model its LSM separately from its LPMs, as for any other control layer. Figure 1.34 shows the LSM that is described for the Internet management layer (i.e., SNMP-based). The model also shows some node concepts defined by SNMP.

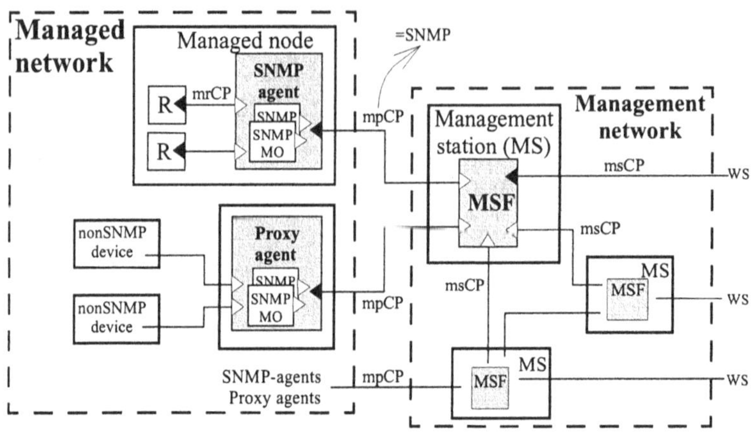

Fig. 1.34 The LSM for the SNMP management layer. MSF=management-station function; WS=work station

1.4.5 Specifying Behavior in AMLn Models

The behavior of an AMLn model is described by separate specifications that are associated to different parts of the model as **attributes**. Almost everything that concerns the behavior can be attributed to the model of Fig. 1.18, which is reproduced here in Fig. 1.35 with some additional information.[5]

The model shows the four primary behavior describing attributes which are all specifications of interfaces:

1. Generic meta-primitives.
2. Generic send and receive operations.
3. A number of abstract protocols that describe interfaces of the LSM.
4. A message protocol specification.

[5] Note that this model represents a control layer. For a connectivity layer, only the LPM part exists. The meta-primitives are replaced by other generic primitives, indicated by another symbol for the actor–agent interface. The actors are not state-machine handlers but actors that terminate, relay, or switch data streams.

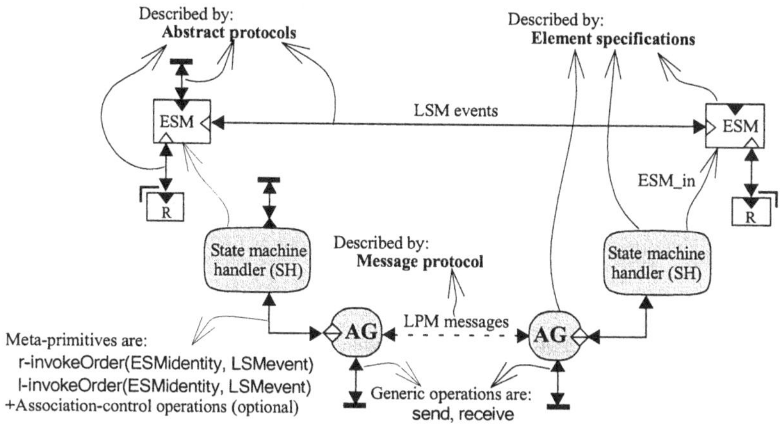

Fig. 1.35 Primary behaviour attributes in network system models

Message protocols for control and connectivity layers will look quite differently:

- *Control layers:* If meta-primitives are specified, the only thing needed is a specification of the translation between meta-primitives and messages, which is the same for all LSM event. The AG–SH interface may also define a number of generic association control primitives (defined by AMLn).
- *Connectivity layers*: The specification looks more like a conventional protocol, describing a number of messages and how their parameters are encoded. In addition, a message-based state-machine description is required. AMLn defines a number of generic primitives for the actor–agent interface (relay, terminate, etc.). The protocol specification defines the translation between these events and messages. It will most likely also include other messages that are used for error and flow control, checking the status of a used connection, etc. These messages have no correspondence in events of the actor–agent interface.

Focusing behavior specification on interfaces is a *universal* principle (at least in AML), which is also the reason that an AMLn model is built by identifying interfaces. After all, we do not want functional elements to be too concerned about other interfaces on which another element depends. However, sooner or later we want to specify how certain elements behave, e.g., for being able to run a model in simulation, or for generating code. We therefore regard an **element specification**, which considers all interfaces on which an element relies, as a secondary behavior description. Such a specification is an attribute of the very functional element. It is designed by specifying the relation between events (and/or messages) in different interfaces that concern the element. This principle is equally applicable on elements of an LPM and an LSM.

We need a specification language for being able to design behavior specifications for an AMLn model. AMLn does not prescribe any particular language, since behavior specifications are just attributes to the model, which implies that any suitable language can be used. We can, however, list a few requirements on such a language, considering what an AMLn model look like:

1. Obviously, the language must allow behavior to be specified in terms of asynchronous events, such as the events of abstract protocols (message communication is, by definition, asynchronous).
2. The language must also allow for asynchronous events to be defined as parts of operations, since that is an expanding form of communication in networks.
3. Furthermore, since sometimes it is convenient to associate operations to objects (e.g., the MOs used by management systems), the language must allow a certain behavior to be associated to an object.
4. Finally, the language must support a behavior to be described in terms of a state machine. This is due to the fact that a major part of all functions in a network system deal with controlling the state of things, e.g., the services given to users and the state of resources used.

The author developed such a language more then 10 years ago, as the first part of AML. We now call it AMLs. AMLs is a language that you may use to describe everything about an LSM. It is a common-value language, i.e., it can be used for describing behavior of systems in any system domain. It therefore does not use the specific names (LSM, ESM, MO, etc.) of state machines that AMLn introduces, but calls all of them **abstract machines** that may or may not include **abstract objects**. AMLs is described in details in Muth (2001). For readers who do not need all details of AMLs, a brief overview is given in Appendix D.

You also need to specify LPM behavior, which comprises the messages of an agent layer protocol and the behavior of the agents. For specifying actor–agent interfaces and layer interfaces that agents use, you may use AMLs as well. However, for describing message types and the mapping between events in actor–agent interfaces and messages, AMLs is of no use. The author therefore developed a method some years ago for generating such mapping descriptions, based on constraints defined by a protocol designer. We now call that method AMLp. AMLp can be regarded as a method for LPM specification.

Since messages as well as parameters of events are data structures, AMLp must rely on some data typing language. An easy way is to use a standardized language (ASN.1, types defined in IDL, or C++), since then you also have access to all encoding algorithms you need to implement in agents. This is not always possible, however, since such methods do create quite a few overhead bits. This may not be acceptable for all protocols, especially not for some switching, transport, and transmission protocols. AMLp therefore also suggests other data types that provide less overhead. AMLp is described in details in Muth (2001). For readers who do not need all details of AMLp, a brief overview is included in Appendix D.

As already indicated, a number of other languages could serve the purpose of AMLs and AMLp as well. For example, both UML and SDL are (in some sense) similar common-value languages, and IDL (the CORBA interface language) can be used for interface specification. However, all these languages are initially developed for software system modeling. They therefore fulfill the general four requirements that were listed previously to various degrees. It is beyond the scope of this book to discuss this matter, however.

1.4.6
The Modeling Dimensions

In AMLn, the complexity of models is handled by defining **modeling dimensions**. These are not just an arbitrary way of separating properties of network systems, but represent important generic architectural structures as well. The purpose with these dimensions is threefold:

1. It gives AMLn users a mental high level image of what network systems look like.
2. It defines model types that are relatively simple, each described by using very few symbols.
3. It defines major partitions of a model in a database.

Model dimensions are defined on two levels. The top level dimension is the management dimension (M) in which we model the relations between two separate planes: traffic plane and management plane (see Figs. 1.32 and 1.33). The M dimension defines two types of relations: managed resource control points (mrCP) and connectivity service points (CSPM) for autonomous management functions (AMF). The planes define functions of the traffic system and management system which are described separately. The "values" associated to the layer dimension (L) and the node dimension (N) denote some kind of elements or interfaces (...CP and...SP). The L and N dimensions are applicable on all types of layers, including OSI layers.

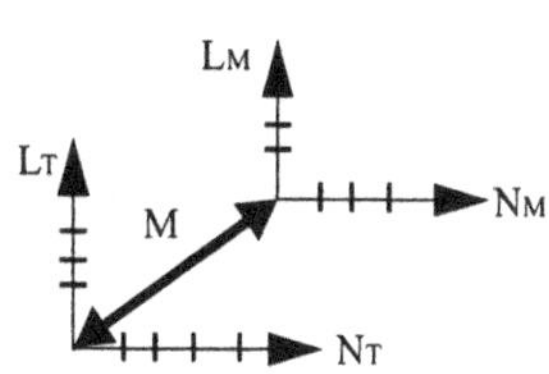

Fig. 1.36 Main modeling dimensions. L=layer dimension; M=management dimension; N=node dimension

When we work on an AMLn model we normally work only in one or sometimes in two dimensions. For example, taking the OSI RM model as a reference, we can either model it separately in the L and N dimensions, or apply a node structure on the L dimension and create a model in the L–N plane, as shown in Fig. 1.37.

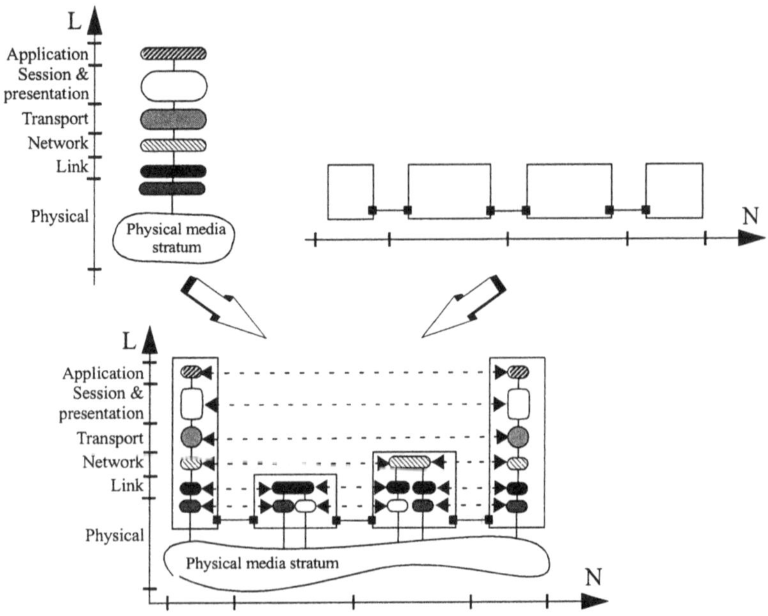

Fig. 1.37 Creating models in two dimensions

Through the AMLn definitions of the actor–agent separation principle and the definition of logical-network structure, each of the L and N dimensions become separated into two lower level dimensions. A layer structure resolves into two layer dimensions: one for control structures (Lac) and one for connectivity structures (Lco). Figure 1.38 exemplifies this for the ISDN network-system model.

This model shows:

- Control structures consisting of actors that communicate over resource control points.
- Connectivity structures consist of agent layers and connectivity layers.
- Control structures and connectivity structures can be maintained separately if SAGs and SACs are specified.
- A model may define more than one control structure if it is stratified
- A single control structure may rely on several independent connectivity structures.

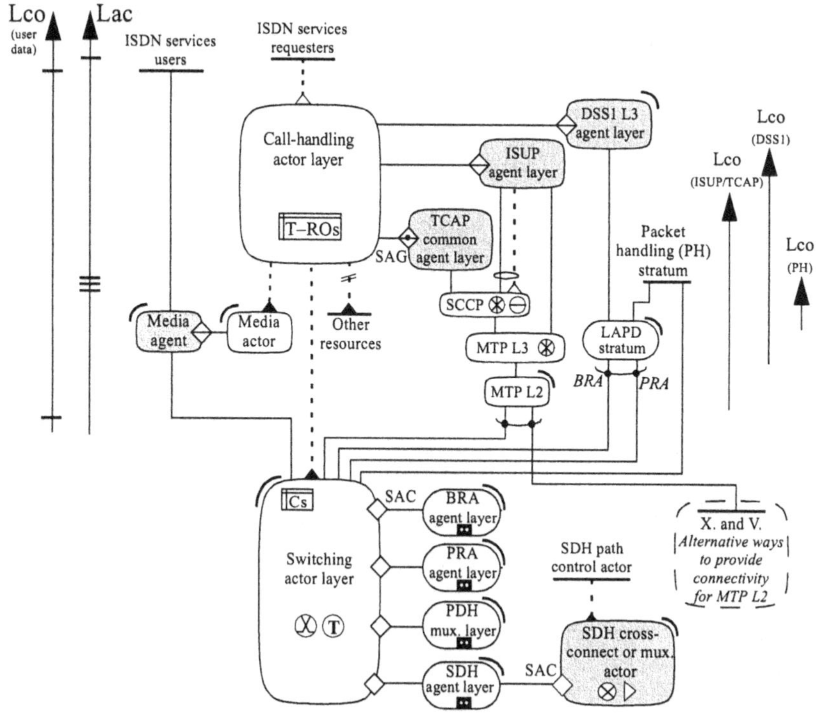

Fig. 1.38 The L dimension consists of several subdimensions, one for control structures (Lac) and several for connectivity structures (Lco)

By applying a structure in the N dimension on the ISDN call-handling layer, for example, we can model its LPM and LSM separately in the L–N plane. Figure 1.39 shows a possible LPM.

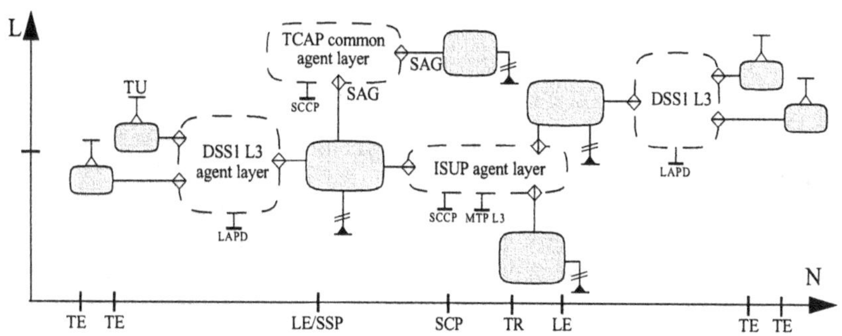

Fig. 1.39 The LPM for an ISDN call-handling layer. LE=local exchange; SCP=service control point; SSP=service switching point; TE=terminal; TR=transit exchange; TU=terminal user

The assumed structure consists of terminals (TE), local exchanges (LE), etc. The corresponding LSM looks as in Fig. 1.40. Note that since it consists of element state machines (ESM) that reside in actors only, this model exists in the LAC–N plane only.

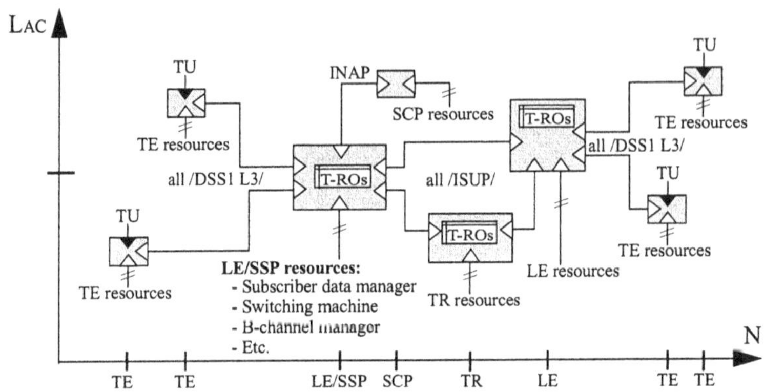

Fig. 1.40 the LSM that corresponds to Fig. 1.39. T-ROs=routing table

The node dimension (N) resolves into a logical node dimension (NO) and a logical network dimension (NW) when a network system is horizontally partitioned into a logical-network structure. For example, if we choose *not* to do so in the IP-network case discussed in Sect 1.4.3, we would describe the system as a single logical network, as depicted in Fig. 1.41.

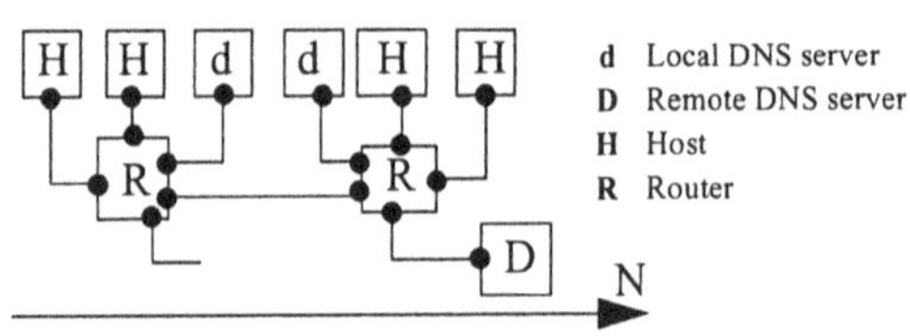

Fig. 1.41 A network described as a single logical network exist in the N dimension

The nodes in this model are derived nodes, including all kinds of layers. For example, hosts will contain layer elements of all of IP, UDP, TCP, DNS (the resolver part), and application layers. Routers will contain IP, UDP, and TCP only while DNS servers will contain IP, UDP, TCP, and DNS. If we introduce the logical-network structure of Fig. 1.30, the N dimension resolves into a logical network dimension (NW) and a separate logical node dimension (NO) for each logical network (see Fig. 1.42).

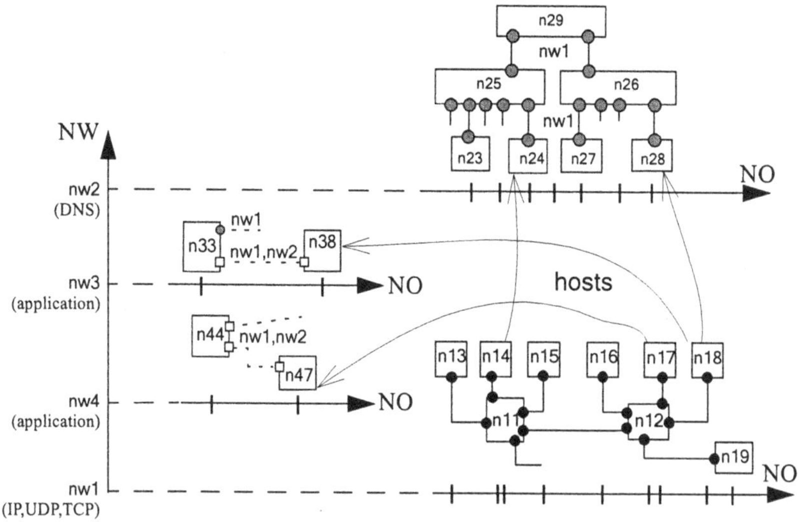

Fig. 1.42 Logical network structures implies that the N dimension expands into an NW–NO plane

Nodes of logical network nw1, for example, are now no longer hosts, routers, etc., but logical nodes of the IP/TCP/UDP switching logical network. Nodes of nw2 are nodes of the DNS logical network only, and so on. Each of these logical networks are considerably simpler to deal with than with a single logical network for the whole network system.

In summary:

- The logical dimensions that are defined by AMLn have evolved from the dimensions originally defined by the OSI RM, which are the layer dimension (L) and the node dimension (N).
- By introducing the separation between management and managed system, we have defined the management dimension (M).
- By introducing the logical network concept we have expanded the N dimension to a plane defined by the logical network dimension (NW) and the logical node dimension (NO).
- By introducing the actor-agent separation, we have separated the L dimension into a layer dimension for control structures (Lac) and a layer dimension for connectivity structures (Lco).

Figure 1.43 summarizes these refinements.

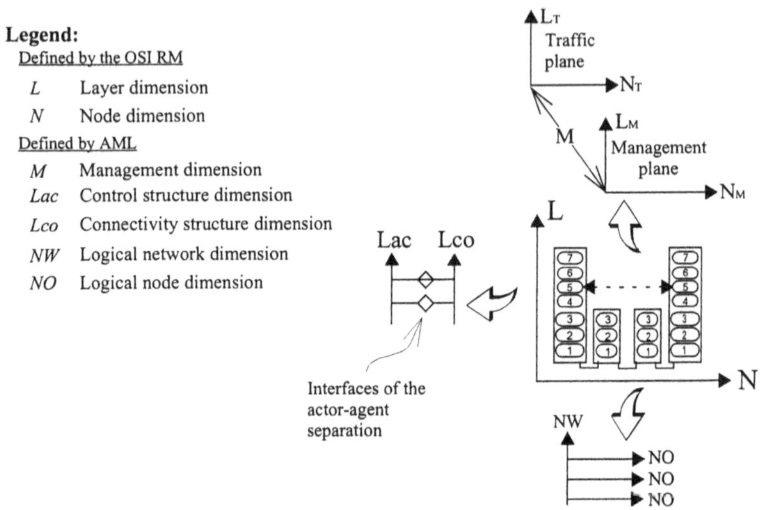

Fig. 1.43 Logical dimensions defined by AML

1.4.7 Views

A **view** is, by definition, whatever part of an AMLn model to which a user has access, and that the modeling tool can display. The modeling dimensions and the structures defined through substructuring (such as stratum structures and the structure of network levels) are often enough for view selection.

Views may be restricted to read and write accesses, however, and to which attributes a particular user has access. Furthermore, it should also be possible to select views that are smaller fragments of the network system model than those defined by modeling dimensions and substructures. For example, a user may want to analyze the behavior with only some of the nodes of a particular logical network involved (typical for when so called "protocol architectures" are designed). The smallest possible view is a single attribute of a single logical node, an agent, or an actor (e.g., the state diagram describing the behavior of a controlling actor). Which views are actually supported is, however, a matter for tool designers to consider.

2 Layer Structures

2.1 Concepts Based on the OSI RM

2.1.1 Layers and Layer Structures

A common appreciation of the concept "network" is that it consists of nodes (machines, such as switches and terminals) that are interconnected by *physical media* (wires, fibres, radio channels, etc.). Inside a node exists a lot of hardware (i.e., electronic circuits) and software elements, the latter performing most of the functions of the network when executed by the node's processing system. Some elements communicate between nodes by means of **message** exchange. When this exchange is described, we normally call that description (message-) **protocol specification** (the bulk of network system standards consists of these types of document).

Such a model of a network is very implementation-oriented, however. It depends on the type of processing system and the technologies that are used to implement the hardware and the software. Since it must be possible to build networks by components (nodes and parts of nodes) that are supplied by different implementors, and since network systems must be open to technology changes, an implementation-oriented view, such as the one indicated, is out of the question as a productive way to create network system models.

An important objective with the OSI RM is to devise a method for describing networks in a way that leaves suppliers of network products to solve all implementation issues. The OSI RM, therefore, makes the concept **layer** known. While a software element or software system, for example, must be confined to a particular node and processing system, a layer is a function that is *network-wide*. This means that it consists of functional parts that are distributed to different nodes (at least two) and communicate by means of message passing. The OSI RM also defines "layer" as (potentially) *a part of* the total functionality of a network, by saying that a layer may rely on services that are provided by other layers over **layer interfaces**, thereby constructing the picture of a layered system, such as the OSI RM seven-**layer structure**. A layer interface connects two layer elements of *different* layers to each other. Layer interfaces are normally regarded as interfaces inside nodes (we will, however, modify this assumption later).

The parts of a layer exist in nodes and are called **layer elements**. Layer elements become fully defined by specifying the messages and rules for exchange of messages (in a protocol specification) within the layer, as well as how they access the services of other layers (specified in **layer–interface specifications**, also called "service definition" in ITU–T standards[1]). This implies that a model created by this method leaves the question of how to implement layer elements completely open. The layer element, therefore, becomes the target for suppliers' implementation efforts.

This is a short recapitulation of what we trust is more or less common knowledge for most readers.[2] We will use the layer in Fig. 2.1 as the starting point for presenting some basic concepts that are associated to OSI RM layers.

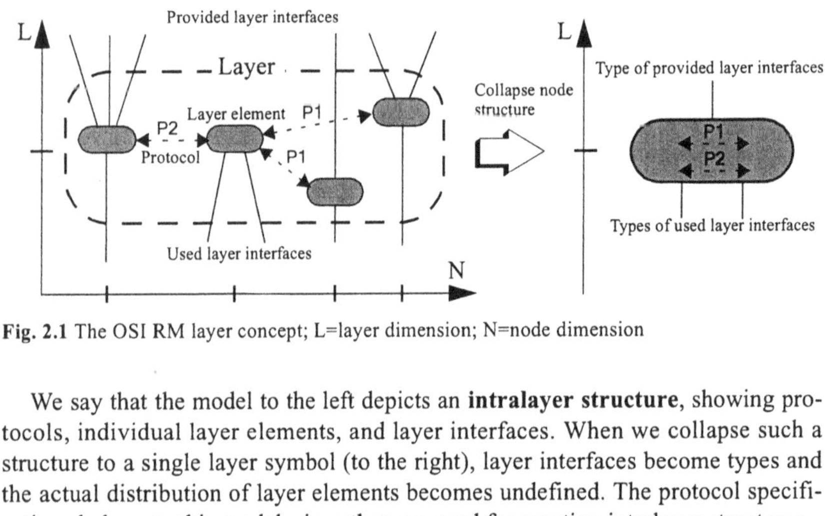

Fig. 2.1 The OSI RM layer concept; L=layer dimension; N=node dimension

We say that the model to the left depicts an **intralayer structure**, showing protocols, individual layer elements, and layer interfaces. When we collapse such a structure to a single layer symbol (to the right), layer interfaces become types and the actual distribution of layer elements becomes undefined. The protocol specifications belong to this model, since they are used for creating intralayer structures.

Through the OSI RM we have learnt that a layer interface is what connects a layer element of one layer to a layer element of another. We do not intend to change this simple principle for identifying what belongs to different layers. Thus, by definition, two layer elements that are interconnected over a mutual layer interface belong to different layers.

A layer interface is, similar to protocols between machines in a network, an interface used for communication. In the OSI RM tradition, however, while protocols define the communication between nodes, elements that are interconnected

---

[1] ITU–T calls its documents "recommendations." We will however use the term "standard" for all documents published by standardization bodies.

[2] If not, look for any text book on data communication. With few exceptions, they all recapitulate OSI RM on some level of details. The ITU–T X.200 recommendation and the ISO 7498 standard are official descriptions of the OSI RM.

over a layer interface always exist in the same node. We will later change this assumption from "always" to "for the most part," since there are an increasing number of layer interfaces that are implemented as protocols as well. However, sticking to the OSI RM for the time being, we can use this assumption to specify layer interfaces in an *abstract* way. This means that we specify communication over layer interfaces in a simplified way, using **abstract service primitives** (ASP, a kind of "function call," or "operation call") as the interaction element between layers.

For example, if a layer element requests a connection to another layer element (of its own layer) from a layer "below", we describe that event as something like connectRequest(otherLayerElementIdentity).[3] The requesting and performing layer elements may be implemented as two different processes in the same machine or in different machines. In any way they need to be connected to some kind of communication mechanism for "intra- or inter-process communication." This is viewed as a problem for suppliers/implementors of the node to solve, however, and therefore disregarded when communication over a layer interface is specified.

Through the use of ASPs, a layer-interface specification shows what the used layer can do for a using layer, normally expressed in terms of the used layer offering some kind of *service* to using layers, the latter accessing the service through ASPs. Besides describing all ASPs and how to use them, layer–interface specifications also describe service characteristics, such as transfer rate for connectivity services and diverse quality of service (QoS) properties.

A layer interface is supposed to offer a single service. There is, however, no clear definition, neither in the OSI RM nor in existing standards, of a "single service" (other than as "a layer–interface specification describes a single service"). An analysis of existing layer–interface specifications, will show that many of them describe a clean cut **connectivity service**, e.g., the IP layer. Others define a single **control service** (such as the call handling layer in ISDN), or a single **information service** (such as DNS[4] and some of the "application service elements" that OSI RM places in the lower part of its application layer). There exists, however, many other layers that define a mix of several types of services, such as TCP that defines both a control and a connectivity service, or the SCCP layer in the Signaling System No. 7 (SS7) that, besides a control service, offers a whole range of different connectivity services, including several methods for addressing.

The fact that a layer interface (from a service point of view) can comprise almost anything makes models of layered systems less valuable, unless layer interfaces are not drawn with some kind of service-type indicating symbol. For example, an important distinction must be made between layer interfaces that offer

[3] We call such an expression not just "primitive" as in most languages but "abstract service primitive" in order to emphasize that, in implementation models, these primitives must be translated to the actual design language used.

[4] The Domain Name System in the Internet architecture, which translates application system names to TCP/UDP "ports" and IP addresses.

connectivity services and control services (e.g., for distinguishing the parts that ISDN allocates to its user and control planes). In AMLn we deal with this problem as follows: since a layer interface can in reality comprise any type and mix of services, we define a small set of symbols for **connectivity service points** (CSP) and **control points** (CP).[5] These symbols are used for drawing layer interfaces in models. Each symbol/point characterizes a major class of services.

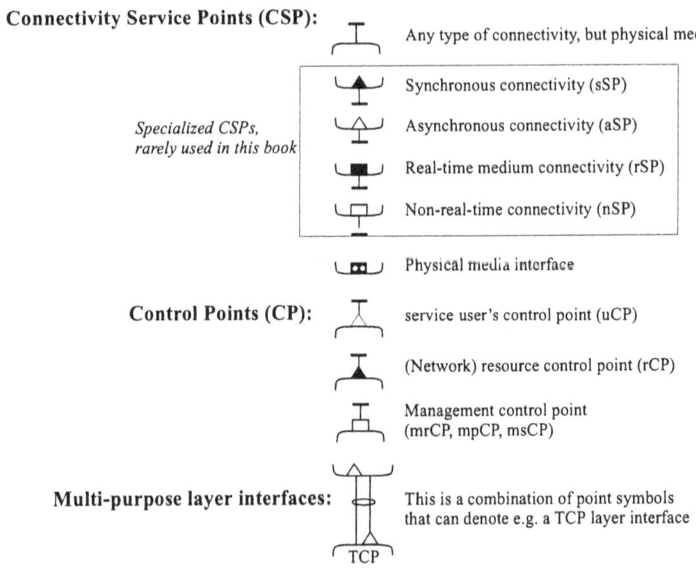

Fig. 2.2 Symbols for connectivity service points (CSP) and control points (...CP)

Some important comments on these symbols:

- The author has experienced that too many symbol types in a model do not help us understand the model. Therefore, we do not use the specialized *CSP* symbols in this book, only the general symbol at the top of the list. Note, however, that this symbol cannot tell which part is the user or which is the provider. In this book we use primarily the OSI RM convention to always place the user of a layer's services above the provider.
- Also, to keep models as clean as possible, we do not suggest any special symbol for information services. The *uCP* symbol is used both as a control point and as

---

[5] A CSP is an interface where the only thing a user can do is to send and/or receive data. CPs are interfaces where a user might control or access any type of service, *except* sending and receiving data. This implies that, for example, a layer interface that provides a connection-oriented package service must be modeled as both a CSP and a CP (see the TCP in Fig. 2.2).

a point where an information service is accessed (specification methods are the same in both cases).

- An *rCP* is an internal control point in a layer structure that is used to denote control of **resources** that are used to provide on-demand services. For example, when an external user requests a connection through a switching network, the call-handling layer must use switching elements. Layer interfaces between a call-handling layer and a switching layer are then modeled as rCPs.
- Management control points *(mrCP, mpCP, msCP)* are used for managing resources in network systems. For example, in circuit-switching networks, a management system maintains routing tables that are used by the call-handling layer. The interface between the management layer and the routing table is then modeled as a management control point.
  We make a distinction between control points that exist in the managed system (mrCP), in between the managing and the traffic system (mpCP) and inside the managing system only or between cooperating management systems (msCP). The same symbol is used in all cases, however (see Chap. 5 about modeling management functions and systems).
  Note also that management control points are associated with layer models only. When we model management functionality, we reduce the management layer to its **layer state machine** (LSM) representation (as we do with all types of control layers). LSMs are modeled by the abstract-machine notation that is described in Muth (2001). A short version is given in Appendix D.
- The distinction between service points (...SP) and control points (...CP) is based on the observation that many layers are of the kind that they provide one or the other. There is, however, no rule that says that they must. The designer of a model may define a layer that offers a whole range of different services over a single layer interface (although we recommend avoiding that if possible). In such cases, symbols for ...SPs and ...CPs may be combined by the notation shown at the bottom of Fig. 2.2 for TCP.

Figure 2.3 gives an example of the use of some of these symbols in an **inter-layer structure** (which is a model in the L dimension only). It also shows how multiplicity is denoted in a layer structures. The exclusive_or symbol denotes that a LAPD transport layer runs on either a BRA or a PRA layer.

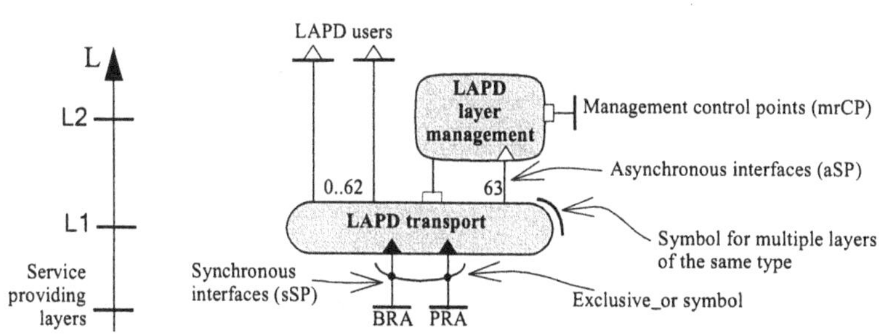

Fig. 2.3 An example of how to use symbols to define types of layer interface. L1,L2=layers

In the connection-oriented OSI RM, all layers except the “application layer” are regarded as layers that offer some kind of “connection” (the abstract primitives connectRequest and dataRequest are used on all the lower six layers). It is therefore important to remember that the AMLn layer definition (also emphasized by the ISDN plane concept and the symbols we have defined) in no way restricts the “layer” concept to layers that offer connectivity only. For example, layer L2 in Fig. 2.3 does not offer any connectivity services. The use of our specialized point symbols helps in identifying the type of layer.

An important feature of many layers is that they support layer interfaces to more than one using layer and use layer interfaces of more than one layer, as depicted in Fig. 2.4. This structure defines six layers.

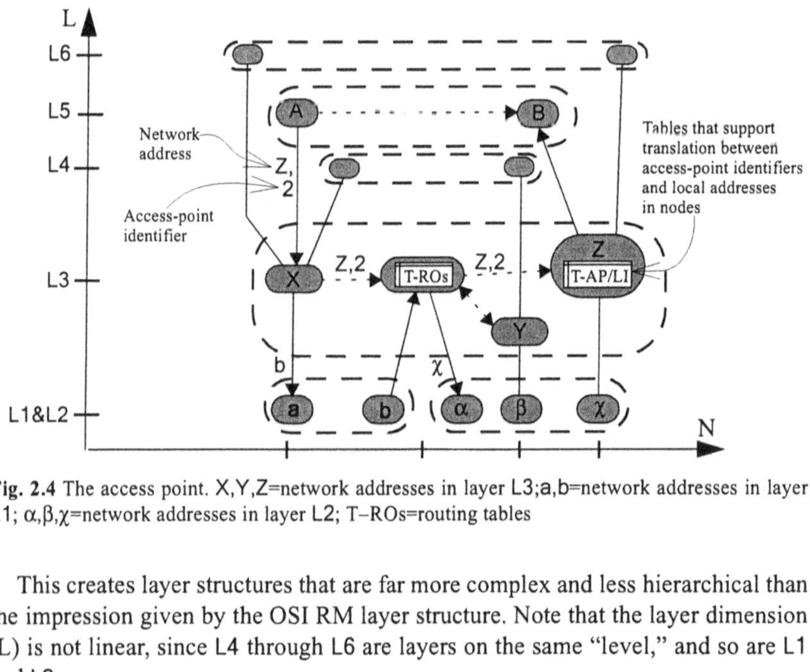

Fig. 2.4 The access point. X,Y,Z=network addresses in layer L3;a,b=network addresses in layer L1; α,β,χ=network addresses in layer L2; T–ROs=routing tables

This creates layer structures that are far more complex and less hierarchical than the impression given by the OSI RM layer structure. Note that the layer dimension (L) is not linear, since L4 through L6 are layers on the same “level,” and so are L1 and L2.

When a layer provides multiple layer interfaces by the same layer element (e.g., X and Z in Fig. 2.4), messages will include an **access-point identifier** (APi)[6] that refers a using layer element. The terminating layer element will translate the APi value to a local address[7] that identifies the using layer element. In this regard, the

[6] Note that there is no consensus in existing standards on this term: OSI RM and LAPD call it “service access point identifier” (SAPI), IP calls it “protocol number,” TCP calls it “port number,” MTP calls it “service indicator” (SI), SCCP calls it “subsystem number” (SSN), etc.

terminating element must manage tables (T-AP/LI) that translate APi values to local addresses. The scenario in Fig. 2.4 tells the following story:

- A layer element A of layer L5 sends a message to layer element B. This part of the scenario is described in the protocol specification of L5.
- In reality, A uses the connectivity services of layer L3 to transport the message data to B. L3 is a layer that provides multiple layer interfaces at its layer elements. A must therefore refer the layer interface to B (identified here by access-point identifier 2), when communicating with its L3 element (X).
- However, access-point identifiers, by definition, are not layer-wide addresses. They are only used by a layer element (in this scenario by Z) to *discriminate* between using layer elements, once the data has arrived at a terminating layer element. The same APi value may be handled in several or all L3 layer elements. In some network systems (e.g., in LAPD, "link access procedure on the D-channel," part of ISDN), the same APi values are assigned to all using layer elements of the same layer, but this is not necessarily a rule.
- Thus, access-point identifiers are not enough to determine the receiving user layer element (B). The layer element (Z) that uses the APi for finding the actual layer interface must also be addressed by a **network address** (Z). Thus, A orders its L3 element (X) to send the message data to network address Z and access-point 2, which together constitute a unique identification of B.

The scenario also indicates how layer L3 (a switching layer) operates:

- The L3 element X has no direct connection to Z. It has, however, access to a layer L1 that can transport data to different L1 network addresses, amongst them b, that is an L1-layer element used by an L3-routing-layer element. Thus X creates an L3 message to the routing element. This message contains, besides the data from A, the L3 network address Z and the access-point identifier 2. X then orders its L1-layer element (a) to send this message data to address b.
- The L3 routing element is a layer element that uses multiple layer interfaces, each interface belonging to a separate layer (L1 and L2). The routing element reads the network address Z and performs a routing analysis, using **routing tables** (T–ROs in Fig. 2.4; the suffix "s" indicates that this is a routing table in a switching layer[8]). The purpose of this analysis is to identify which layer (L1 or L2) to use for further transport, and (in case) which network address and access point of that layer to refer. In this case it selects L2 and orders its L2-layer element (α) to send the message data to network address χ.

---

[7] By "local" we mean an address that exists in the (derived) node where the layer element runs. In such a node, a processing platform exists to support communication between adjacent layer elements through local addresses.

[8] In Chap. 3 and Appendix C, we define a number of AMLn tables that can be used for specifying network configuration properties. The T-ROs, T-AP and T-LI in Fig. 2.4 are such tables.

- When Z receives the data from χ, it interprets the message from the routing element. It first checks the network address. Since it is its own address, it knows that it must terminate the data, i.e., hand it over to one of its users. It reads the access-point identifier 2 and uses some address translation tables (T-AP/LI in Fig. 2.4) to identify the local address of B. It then sends the data to B, which ends this scenario.

Let's summarize these findings:

- The internal structure of a layer is an intralayer structure, showing layer elements, which protocols they handle, and which layer interfaces they provide and use.
- Layer elements of different layers communicate over layer interfaces. There are basically two types of interfaces: interfaces that offer some type of connectivity services (i.e., transporting data between two using layer elements) and interfaces that offer some other type of service (e.g., setting up connections, translating addresses, accessing information, controlling a network resource, etc.). Actual layer interfaces may be one or the other, or a mix of both.
- The functionality of a network system can be described as an interlayer structure. Some layers are connectivity layers, others are control layers.
- A layer element can offer connectivity services to several layer elements by identifying them with different access-point identifiers in messages.
- A routing- and switching-layer element can make use of more than one layer interface that offers connectivity services. The process of transferring a message that has arrived over one used layer interface to another is commonly called *relaying*. If an analysis of a network address in the message is needed, it is commonly called *routing analysis* and the relaying activity *switching*.
- When a layer element of a switching layer receives a message that refers its own network address, it *terminates* the message.

In network system standards, a protocol specification is normally associated with a single layer. According to our understanding of layer interface, if messages of the protocol contains an access-point identifier, those layer elements that are referred to by that identifier belong to a layer other than the one defined by the protocol. However, if we analyze existing ITU–T protocols with respect to access-point identifier, we find that it is more the rule than an exception that a protocol specification comprises several layers. One example is the LAPD protocol, which is described in a single protocol specification. Figure 2.5 shows the interlayer structure that this standard reveals.

From the AMLn point of view, the LAPD protocol specification does not describe a single layer, but a **stratum** that comprises two separate layers (L1 and L2 in the model). The reason is that the LAPD-transport layer performs discrimination among 64 possible layer interfaces, identified by an access-point identifier called SAPI. The protocols of these layers depend only on the properties of the service in the layer interface, i.e., they are independent of how the LAPD transport protocol looks.

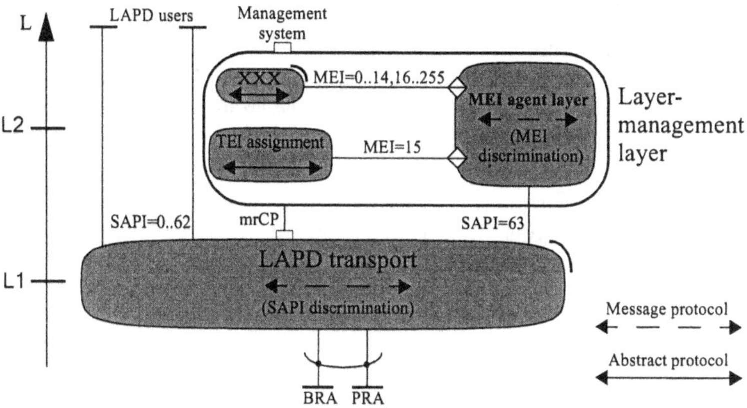

Fig. 2.5 Layers that are identified by access-point identifiers (defined as SAPI in LAPD) and actor-layer identifiers (defined as "management entity identifier" (MEI) in LAPD)

A second level of function separation is performed within the layer-management layer (see Fig. 2.5).[9] This layer defines a single message protocol and up to 256 different management functions that can operate simultaneously, independent of each other. The model separates the MEI agent layer from the management functions themselves. This separation is an example of **actor–agent separation**: all management functions rely on the same message protocol; an **actor-layer identifier** (here MEI) in the layer protocol provides discrimination between management functions (i.e., between **actor layers**). Note that although the LAPD standard defines 256 possible actor layers, so far it only describes one: a function called "TEI assignment," (identified by MEI=15).

In the OSI RM tradition, layer management is regarded as part of the layer that is managed, which is why many ITU–T protocol specifications are very complex and comprise several layers. However, since the protocols of the layer-management layer are specified independently from the LAPD transport protocol, it is better (both from a modelling and design maintenance point of view) to always separate the specification of a managed layer from specifications of the layers that manage its resources. This separation of functions is supported by the **management dimension** (M) and the **plane** concept defined in AMLn. Read more about this in Chap. 5.

Therefore, when we discuss "protocols" in this book, we refer to layers in that sense. Furthermore, we will make a distinction between **message protocols** (such as the ones that describe the LAPD transport layer and the MEI agent layer), and

[9] This layer performs an autonomous management function (AMF) over ISDN accesses. It is therefore part of the ISDN management **plane**. Note, however that it exists in the ISDN traffic **system**.

abstract protocols (such as the ones that describe communication inside the actor layers of the LAPD layer-management layer).

A well-formed layer normally has a single, well-defined purpose. The LAPD transport layer in Fig. 2.5 is an example of such a layer: its only purpose is to transport data frames between other layer elements in ISDN terminals and in ISDN local exchanges. This basic function can be defined by a single protocol. However, there are innumerous situations where a layer must be described by several protocols, in particular for layers that perform some type of control function in a network. A typical layer of the kind is the call-handling layer in ISDN. Figure 2.6 gives an example of the intralayer structure for such a layer.

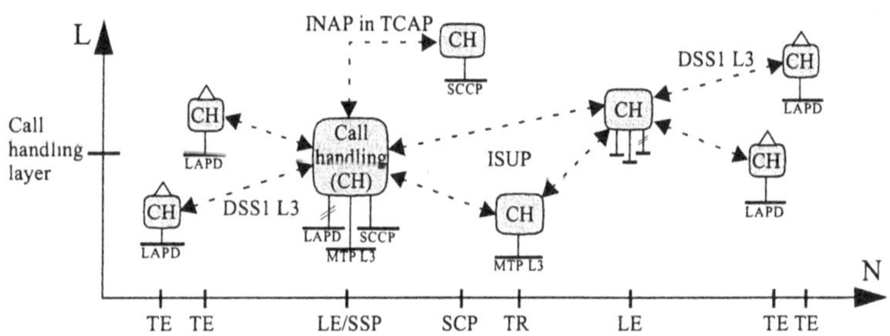

Fig. 2.6 The call-handling layer in an ISDN. LE=local exchange; N=node dimension; SCP,SSP=node names defined in the "intelligent network" (IN) concept; TE=terminal; TR=transit exchange

This layer is defined by at least three different protocol specifications: DSS1 L,3 ISUP (OSI type of message protocols), and INAP (an abstract protocol). Contrary to the LAPD standard, where a single protocol specification defines many layers, in this case we have a single layer that is described by many protocol specifications.

Note, however, that there exists no specification in ITU–T standards of a network-wide "call-handling layer," as depicted in Fig. 2.6. This brings us to the question of how to identify layers out of the mass of protocol specification documents that are published by ITU–T (and other standardization bodies). The key to this and other problems in modeling networks is to distinguish between the *message handling* that a layer element performs and the *actions* it takes as a result of receiving correct messages. This is the actor–agent separation principle (see Sect. 2.3).

2.1.2 Service Types and Layer Types

In the OSI RM as well as in AMLn, services are offered over layer interfaces. The OSI RM is, however, somewhat vague as regards the relation between services and layers. The basic view that OSI RM takes through its seven-layer structure is that a layer provides some kind of connectivity service, which in AMLn terms means that its layer interfaces represent different types of connectivity service points (...SP). A basic distinction has to be made between synchronous data streams (such as streams supported by circuit-switching layers) and asynchronous data streams (such as streams supported by packet-switching layers). More detailed distinctions are the transfer rate, length of packages, and the quality of service (QoS). In AMLn, a connectivity service always defines how to transfer bits in some type of structure: bitstream, octets (or "bytes"), frames, superframes, packages, etc. The OSI RM does, however, also define connectivity services that transfer things such as "abstract data types." In AMLn we handle these types of interfaces differently (see Sect. 2.3).

Users who are connected to connectivity service points can in AMLn only *send* and *receive* data, while in the OSI RM they can do other things as well, such as requesting connections, resetting connections, etc. In the OSI RM-application layer there also exists many functions that do things other than offering connectivity services. For example, the ACSE function provides a user interface that offers the establishment of associations, the FTAM provides a user interface for accessing globally distributed files, and the ITU–T X.500 directory provides a user interface that offers address translations. Since these interfaces exist above layer 6, but are not intended for end-users, the OSI RM sometimes classifies them as "sublayers in layer 7." The OSI RM makes no distinction between these types of services and connectivity services in models, i.e., all layers and sublayers are modeled the same way. Only the names of layers give a hint of what they are doing (if you know what the name implies).

AMLn corrects this, and also eliminates the need for "sublayers," by separating services in two main classes: connectivity services and control services, the former offered over ...SPs, the latter over some type of control points (...CPs). To support this distinction in models, AMLn introduces different graphical notations for connectivity service points and control points, as discussed in the previous chapter.

A result of the AMLn approach is that models will exhibit two types of layers as well: connectivity layers and control layers. In layer-structure models the type of a layer becomes very obvious when the point symbols for these layer-interface types are used. Figure 2.7 compares the OSI RM layer concept with those for AMLn. Note the differences in AMLn model structures when a layer supports...SPs and ...CPs respectively:

- A connectivity layer has a symmetric appearance in models, similar to an OSI RM layer, since there are two or more users who together use a service. The difference is that the ...SP layer interface only comprises send and receive operations while the OSI RM layer interface may also define operations for making the service accessible (by connectRequest operations, etc.). The protocol is characterized as a **connectivity protocol**.
- A control layer has an asymmetric appearance in models. A service is used by a single user over an ...CP. Other layer elements may not have any control points because they represent remote resources or control functions of some kind. The

operations of a control point can be *anything but* send and receive. The protocols are characterized as **control protocols**.

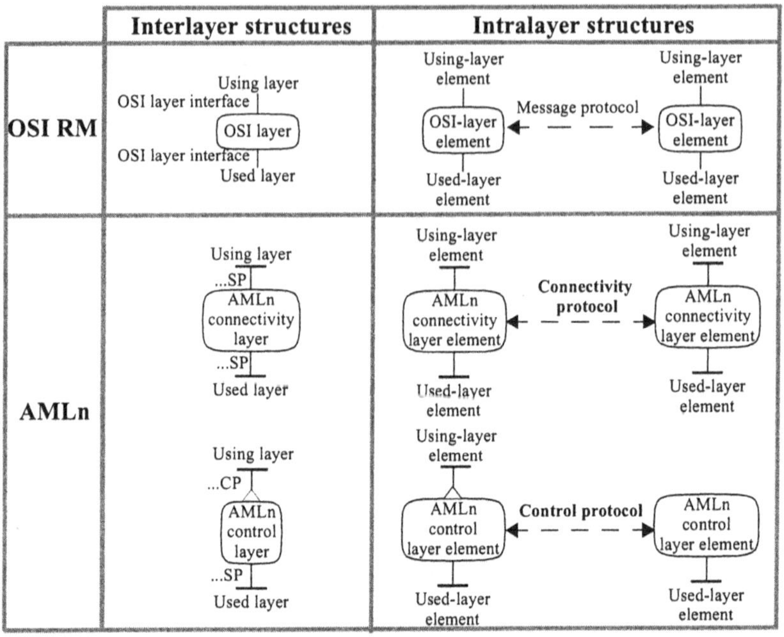

Fig. 2.7 Difference between AMLn layers and OSI RM layers

In case a layer offers both a control point and a service point over the same layer interface (as connection-oriented packet-switching networks do) the combined ...SP and ...CP notation may be used, as defined in Fig. 2.2.

2.2 Discrimination (Connectivity Layers Only)

2.2.1 Introduction

Besides offering send and receive operations and a certain QoS, connectivity-layer elements also have another important function: to *discriminate* between multiple provided layer interfaces. Such discrimination functions must exist in terminating-layer elements that support multiple users.

The OSI RM is defined for packet-switched services only. It defines two principles for discrimination in connectivity layers: either by some kind of **connection endpoint identifier** (CEi) in layer interfaces and a **connection identifier** (Ci) in the protocol, or by some type of network-wide layer-element identifier. Such an identifier can comprise either a **network address** (NA) only, or an access-point identifier (APi) only.

The Internet (another packet-switched network system), on the other hand, combines discrimination by NA and APi in the same connectivity layer (the IP layer). Furthermore, the Internet also supports discrimination by **sockets** (a kind of "connections" offered by TCP and UDP) by combining the NA of a lower layer (IP) with an APi of a higher layer (TCP or UDP). Other forms of discrimination exist as well, as described in the following sections.

2.2.2 Discrimination in the OSI RM

In the OSI RM, the way the other layer element is identified in send and receive operations is used to characterize the class of service as either connection-less (CNL or "datagram") or connection-oriented (CON). The CON-type of service implies that before packets can be sent and received, the network must establish an end-to-end "connection," which in a layer interface is identified by a CEi.[10] Strictly speaking, the CON–CNL classification applies only to packet-switching networks. Although "circuits" in circuit-switched networks have similar properties and establishment procedures as the OSI "connection," layer elements use no identifiers at all at sending and receiving. We will, in general, use the term CON both in the OSI RM sense and to characterize circuit switched connectivity services.

The way OSI-layer elements that use a connectivity service can be structured internally depends on whether a CON or a CNL service is used (see Fig. 2.8).

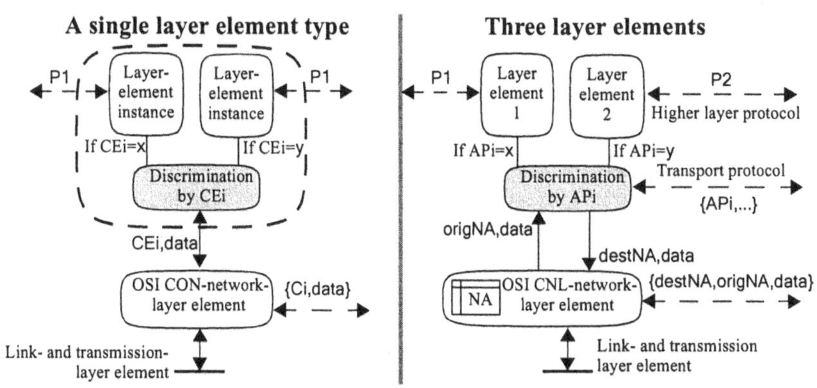

Fig. 2.8 Structuring OSI-layer elements depends on the type of connectivity service. APi=access-point identifier; CEi=connection-endpoint identifier; Ci=connection identifier; CNL=connection-less; CON=connection-oriented; dest...=destination ...; NA=network address; orig...=originating...; P...=protocols

[10]In the telephony world, the expression "on-demand service" is often used when a user must order the connectivity service before he can use it.

The OSI RM network layer that offers CON services (to the left) terminates multiple channels, with each channel identified by a connection identifier (Ci) in the message. The network layer delivers the user data, together with a (local) connection-endpoint identifier (CEi) value, to the layer element, which may use this feature to distribute different streams to *instances* of the layer element (or many streams to the same instance). There must therefore exist a special element that interfaces the CON server and performs **discrimination by CEi**. In case CEis are established on-demand, this element can only support instances, i.e., elements that operate according to *the same protocol* (P1 in the figure). Note that the discrimination element *need not handle any protocol* for being able to perform its task, which creates a semantic contradiction: if the interface between the discrimination element and the OSI CON layer is a layer interface, the discrimination element cannot be part of the layer (by OSI RM definition) since it handles no part of the using layer's protocol. In AMLn, this contradiction is resolved by the actor–agent separation (see Fig. 2.47).

The OSI RM network layer that offers CNL services (to the right in Fig. 2.8) performs no discrimination. It terminates a single stream of messages, where each message includes a *destination* network address (destNA) and an *originating* network address (origNA). Discrimination of using layer elements must therefore rely on APi, as depicted. Since the network layer should not know anything about the layer structure above, there must exist a special element that interfaces the CNL server and performs **discrimination by APi**. Contrary to the CEi discriminator, this element must handle a protocol that transfers access-point identifiers between terminating layer elements. This is done by the transport-layer protocol in the OSI RM. Contrary to the CON case, the using layer elements can operate according *to different protocols* (P1 and P2 in the figure). Thus, these elements can be parts of different layers that can exist simultaneously in the network.

Layers that use a layer offering discrimination by APi can use this service in two principle ways:

1. All layer elements of the same layer are associated with the same APi value. We say that such an APi is a **layer APi**.
2. Layer elements of the same layer can be associated to different APi-value. Consequently, such an APi is a **layer-element APi**.

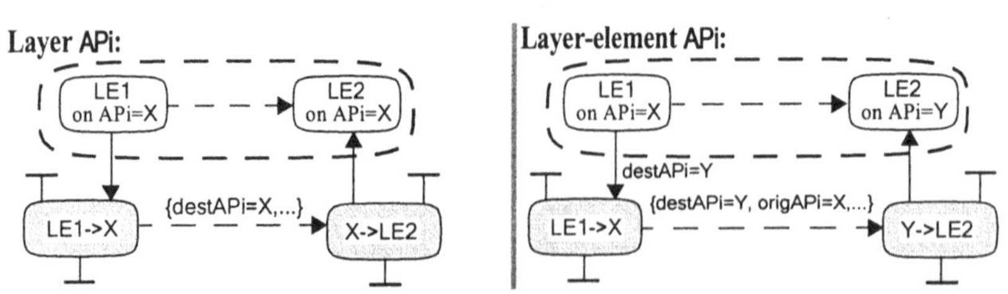

Fig. 2.9 Two approaches of using access-point identifiers (APi). LE...=layer element; X,Y=access-point identifiers

Figure 2.9 models implications of these approaches:

- In case of **layer APi**, the association between a particular layer element and the APi value (by which it is referred to in messages) is a value that is defined by the discriminating layer. This approach can be used in network systems where there are few using layers which all are part of the same network system. Examples of such discriminating layers are the IP layer, the SCCP layer (in SS7) and LAPD. Discriminating layers that support layer APi are recognized by the fact that they define APi values in their protocol specifications, and that the APi value-range is rather small.
  A benefit with this approach is that a using layer element need not refer the APi of the destination element when requesting the discriminating layer to send something. This is because the association between a layer element and its APi value is set in every node when the actual using layer is first installed in the network. Furthermore, the discrimination protocol need not identify sender's APi.
- In case of **layer-element APi**, only the range of the APi parameter is defined by the discriminating layer. The actual APi values are defined by a separate specification that belongs to the network system as a whole (e.g., for the Internet, the IANA publishes a separate list of "port numbers," which is the APi that TCP/UDP uses for discrimination). This approach is used when there can be a large and variable number of using layers that are regarded as applications of a network system. Layer elements of the same using layer can be installed at different points in time, and can therefore be associated to any idle APi value. Examples of such discriminating layers are TCP and UDP. SCCP also allows this mode of operation.
  The layer-element APi approach is more complex than the layer APi method in that a sending using element must know the destination APi, and the discrimination protocol must include both the originating and the destination APi.

2.2.3 Discrimination in the Internet

The IP layer (the Internet "network service") offers a datagram service that performs discrimination according to the layer-APi method, based on an APi that is called "protocol number" (PRN). The IP layer can therefore, contrary to the OSI RM network layer, support multiple "transport layers." Most well-known are the TCP (a CON service) and UDP (a CNL service).

Furthermore, the Internet layer structure supports discrimination by APi recursively since TCP/UDP also performs discrimination, according to the layer-element APi method, however. The APi is called "port number" (PON). Application layers (e.g., the "simple message transfer protocol" layer (SMTP)) can use TCP and UDP services by the same port numbers (see Fig. 2.10).

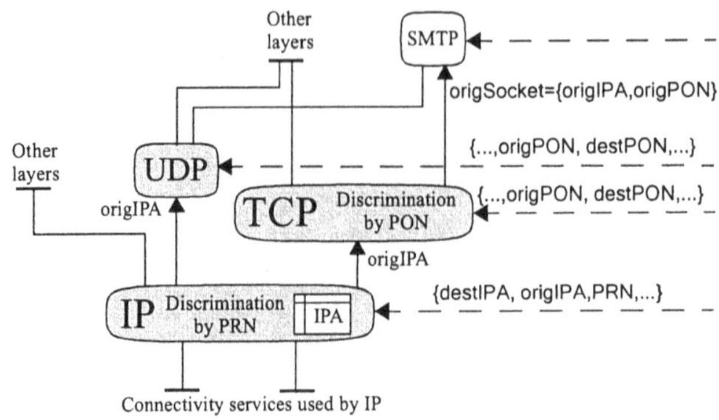

Fig. 2.10 Discrimination offered by TCP,UDP/IP. dest...=destination...; IPA=IP address; orig...=originating...; PON=port number; PRN=protocol number

TCP offers a connection-oriented service. The TCP correspondence to the OSI CEi is called "socket." A socket is identified by an IP network address (IPA) and a TCP/UDP port number, which implies that users of different sockets can handle different protocols, i.e., they are not instances, as in the CON case of the OSI RM. We call this **discrimination by sockets**.

2.2.4 Discrimination by Multiple Network Addresses

Network addresses (NA) are used inside connectivity layers primarily for routing. That implies that when a message has reached its terminating layer element in the layer (which is a layer element that listens to a particular destination NA), the destination NA has normally ceased to play a role (see the scenario in Fig. 2.4).

Thus, destination NAs are normally not used for discrimination (neither in the OSI RM nor in the Internet). In circuit-switching networks, terminating nodes are sometimes assigned multiple NAs. For example, in PSTN and ISDN, terminating nodes (called "end systems") of the "private automatic branch exchange" type (PABX) may support hundreds to thousands of terminals, each having its own ISDN network address. How these are connected to the ISDN is of no concern to the ISDN. Since these NAs are also used for routing inside the ISDN, this is not a discrimination case, but a case of *interworking* between two separate networks (an ISDN and a PABX) within the same NA numbering system (see Fig. 2.11). NAs within a particular PABX are assigned within a particular series number, which is the part of an NA used for routing inside the ISDN.

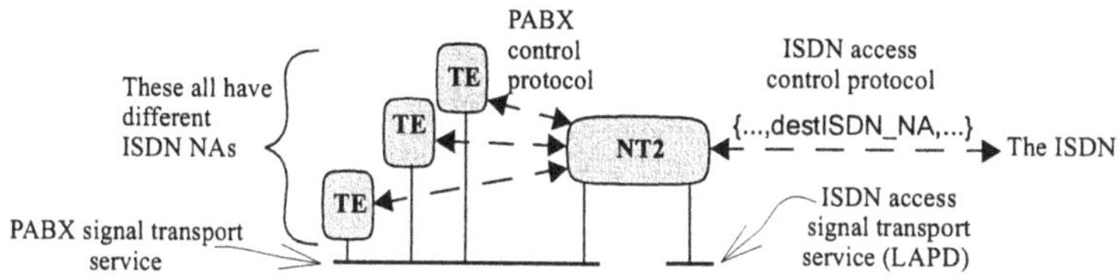

Fig. 2.11 Supporting multiple network addresses through interworking

Nevertheless, individual NAs could be used for discrimination as well. The hypothetical cases of assigning multiple NAs to terminating layer elements in IP and X.25 networks (in order to use them for discrimination) are depicted in Fig. 2.12.

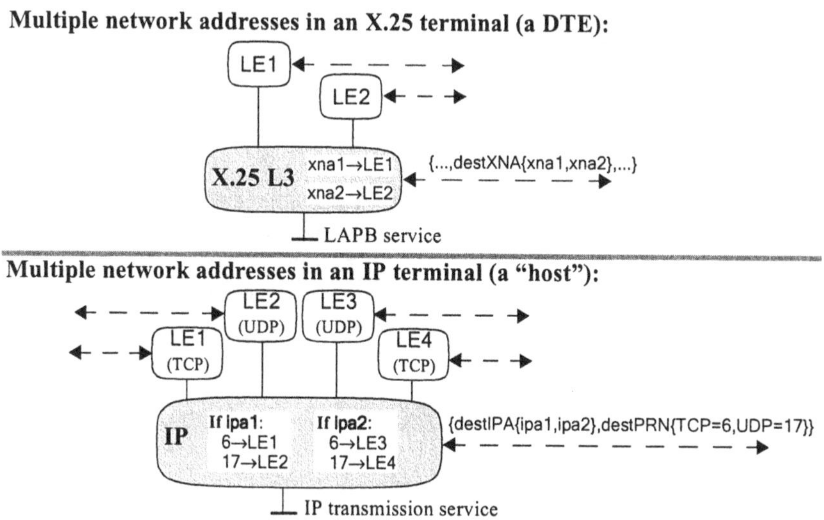

Fig. 2.12 Discrimination by multiple network addresses. XNA=X.25 network address; IPA=IP network address

Note, however, that from all aspects except for routing, switching, and transmission, layer elements in X.25 terminals and IP hosts that rely on different NAs belong to completely separated networks.

2.2.5 Discrimination in Circuit-Switching Layers

Discrimination by a terminating layer element can also be provided without using any particular identifier. This is specially so in circuit-switching layers. Figure 2.13 depicts how termination is performed in an ISDN terminal of the BRA type.

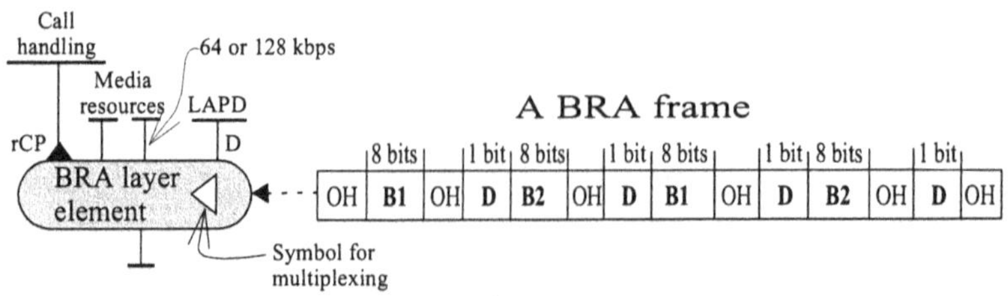

Fig. 2.13 Discrimination in the ISDN circuit-switching layer. B1,B2=bits of B-channels; D=bits of the D-channel; OH=overhead bits

Synchronous frames, as depicted in Fig 2.13, arrive continuously from the network. The bits in this frame represent three channels: D (16 kbps), B1, and B2 (each 64 kbps). Since the number of bits of respective channels is the same in all frames, and always have the same position, there is no need for any special parameter to identify each channel. We call this **discrimination by position**.

The D channel terminates in the LAPD-layer element, while B channels are used primarily for creating media services, and are therefore terminated in diverse media layer elements (e.g., a voice A/D, which is an analog↔digital voice converter that requires 64 kbps (B1 or B2 channel)).

Since a transmission rate of 128 kbps is available, media services that require more than 64 kbps can be supported as well. The discrimination function in the layer element is therefore dynamically controlled from the call-handling layer over a resource control point (rCP). Over this control point, the call-handling layer tells the BRA if 64 or 128 kbps is required, and over which media layer interface to terminate the bits. If 128 kbps is required, the layer element multiplexes both B channels over the actual layer interface.

The same principle is used for discrimination in ISDN terminals of the PRA type. The difference is that there are 30 B channels in that case, which can support media services that require up to 2 Mbps by multiplexing in the PRA layer element.

2.2.6 Summary

The following five forms of discrimination have been discussed:

1. **Discrimination by CEi and Ci,** which (in the OSI RM) supports instantiation on the network level and requires no extra protocol.
2. **Discrimination by APi,** which supports concurrent layers but requires an extra protocol. Two ways of using APi for discrimination have been identified: the **layer APi method** and the **layer-element APi method**.
3. **Discrimination by NA,** which is feasible but rare and does not support instantiation.
4. **Discrimination by sockets,** which (in case of TCP) supports both CON-service properties and concurrent layers, but not instantiation.

5. **Discrimination by position**, applied primarily in circuit-switching layers.

Be aware that, in standardized network systems, the distinction between these forms of discrimination is not always explicitly defined. Furthermore, terminating elements may handle other forms as well. Sometimes a single, very complex identifier is used for several forms of discrimination. An exceptional example is the parameter "payload type identification parameter" (PTI) in ATM messages, described in ITU–T I.361 (see Sect. 6.4.3 for an explanation). This parameter defines four types of discrimination in a single parameter of three bits. Such smart-coded parameters may save some bits in transmission, but they also cause severe interpretation problems on the part of the human interpreter.

2.3 Agents and Actors

2.3.1 Introduction

As a general principle (in AMLn as well in most supplier solutions), a layer element can always be separated into one or several **actor** elements and one or several (protocol) **agent** elements. In Fig. 2.14 a layer element of a simple control layer is depicted in this fashion.

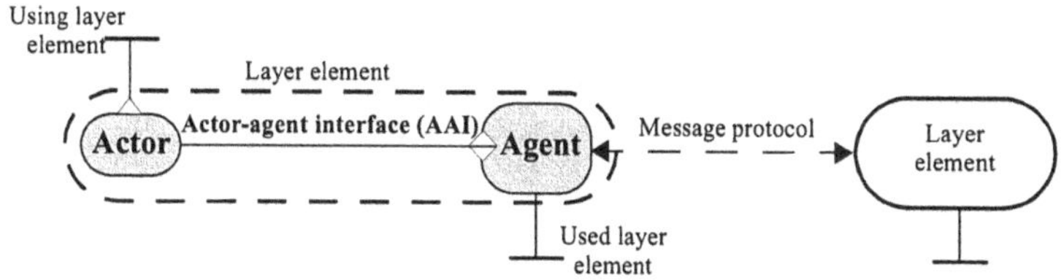

Fig. 2.14 Every layer element can be separated into one (or several) agents and one (or several) actors

The interface between an actor and an agent is called **actor–agent interface** (AAI) and is invisible outside the very layer element. We use a special graphical symbol for this interface in order to distinguish it from layer interfaces.

An agent element deals only with message reception and sending, which primarily implies formatting and encoding. Consequently, agents are the elements that use layer interfaces offering connectivity services. An agent may also deal with other message-oriented functions, such as message error control, segmentation, flow control, etc. Beyond that, an agent does nothing but informing its actor (over the AAI) about what a received message tells (in case of control layers) or hand some user data to the actor (in case of connectivity layers). Since this interface is local within the layer element, any suitable language can be used inside a layer element. Thus, *which language actors and agents use over an AAI is independent of the format and encoding of a received or sent message*, i.e., actors are not concerned with what messages look like.

- An actor in a connectivity layer gets user data over the AAI. It does not interpret the data. It only decides on whether to terminate the data or relay it (over the same or another AAI).
- An actor in a control layer takes decisions (based on information in received messages) about whether to accept the information, perform some action, and (in case) inform its peer layer elements about its actions through information in sent messages. This implies that such actors are the *elements that can change the state* of the layer (agents cannot, provided that they behave properly).

These are the most important reasons for separating actors from agents. An actor may also handle layer interfaces to layer service users (if there are any).

In most cases, standards do not define actor–agent interfaces. For example, every implementor designs his own solutions to actor–agent interfaces for TUP, ISUP, and DSS1 L3. One outstanding exception is the use of TCAP[11] agents. Figure 2.15 shows a scenario of how TCAP is used in PSTN/ISDN for the IN actors SCP and SSP. Note that we call an actor that generates an event **generator** and the actor that handles the event **detector**. The identity of generators and detectors must be separated from other identifications (such as network addresses and access-point identifiers), since a layer element may include more than one actor.

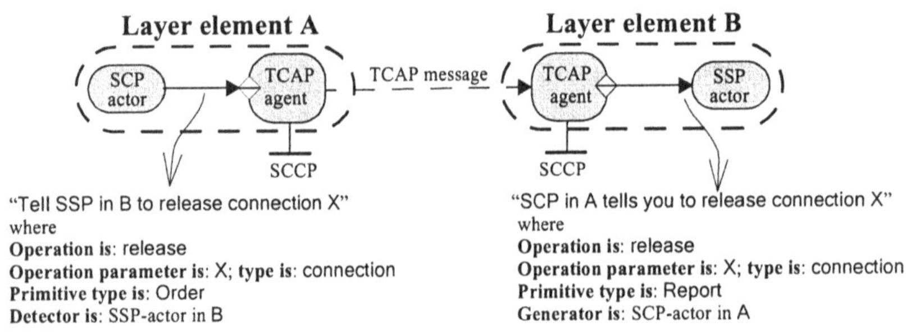

Fig. 2.15 The use of TCAP in PSTN/ISDN. SCP and SSP are (derived) nodes of the IN architecture, not to be confused with the AMLn "control point" and "service point"

The scenario shows the SCP actor *ordering* its agent to "tell SSP in B to release connection X" in any suitable language used in layer element A. Its TCAP agent translates the order to information in a TCAP message. On the SSP side, the TCAP agent interprets the received message and translates it to a *report* to its SSP actor that says "SCP in A tells you to release the connection X," using any suitable lan-

[11] "Transaction capabilities application part". TCAP has evolved from the ACSE, ROSE, and ASN.1 standards of the OSI application layer. TCAP relies on SCCP for connectivity. SCCP is a kind of transport layer in SS7.

guage in B. The scenario also indicates some of the parameters that must be included in the TCAP message for this scenario to succeed: an **operation name**, relevant **operation parameters**, a **primitive type** parameter, a **detecting actor address** and possibly a **generating actor address** (we also say that an operation name concatenated with a primitive type is a **primitive name**).

The actor–agent separation principles offers an alternative to how to specify protocols:

- The method applied in most existing standards is to produce a single protocol specification that describes how an observer who stands in between the layer elements would interpret the messages, and which conclusions he would draw about what the actors do. Such protocol specifications always tend to be very voluminous and complex, and therefore difficult to interpret.
- The actor–agent separation principle can be applied for all protocols. It separates a conventional control protocol specification into one part that describes the communication between actors, one part that describes the communication between agents, and one (language-dependent) part that describes how the events in an actor–agent interface are mapped on messages. For example, the communication between actors in the above scenario says simply "release connection X." We regard such an expression as an event in the abstract protocol between actors. The corresponding TCAP message (part of the TCAP message protocol) is obviously considerably more complex.[12]

Note that the effect of any separation of layer elements into actors and agents is that actor–actor communication becomes *transparent* to all communication problems, since agents handle both the messages and the layer interfaces to used connectivity services. The only thing a generating actor must do is to deliver the specification of an actor–actor event and a detector address to its agent. Thus, such an event becomes mapped from one AAI to another, as depicted in Fig. 2.16.

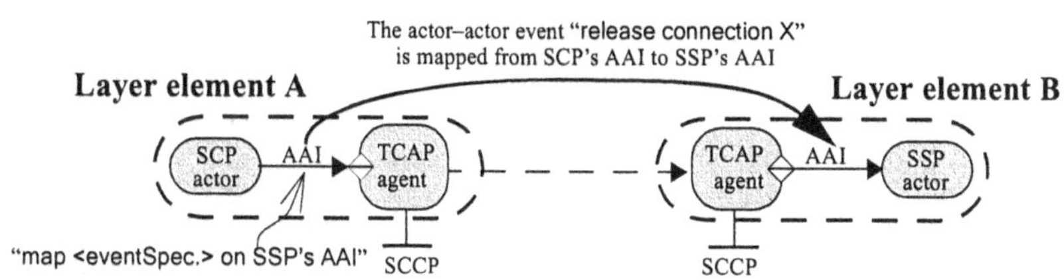

Fig. 2.16 The role of agents

[12]Some abstract protocols that rely on TCAP are INAP (the abstract protocol between IN actors), MAP (an abstract protocol in GSM), and OMAP (an abstract management protocol in PSTN/ISDN).

Note that agents do not need to know anything of what the actor–actor event is about. Therefore, the actor–agent event (in an AAI) can be conceptually expressed as "map <eventSpecification> on the AAI of <detecting actor address>", as described in Fig. 2.16. This expression is obviously valid for any type of actor–actor event, as long as the actor and the agent have a common language for specifying such events and a common system for identifying actors. This fact gives the opportunity to separate the protocol specification between A and B into the three specification components discussed previously. We will discuss these components in detail in the following sections, but separated into connectivity and control layers since there are essential differences as to how AAIs look like and what actors do in these layer types.

2.3.2 Agents and Actors in Control Layers

2.3.2.1 A Case Study

We will base this discussion on an analysis of a small part of the DSS1 L3 protocol (part of the call-handling layer depicted in Fig. 2.6), to see how an actor–agent separation affects an OSI RM protocol specification.[13]

A protocol defines a number of **message types** as parts of the specification. A message type is a structured data type, which means that it defines a structure of parameters and their types, value range, and how to encode them into bits. One of the messages in DSS1 L3 is called SETUP. This message is the first that is sent between an ISDN terminal (TE) and a local exchange (LE) of the network when either the terminal requests a service from the local exchange (normally an "outgoing call"), or the local exchange offers a service (e.g., an "incoming call") from another terminal to (one of) its own terminals. The existing protocol specification defines both the message type and the actions that are expected from the LE or the terminal when receiving a SETUP. Figure 2.17 shows the reception of a SETUP at an LE and a very small part of what the DSS1 L3 specification says about that (also slightly transcribed by the author).

We are only scratching the surface of the message type specifications of DSS1 L3 here (more than 400 pages). As a matter of fact, the bulk of the DSS1 L3 specification deals with specifying all the details of messages. The actions associated to the SETUP message in Fig. 2.17 was constructed by two text fragments that were found (through tedious readings) in different places in the protocol specification.

[13]Since the OSI RM does not define the actor–agent separation in layer elements, we regard as "OSI layer" any layer that does *not* define this distinction.

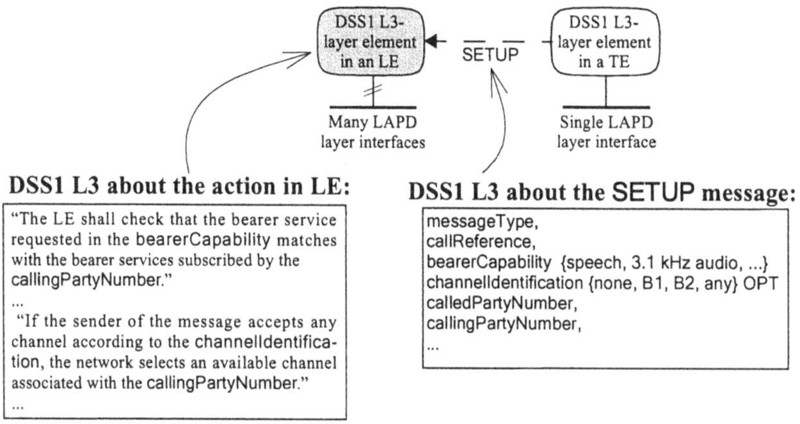

Fig. 2.17 A small part of the DSS1 L3 protocol specification

The action part obviously describes what the actor in the layer element does, while the message description defines implicitly what the agent must do, as modeled in Fig. 2.18.

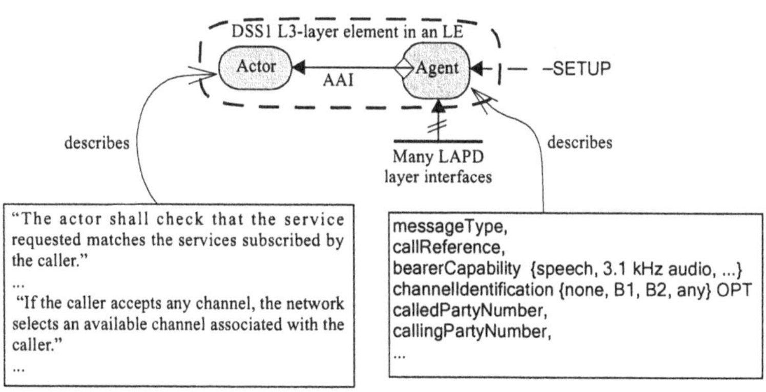

Fig. 2.18 The actor and the agent in the DSS1 L3-layer element in a local exchange (LE)

The scenario for handling a SETUP reception in this model is divided into two description parts:

1. The bits of the message arrive at the agent over a layer interface of a LAPD layer. The agent must identify all the parameters of the SETUP, using the encoding rules defined by the protocol specification. Each value will also be represented as a local data-type value, according to the local processing environment of the layer element. The agent must then check the validity of the message. For example, some parameters are mandatory, some are optional, some may have a

default value associated. Other things to check is that all values are within defined value ranges. Any errors or inconsistencies in these respects will not start any other action in the layer element than a negative response message (DISCONNECT) to the calling terminal, i.e., the actor will not at all be involved.
2. If the message is valid, the agent delivers parameters of the actor–actor event to the actor, which takes actions according to the description to the left.

These actions depend of course on the values that were carried by the SETUP, but in no way on parameter names in messages, the message format and the encoding of parameters, and how the message-handling process deals with its problem. In Fig. 2.18 we have used this insight to simplify the actor description, primarily by removing references to names of SETUP parameters from the description.

Note that through the actor–agent separation, the actor part description holds for a variety of modifications of the message protocol (handled by the agent). For example, other parameter names or another set of encoding rules would do as well. If it was necessary to separate SETUP messages into several submessages, it would not bother the actor, as long as the agent reassembles submessages to a complete actor–actor event before contacting the actor.

If we read the actor description carefully, it implies that the actor has access to **resources** that are of no concern to the agent. For example, the actor description implies that it uses some kind of register that defines which services subscribers are subscribed to. It also implies that the actor relies on a register of idle B channels.[14] Other resources that are controlled by the actor in an LE are switching elements, for example. Note, however, that how we describe these resource in a model in no way affects the description in the DSS1 L3 protocol specification of what the actor does.

In the OSI RM, resources are not explicitly modeled. In general, however, resources on which an actor relies may be distributed or exist remotely to the actor because they may be used by many other layer elements and layers as well. There must therefore exist special agents and protocols for resource access. For example, let's assume that we allocate subscriber data in a remote database, common for all DSS1 L3 actors in the network. The DSS1 L3 actor may get access to such data over a special resource access protocol, as depicted in Fig. 2.19.

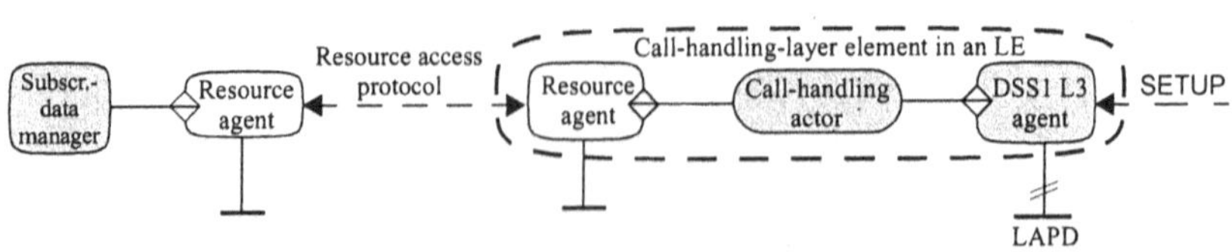

Fig. 2.19 One method for providing remote resource access to actors in control layers

[14]These are channels that carry user data between a terminal and its LE, once a path is established between the called and calling parties.

Note that this is an example where a layer may define several protocols for different purposes. The depicted actor is now more than just a "DSS1 L3 actor." It is an actor of the call-handling layer, which is a layer that comprises DSS1 L3 agents, resource agents, call-handling actors, and a remote subscriber-data manager. This emphasizes the importance of separating actors from agents.

There are several disadvantages of making resources integrated parts of a layer as in Fig. 2.19. Among others, the using actor must know if a particular resource is available locally or remotely. Resource configuration also becomes an issue for the very layer since a resource using actor must know addresses to remote resources. In addition, if the actor–agent interface is not standardized, every modification of the access protocol will affect the call-handling layer element.

A more general solution to resource access is therefore to provide local interfaces to all kinds of resources. We model such interfaces as layer interfaces of the type **resource control point** (rCP). In Fig. 2.20, the previous model is redrawn with all resources that the actor needs accessible over rCPs.

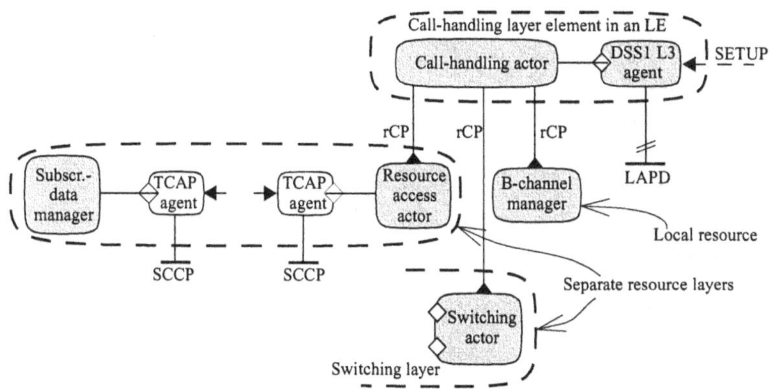

Fig. 2.20 Transparent resource modeling using resource control points (rCP)

The protocol for accessing subscriber data needs an asynchronous connectivity service that is accessible in all nodes of an ISDN. We have therefore assumed that TCAP is used (over SCCP). The DSS1 L3 agent is the part that interfaces connectivity layer interfaces (to different LAPD layers here), which is the only resource this agent needs for its operation. The actor interfaces all other types of (local and remote) resources that are needed for call handling.

By using rCPs, the call-handling actor is (in principle) completely independent of if resources exist locally or remote, provided that rCPs are standardized and preserved. The rCP notation supports modeling both local and remote access. For example, the B-channel manager is drawn as a single functional element that has no associated intralayer structure because we define it as local resource, i.e., it must exist in the same node as the call-handling actor. We may later add an intralayer structure to this element which (in principle) is transparent to the this actor. The other resources are drawn as parts of resource layers, however. Such layers will show up in the layer structure of the network system as layers separated from (in this case) the ISDN call-handling layer.

Let's now turn to the behavior and functions of the actors. As we have seen, the agents perform a generic functions that is independent of what the actors do, i.e., of the semantics of the actor–actor communication. The function of actors in a control layer can generally be described as *controlling the state of the layer*, which is related to the resources that are implied. For example, before the SETUP arrived, there was no call established through the network. This state was represented by the fact that the corresponding B channels managed by the B-channel manager were idle, and that no paths existed through the switching actor for the call requested by the SETUP. As a result of SETUP reception, the call-handling actor will change this situation: a B channel is now made busy and a path through the switching actor is established. This is interpreted as a new **layer state** that was realized by, and will be supervised by, the actor of the call-handling layer element.

A state machine is a way of describing a behavior.[15] We therefore say that an actor of a control layer handles an **element state machine** (ESM) that is virtually contained in the actor. Actor–actor communication can therefore be described as communication between ESMs of the layer. We say that the system of such state machines in a layer is a **layer state machine** (LSM) (see Fig. 2.21 for our DSS1 L3 example).

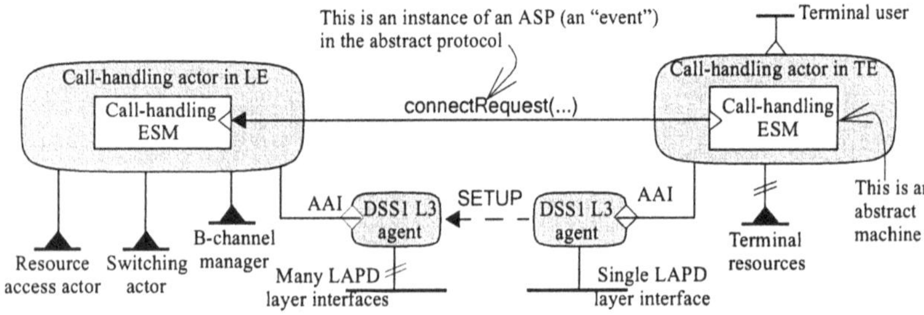

Fig. 2.21 Actors contain element state machines (ESM) which communicate by instances of abstract service primitives (events)

Note that since the actor also has functions that can handle actor–agent interfaces, we must make a distinction between the actor and the state machine it "contains." Any system of state machines can, by using AMLs (see Appendix D or Muth (2001)), be described as an **abstract machine system** that defines state machines as **abstract machines** which communicate by instances of abstract service primitives (ASP; we frequently call an instance of an ASP **event**). When we

[15]We assume that the reader has a general understanding of what a state machine is. If not, see Muth (2001) that gives an in-depth presentation of the state-machine concept in AMLs.

describe the communication between abstract machine in terms of ASPs, we describe an **abstract protocol**.

The ESMs are also the elements that handle resources. Since communication over rCPs is independent of DSS1 L3 agents (and the AAIs) as well as how resources are distributed, an ESM sees resources as (local) abstract machines only, as depicted in Fig. 2.22. Note that the ESM in a TE sees the terminal user as an abstract machine as well

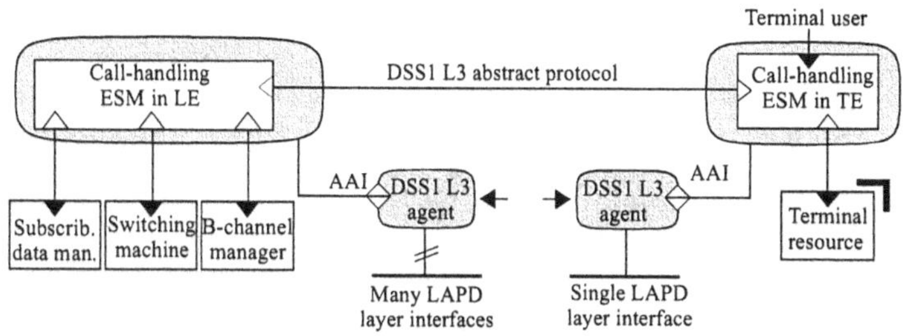

Fig. 2.22 An ESM sees resources as abstract machines

By separating an ESM from its "containment" in an actor, as depicted in Fig. 2.23, we actually define a layer in two separate models: an LSM and a **layer protocol machine** (LPM), related through containment relations that we call **ESM_in**.

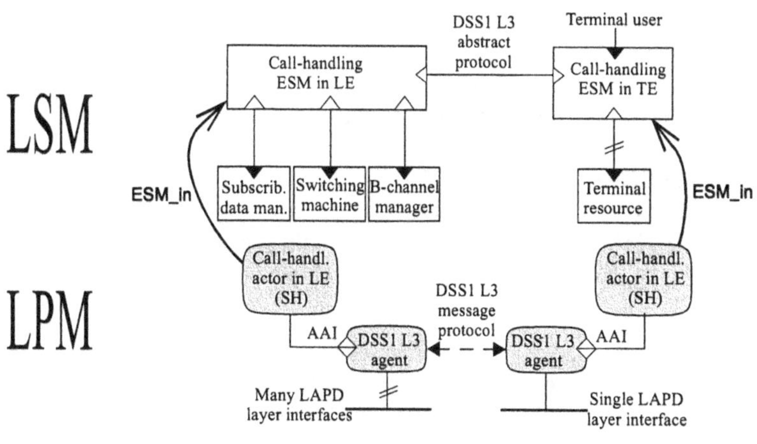

Fig. 2.23 Layer state machines (LSM, an abstract machine system) are contained in layer protocol machines (LPM). SH=state-machine handler

The DSS1 L3 LPM is a machine that supports distribution of the DSS1 L3 LSM in the network. The LPM can be designed so that it becomes independent of what the LSM is about, i.e., all functions that are unique to the DSS1 L3 layer are then defined by the LSM.

Let's therefore analyse how the DSS1 L3 protocol can be described on the (much simpler) LSM level, where interactions are in terms of events instead of message exchange. In Fig. 2.24 we compare the message exchange between DSS1 L3 agents with the events between DSS1 L3 ESMs for the most common messages that are used in the setup procedure for an originating call.

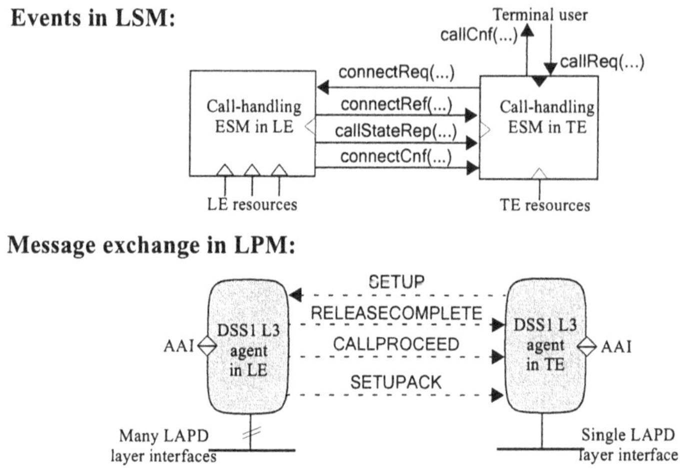

Fig. 2.24 The translation between LSM–ASPs and messages in DSS1 L3

These messages are:

- SETUP from the TE that requests a call. A lot of message parameters are included, e.g., the network address of the called party (called calledPartyNumber, see Fig. 2.18).
- RELEASECOMPLETE from the LE that informs the TE that the request cannot be fulfilled (parameters are included that tell why)
- CALLPROCEED from the LE that informs the TE that the request has been accepted but the call is not yet established end-to-end.
- SETUPACK from the LE that informs the TE that a call connection now exist end-to-end, i.e., users can start talking.

In general, there is no guarantee that a message type corresponds exactly to an LSM–ASP (which is one important reason to separate LSMs from LPMs). However, in the DSS1 L3 protocol there is (more or less) a one-to-one correspondence between messages and ASPs.

There exist a number of ASP notations but none are standardized. We therefore use the one defined by AMLs. In AMLs, ASPs that belong to the same operation are named with the same operation name (connect here) and given a primitive type denotation that tell whether it is a request (e.g., connectReq), a confirmation of that the request is performed (e.g., connectCnf) or a refusal to perform the request (e.g., connectRef). These ASPs correspond to the messages SETUP, SETUPACK and RELEASECOMPLETE. The CALLPROCEED corresponds to an ASP that reports the actual state of the call, and is therefore denoted as the primitive type of Report of the operation callState, i.e., callStateRep.

A specification of the LSM for a layer is considerably simpler, more intelligible and more valuable than a conventional protocol specification. This is because all message-handling functions are excluded, and because it gives a clear picture of the essential states and state transitions of a layer, i.e., the behavior in the layer. This picture is not blurred by the often large number of LPM states, since these are not visible to the LSM (if the LPM is correctly designed). Figure 2.25 gives an indication of how the abstract protocol for the LSM in Fig. 2.24 can be described.

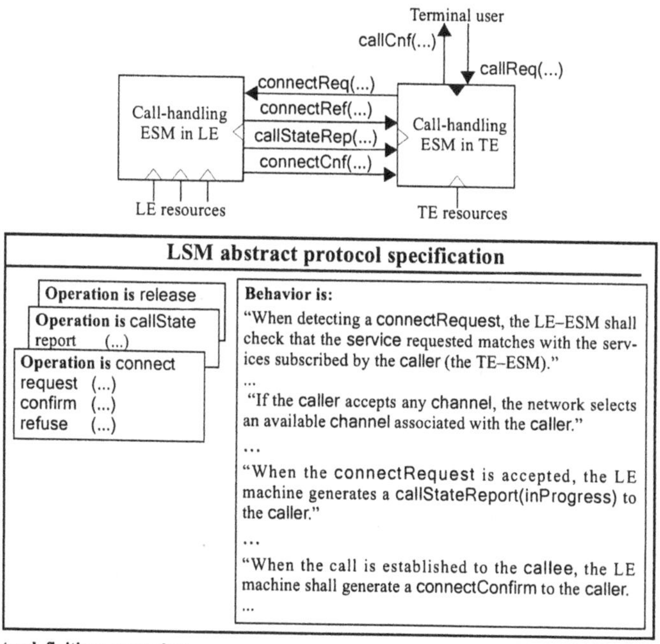

Parameter definitions, examples:

connectRequest(call TYPE, service TYPE, caller TYPE, callee TYPE, channel TYPE, ...)
connectConfirm(call TYPE, callState{inCall(1)} TYPE)
callStateReport(call TYPE, callState{inProgress (2)} TYPE)

Fig. 2.25 The abstract protocol is described by an AMLs model (an abstract-machine system), a number of operation specifications, and a behavior specification (textual here, or graphical as a state diagram)

The abstract protocol specification consists of, besides the model, a number of **operation specifications** and a **behavior specification**:

- Each **operation specification** defines one or several ASPs and their parameters. Each parameter has a name (that is not necessarily the same as the corresponding message–parameter names) and a value range. It also has a type designation which defines how it shall be encoded in messages. Different type and encoding systems may be used, e.g. ASN.1 ("abstract syntax notation no 1") and BER ("basic encoding rules"). You can read more about operation specifications in Muth (2001).
- The **behavior specification** in Fig. 2.25 is extracted by textual fragments from the DSS1 L3 standard. It is also transcribed so that it refers ASPs and ASP parameters instead of messages and message parameters. This makes it impossible to confuse and mix LSM behavior with LPM behavior (as in conventional standards).
  In this case study we have (more or less) used the existing standard text fragments. In reality we would have produced a state diagram representation instead, possibly annotated with text fragments where needed since behavior descriptions on the LSM level are considerably simpler than when LSM and LPM behavior are mixed (in which case the state diagram often becomes too complex to be of any value).
  Within the context of a standard, resource control points (rCP) are normally not specified.[16] Implementors can, however, use this model as a basis for their implementations by adding rCP specifications and refine the behavior specification into complete ESM specifications, from which implementations of ESMs can be generated. You can read more about this in Muth (2001).

Let's now turn to the LPM specification and its relation to the LSM. Assuming that the LSM abstract protocol is specified as described, the only thing the LPM has to do is to create messages for the ASPs defined by the abstract protocol, and to interface the connectivity service, which is the LAPD service. Since all ASPs are (presumably) built in the same way, a general format for messages that carry ASP information may look like Fig. 2.26. In general, an LPM may also perform other functions except mapping ASPs on messages, such as message error control, fragmentation, and flow control (but not in DSS1 L3). There may therefore exist messages as well that do not carry ASPs.

The translation (or "mapping") between the ASPs of the LSM and the messages of the LPM is performed by the DSS1 L3 agents. It must therefore be described as part of the LPM specification. Compared to the existing standard, the LPM specification becomes rather simple because all ASP parameters are defined and typed in the LSM specification. The LPM specification need therefore only comprise the depicted formats, encoding algorithms and LPM-specific functions and behaviors.

[16]The exception is when a resource is represented by a resource layer, e.g., as directory resources are in the Internet architecture (DNS) and in OSI networks (the X.500 directory).

Messages that carry ASP information:

Message type	Detector reference (optional)	Generator reference (optional)	Object identifier (optional)	Event identifier (optional)	LSM event information			LPM-specific parameters (if any)
ASP					Operation name	Primitive type	ASP parameters	

Example: {...,...,...,...,...., connect(encoded), Request (encoded), 3.1.kHz voice (encoded), 1234566 (calledPartyNumber encoded), etc., ...}

Messages that carry LPM information only:

Message Type	LPM-specific parameters for the message type
notASP	

Fig. 2.26 General formats for messages when a layer is separated into an LSM and an LPM

Since the existing DSS1 L3 standard does not define the LSM, no general approach to message formatting, data typing, and encoding has been used. Message description and encoding to the bit-level is therefore defined uniquely for a large number of message types to a great length. Also, as a result of the absence of an LSM description, there exist message types that represent several ASPs, depending on some parameter value in the message, or on who sends the message (the TE or the LE). Such a mapping dissonance makes it harder to understand and maintain the standard and complicates agent functions.

Let's now take a closer look at the actor–agent interface (AAI) for control layers. We saw that an LSM can have an existence only if it is contained in an LPM and if there exist an AAI in the actual layer element. We also saw that, in natural language, events in an AAI could be expressed as map <LSMeventSpecification> on the AAI of <detecting actor identifier>. An obvious observation is therefore that *AAI-event types are ASPs that carry the ASPs of the LSM*. We say that ASPs in such AAIs are **meta-primitives**, i.e., primitives that carry other primitives (a common term for meta-primitives in some standards is invoke). In AMLn only two such meta-primitives are needed: the **r-invokeOrder** and the **l-invokeReport** (**r-invOrd** and **l-invRep** for short), see Fig. 2.27. Since these primitives are associated with control layers only, we refer to them as the primitives of a **specified agent interface** (SAG).

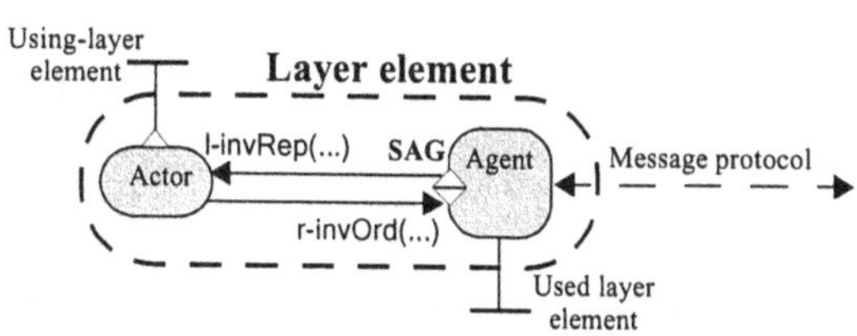

Fig. 2.27 Meta-primitives in a specified agent interfaces (SAG) carry LSM primitive information

The r-invokeOrder(...) is used by an actor to invoke its agent in order to invoke a remote actor with an LSM-event. The l-invokeOrder(...) is used by an agent to invoke a local actor in order to make it act on an LSM event that was generated by a remote actor. The parameter field of these meta-primitives are identical and are (more or less) given by the general message format in Fig. 2.26, which means it consists of all or some of the following parameters: **detector reference** (optional), **invocation identifier** (optional), **object identifier** (optional), **operation name**, **primitive type** and **ASP parameters**. A more detailed and formal specification of these meta-primitives is given in Appendix B.

To model a SAG when specifying control layer protocols is always an excellent conceptual tool for making such specifications reasonably small and intelligible. It does not, however, mean that SAGs must be standardized. It all depends on what purpose we intend to use it for:

1. An obvious and very important purpose (considering how standards are written today) is to use it for simplifying the specification of an OSI layer, as described previously. If that is the only objective there is no need to standardize SAG primitives, which implies that they will be realized by each implementor in his or her own way.
   An implementor who separates actors and agents in different modules may choose to use some local message passing mechanism for communication over SAGs. In that case, different local messages will be defined to represent the two invoke types, or an operational approach may be used that for each invoke type defines a library procedure.
2. In any case, such a SAG is not reusable for any other actor than the one it was designed for. Furthermore, it is also not possible to simulate layer behavior, unless implemented agents are available. An alternative is therefore to define a standardized set of SAG primitives that can be used for all control layers, such as the AMLn r-invOrd and l-invRep.

SAG primitives have been defined only for a few control-layer applications. One (the ROSE, an OSI RM "sublayer") is used for OSI RM applications only, another (the TCAP, part of SS7) is used for some PSTN/ISDN control layers only (e.g., INAP, MAP, and OMAP), and again another (the GIOP) is used for distributed object systems on CORBA platforms. ROSE, TCAP, and GIOP are not compatible, however. Thus, there exist no standardized definition of SAG primitives that can be used for all network systems.

As we understand by now, a prerequisite for such a standard is that all control layers define the same set of parameters for LSM events (which of course implies that they specify the LSM of the layer in the first place). ROSE and TCAP define a kind of invokes that contain similar parameters as the AMLn r-invOrd and l-invRep, with one important exception, however: the AMLn parameter *primitive type* is not a parameter in the invokes, but used to create a specific invoke for every possible ROSE/TCAP-defined primitive type (as a matter of fact, ROSE defines nine SAG primitives, TCAP defines 17 through its rather strange construction, and AMLn copes with two). This is because ROSE/TCAP confuses LSM primitives with SAG primitives, thereby creating an unfortunate and unnecessary tight coupling between an LSM and the LPM on which it relies (in addition to unnecessarily complex SAGs).

2.3.2.2
Agent Layers and Actor Layers

The distinction between the LSM and the LPM representation of a control layer offers an excellent possibility to separate control structures from connectivity structures in layer structures. The key to this modeling approach is to focus on the actor-agent interfaces instead of on layer interfaces. This allows us to define **agent layers** separately from **actor layers**, as depicted in Fig. 2.28 for the DSS1 L3 layer. We define an agent layer as a system of agents that communicate by a message protocol, and interface one or several connectivity strata (in this case LAPD strata).

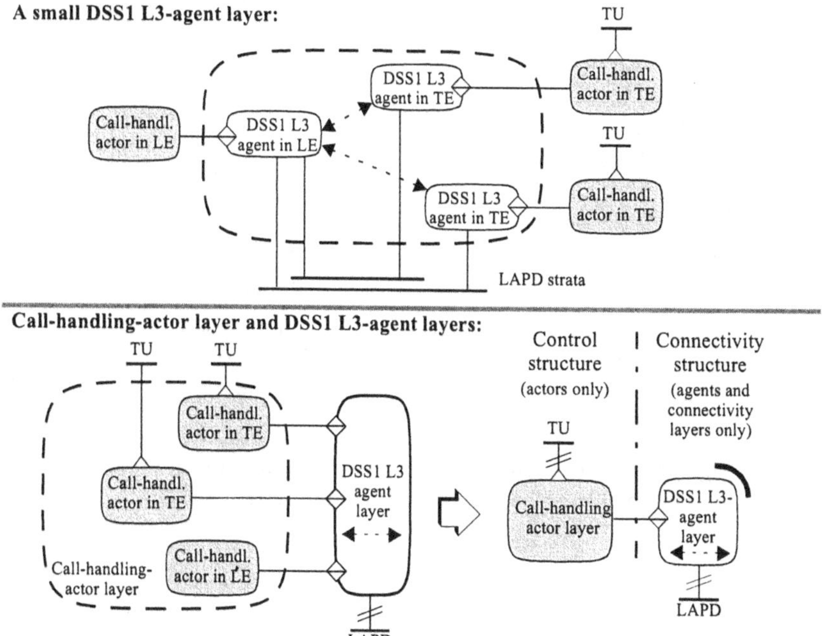

Fig. 2.28 The actor layer and agent layer representation of DSS1 L3. TU=terminal user

The three DSS1 L3 agents together constitute a DSS L3 agent layer in this case. The set of actors belong to the actor layer. To the lower right a model that separates the call-handling layer into its control structure and connectivity structures. This model hides the distribution of agents and actors completely, and is therefore valid for any node structure of an ISDN.

The benefit of the agent layer notation is that all LPM aspects (i.e., message protocol) are confined to an agent layer, and all LSM aspects to an actor layer. The relation between the actor layer and the agent layers on which it relies are fully defined by defining AAIs as SAGs. As we know, this implies that modifications of the agent protocol, as well as other changes of the layer structure on the connectivity-structure side are transparent to an actor layer, as long as the SAG remains the same.

The separation of actor layers from agent layers shows a property for control layers that cannot be made visible in a layer structure that rely on the OSI RM layer concept: the distinction between a control structure and a connectivity structure model. We can define the control structure that is realized by an actor layer by extracting the LSM out of the call-handling-actor layer, as we have done in Fig. 2.29.

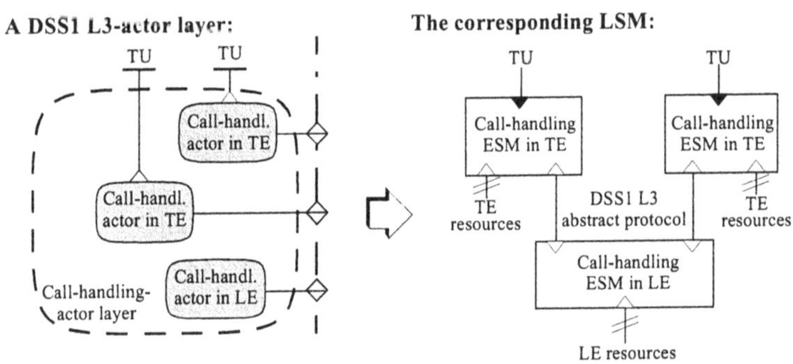

Fig. 2.29 The actor layer and its LSM representation for a small DSS1 L3 layer

Note that DSS1 L3 defines a single abstract protocol. In general, however, an LSM model may include several abstract protocols. As a matter of fact, the LSM model in Fig. 2.29 is just a small part of the complete LSM model for call handling in ISDN. A more complete model is discussed in Sect. 2.3.2.4 (see Fig. 2.37 in particular).

2.3.2.3
Common Agent Layers

The DSS1 L3 standard does not define a SAG for the AAI between its actors and agent layer, since the agent layer is not designed for being used by anything other than a single DSS1 L3-actor layer. In case we want to design an agent layer that can support multiple actor layers, we call it **common agent layer** (which is a system of **common agents**). One prerequisite for such a layer is that it specifies a SAG, as for ROSE, TCAP, and GIOP. Another is that actors can identify each other over the network in which the common agent layer runs. In the DSS1 L3 case, this was not an issue (see the upper model in Fig. 2.28): every TE is connected semi-permanently over its own LAPD stratum to the LE and therefore need not use any kind of identifier. An LE just chooses the LAPD stratum that also serves the actual TE. The same goes for all kinds of link layers.

For a common agent layer, actor identification is an issue, however. This is due to:

1. Common agents will most likely run on switching networks, which requires that an agent must know the network address of the agent that handles the AAI to the wanted actor. If one wants to avoid making actors dependent on where they are allocated in a network, an actor identification system is needed.
2. Most likely we want a common agent layer to be able to support many independent actor layers simultaneously. In that case we do not want an actor to communicate with an actor that belongs to another actor layer. Obviously, this requires actors to be identified with respect to which actor layer they belong.
3. A common agent layer may be used for simple request–reply types of interactions between actors. Often, however, actors may be engaged in longer sessions, in which case it is favorable (for many reasons, authentication being one) if they can refer to some type of session identifier, as soon as a session is established. We use a terminology inherited from the OSI RM that calls such sessions **associations**, identified by **association identifiers** (ASi).

Let's take an example: we want to design an OSI layer that supports multiple actor layers. Figure 2.30 shows the common agent layer and two actor layers, ALi=1 and ALi=2 in their LSM representation.

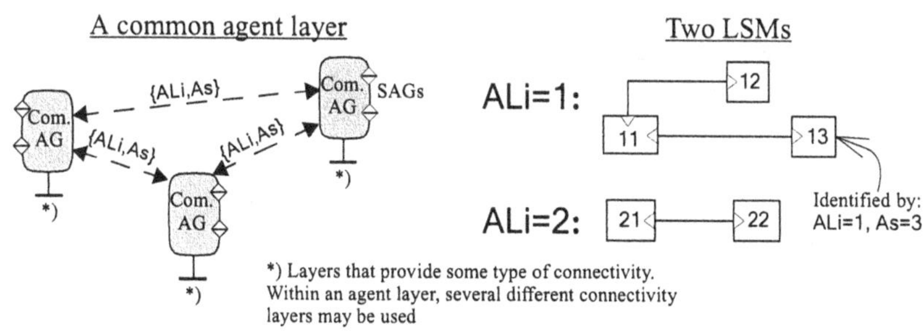

Fig. 2.30 An OSI layer can contain many actor layers if based on a common agent layer. ALi=actor layer identifier, As=actor suffix (identifies an actor within an actor layer)

We expect that the agent layer can support a large number of actor layers and protect actors of different layers to accidentally accessing each other. We also assume so far that the agent layer does not support associations (i.e., actors can communicate by request–reply only).

Thus, every actor in this layer must have a unique identity, which must also help the common agent layer prevent actors from communicating over actor layer boundaries. An **actor address** (AA) will therefore be built by two parameters: an **actor-layer identifier** (ALi) and an **actor suffix** (As). By explicitly defining actor-layer identifiers, actor suffixes can be assigned individually for each actor layer.

We can now model the OSI layer that is built by the common agent layer and the two actor layers (see Fig. 2.31).

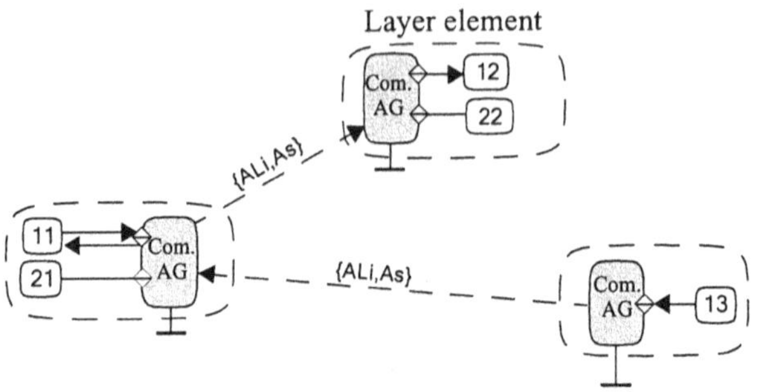

Fig. 2.31 An OSI layer built by a common agent layer that supports several separate actor layers

The way a common agent operates under these circumstances is depicted by the scenario in Fig. 2.32. Here we have also made the reasonable assumption that there exist only one actor per actor layer in each layer element (if there are more than one, the actor suffixes of detectors and generators must be given in messages as well).

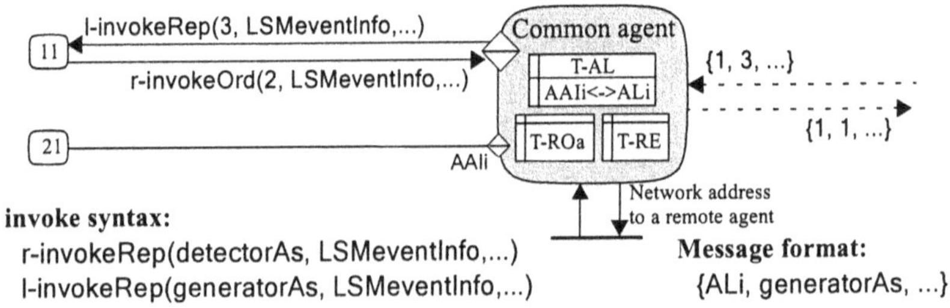

Fig. 2.32 Identifying actors when a common agent layer is used

The model shows one of the common agents from Fig. 2.31, and the actors it connects. We assume that when an actor is initially configurated to an actor–agent interface, the agent relates the identity of this interface (a local identifier called **actor-agent interface identifier** (AAli)) to the name of the actor layer to which the actor belongs. This is done in a table (called T-AL) that associates an AAli to an ALi. Another part of the configuration activity may be to store information about which remote agents will be addressed for each remote actor of interest to the agent's own actors. By storing this information, actors do not need to know network addresses to remote agents, which protects actors from configuration changes on the network level. This information is stored in two tables, called T-ROa and T-RE in models, where T-ROa is used to translate an actor address to a **route endpoint identifier**

(REi), and T–RE to translate an REi to a network address. You can read more about these identifiers and tables in Chap. 3 and Appendix C.

The *message-reception* scenario of Fig. 2.32 is as follows:

- The agent receives a message from actor 13.
- The agent looks up its T-AL table and finds the AAI in its own layer element that connects an actor of actor layer 1 (which is the actor 11 in the model). It generates an l-invokeRep in that AAI that includes the generating actor suffix 3.

The *message-sending* scenario is as follows:

- As a result of detecting the LSM event from actor 13, actor 11 decides to request something from actor 12. It generates an r-invokeOrd, that refers the actor suffix 2 as the detecting actors identity.
- The agent looks up its T-AL table and finds that the actor layer identifier of 11 is 1, and therefore also the actor layer to which 2 belongs.
- The agent then takes the actor address 12 and finds, through the T-ROa and T-RE tables, how to address the remote agent. The outgoing message contains the generating actor identity, i.e. 1.

The operations of this common agent layer is rather simple, mainly because it allows only one actor per actor layer in a layer element and because it supports only simple request–reply interactions between actors. This works fine as long as all actors are always available, honest, work for free, and operate according to mutually agreed abstract protocols. In case this cannot be assumed, the common agent layer might require that an **association** (in TCAP referred to as a "dialogue") is first established between the actors, before they can start using r-invokes and l-invokes in their AAIs. An "association" is a communication relation between two actors that can imply a number of things:

1. The called actor will be responsive during the whole session.
2. The common-agent layer guarantees that the called actor is the one called, i.e., it belongs to the same actor layer and has the actor suffix given by the calling actor.
3. The calling actor is the one it claims to be (authentication).
4. The two actors are functionally compatible, i.e., the abstract protocol that the initiating ESM handles is supported by the other ESM as well.
5. A charging procedure has started.

Associations rely on "connections" that connect common agents to each other. Associations must, however, not be confused with such connections:

- Associations connect actors, connections connect agents
- A connection between two common agents may be used for several associations simultaneously (implied by the model in Fig. 2.31)
- Connections are used for message transfer (an LPM activity), associations for LSM event generation an detection

If a common agent layer will support associations, we must add additional ASPs to the specified agent interface (SAG). These ASPs are used to request, manage, and release associations through the common agent layer. They are not meta-primitives since they do not carry any LSM events. Since an association ties up resources both in the agent layer and in the actors, association handling comprises an establishment phase and a release phase, similar to connection handling.

In order to understand which operations are needed in a SAG to support associations, we show in Fig. 2.33 a scenario in terms of a sequence diagram, where an association is used between actors 11 and 12. An association is identified by an association identifier (ASi). In this scenario we only included requirements according to bullets 1 and 2 above. Note that layer models should indicate if a SAG includes association handling or not. We therefore insert a black dot in the actor–agent interface symbol to indicate that the SAG, besides invokes, also defines association-handling operation.

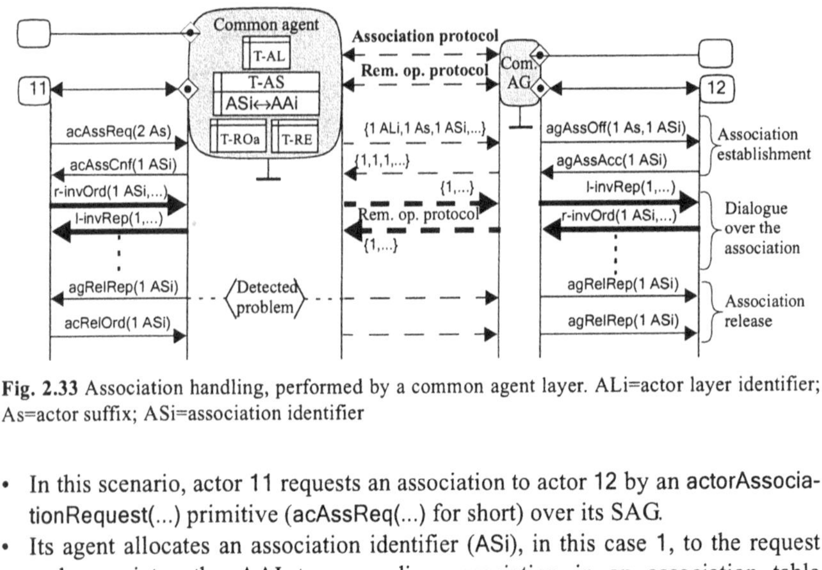

Fig. 2.33 Association handling, performed by a common agent layer. ALi=actor layer identifier; As=actor suffix; ASi=association identifier

- In this scenario, actor 11 requests an association to actor 12 by an actorAssociationRequest(...) primitive (acAssReq(...) for short) over its SAG.
- Its agent allocates an association identifier (ASi), in this case 1, to the request and associates the AAI to a pending association in an association table (ASi↔AAIi). It then sends an association request message to the remote agent that manages actor 12. This message is defined in the **association protocol** part of the agent layer protocol.
- The remote agent looks up (in its T-AL table) which AAI connects the element 12 (if any) and offers this actor, in an agentAssociationOffer(1 As,1 ASi) primitive (agAssOff(...) for short), the association to actor 11. If actor 12 accepts the offer it answers with an agAssAcc(1 ASi) primitive. The establishment phase then ends with an acAssCnf(1 ASi) primitive to actor 11.

- From here on, actors 11 and 12 invoke each other by r-invokeOrd(1 ASi,...) and l-invokeRep(1 ASi,...) which are mapped on messages of the **remote operation protocol,** which is the other part of the agent layer protocol.
- Before the actors end the dialogue voluntarily, an agent may detect problems that forces it to release the association. These problems could be related to the connectivity layer it uses, local processing problems, actors creating invalid invokes, or a remote agent sending invalid messages. In any case, the agent informs its own actor by an agentReleaseReport(1 ASi) primitive (agRelRep(...) for short) and sends a corresponding message to the remote agent that does the same to its actor.
- Normally (hopefully) the actors release their association voluntarily. The initiating actor does that by an actorReleaseOrder(1 ASi,...) primitive (acRelOrd(...) for short). Its agent sends a message to the remote agent, that informs its actor by an agRelRepReport(1 ASi,...) primitive that the other actor has released the association.

By this scenario we have identified a minimum set of four operations that are needed in a SAG to support associations: acAss (operation class is REQUEST), agAss (operation class is OFFER), agRel (operation class is REPORT), acRel (operation class is ORDER)[17]. The parameters of these operations consists of actor suffixes and association identifiers, as indicated in the scenario. In addition, the acAssReq and agAssOff primitives may include other parameters for negotiation about the abstract protocol version to be used, authentication, and charging of used services. Release primitives may include parameters that explain the reason for releasing the association.

The number and the actual definition of association operations in standards differ a lot, depending on the context. In open network systems, such as the Internet, authentication and charging may be important issues while it is not in closed systems (such as the use of TCAP in ISDN).

2.3.2.4
Modeling the Actor Layers of an OSI Layer

Since almost no network system standard applies the actor–agent separation, we will show how actors and actor layers can be identified out of an OSI-control-layer standard. Let's therefore return to the intralayer structure of the ISDN call-handling layer, modelled as an OSI layer (see Fig. 2.34 that is identical to Fig. 2.6).

[17] REQUEST, OFFER, etc., are names of AMLs operation classes. An operation class defines a particular set of primitive types. For example, REQUEST defines Request, Confirm, and Refuse. Read more about this in Appendix D.

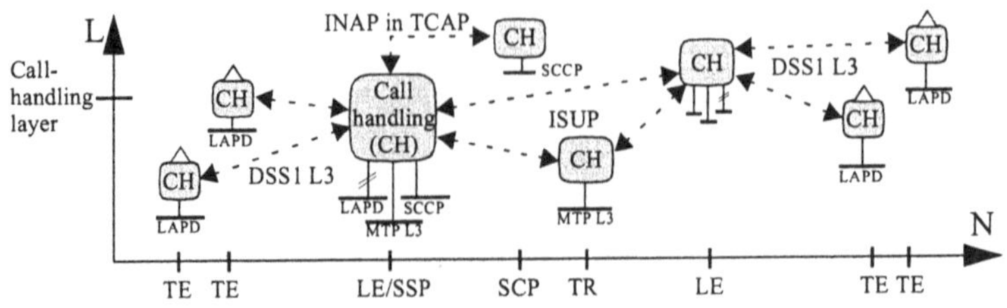

Fig. 2.34 The call-handling layer in an ISDN

Let's now see how we can identify the actor of the layer element in a combined LE/SSP. This layer element handles three different protocols (more than that in reality). By separating the agent part from the actor part for each of them, we can identify the call-handling actor of an LE/SSP (see Fig. 2.35).

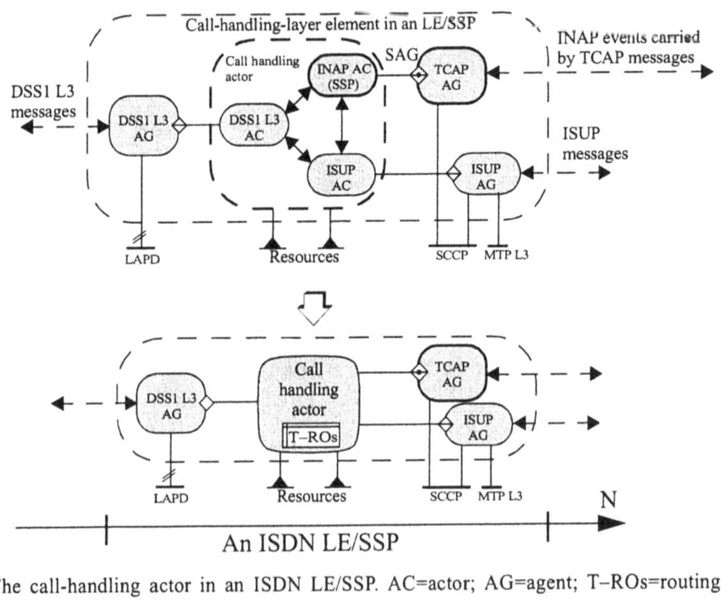

Fig. 2.35 The call-handling actor in an ISDN LE/SSP. AC=actor; AG=agent; T–ROs=routing table for a switching layer

The DSS1 L3 standard describes the DSS1 L3 agent and the call-handling actor as seen in the network interface between a terminal (TE) and its LE. The ISUP standard gives the corresponding view of this actor in the network interface between the LE/SSP and another LE or transit node (TR), as is the general rule in standards. The INAP and TCAP standards are the exceptions of the rule, however. TCAP is a protocol that defines a common agent layer. INAP is an abstract protocol that defines an LSM, called "the intelligent network" (IN) in standards. The depicted INAP actor in Fig. 2.35 handles the SSP function defined in IN, and is an added part of the ISDN call-handling layer. INAP is the abstract protocol that this actor handles in communication with other IN actors of the SCP type.

If we analyse the relations between the three parts (DSS1 L3, ISUP, and INAP/TCAP) of the layer element, we will find that a message that is received by anyone of the agents frequently results in a message sent out by one or several others of the agents. Thus the DSS1 L3, ISUP, and INAP actors are not isolated from each other. On the contrary, they are all fictitious parts of one and the same call-handling actor that takes all the decisions in the layer element that regards the state of the layer. The difference between these parts is only that the call-handling actor uses a standardized SAG (TCAP **invokes**, etc.) when communicating with the TCAP agent, but means that are defined by the implementor when communicating with the other two agents.

If we refine every layer element of Fig. 2.34 into actors and agents, we can draw the model of the call-handling actor layer in two ways, as discussed before: as a layer protocol machine (LPM) and as a layer state machine (LSM). Figure 2.36 shows the LPM.

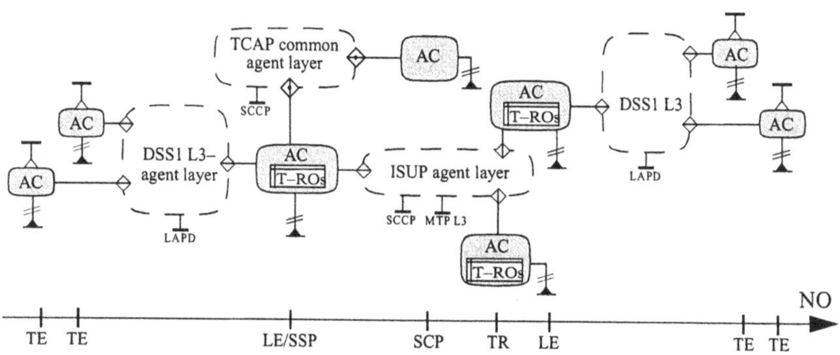

Fig. 2.36 The call-handling LPM in an ISDN

In this model, all agent layers are drawn as functional elements of their own, interconnecting actors over actor–agent interfaces. Only TCAP does define a standardized SAG, however.

The LSM representation of the layer (an abstract machine system) is shown in Fig. 2.37 (compare with Fig. 2.29). We added the NO dimension for this representation as well to show in which node the element state machines (ESM) reside. Note that when a standard does not apply the actor–agent separation, there are no abstract protocol names to refer. We therefore name abstract protocols by taking the message protocol name (e.g., DSS1 L3) and put it in between slashes, e.g., /DSS1 L3/.

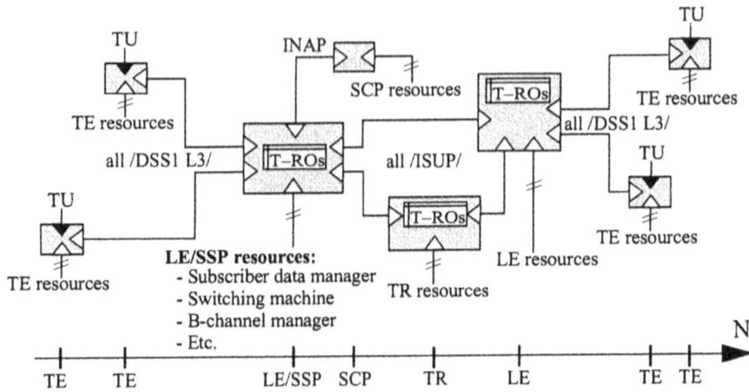

Fig. 2.37 The call-handling LSM in an ISDN

The designers of the ISDN call-handling layer knew of course exactly how the layer was supposed to operate, but they fragmented their knowledge into (at least) three different message protocol specifications. Since actors are almost always separated from agents in implementations, it became a considerable burden on the part of interpreters of standards (implementors's R&D departments and students of ISDN) to find out how the LSM was supposed to work. A lot of effort has been spent by implementors over the years to create simpler and more intelligible descriptions of call handling than what is represented by the around 2000 pages of the message protocol standards. Implementor's documents stay however as proprietary documents[18]. All this effort could have been avoided if the standard, instead of fragmenting the knowledge, had provided a description of the ISDN call handling layer in terms of its LPM and LSM models instead.

The exception to the rule in ITU–T standards is the TCAP common agent layer. Although not as general as the AMLn common agent layer, the TCAP layer can be used by a number of ISDN actor layers simultaneously, as long as they define their abstract protocols according to the operation classes that ROSE defines, and use the SAG defined by TCAP. This implies that TCAP agents should not be drawn in an OSI layer-structure model as if they were a part of an ISDN call-handling layer element. A common agent layer must always be drawn as a separate layer, related by actor-agent interfaces to other layers, as illustrated in Fig. 2.38.

[18]The author's most significant personal experience in this respect was when working on a large mobile network system project. As usual, designers had a very fragmented picture of how call handling was supposed to work in the system, due to the way standards were written. One designer had the ambition to understand how all fragments were related, and acquired the deep and overall knowledge of call handling. By investing man-years of efforts, he developed and gave a course that described call handling as (the correspondence to) an LSM model. This course always sold out. It became the most popular course by many hundreds of designers and testers, and a major contributor to the overall success of the project.

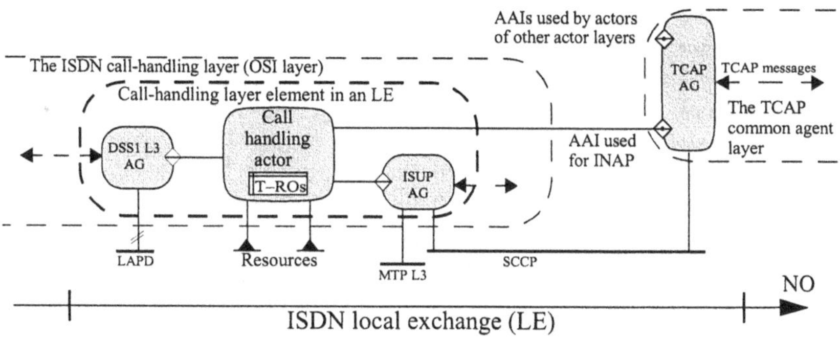

Fig. 2.38 Common agent layers, such as TCAP, must be described as separate layers

2.3.3 Agents and Actors in Connectivity Layers

2.3.3.1 General Modeling Principles

A completely different type of actors exist in connectivity layers. We refer to any type of functional element that transports data between some kind of inlets and outlets, without changing anything in the data stream. The type of functions performed by the actors varies over a wide range, but can be classified as either some type of **relaying** (multiplexing, cross-connecting, switching) or just **termination**. The term "relaying" is also used to denote elements that relay data between just one inlet and one outlet, such as amplifiers and regenerators, which we will, however, disregard in the present context.

A relaying function handles no message protocol. It only takes bits of an inlet and relays them to an outlet. A relaying function may take bit streams from several inlets and relay them to a common outlet (which means it performs a multiplexing function). This still does not change anything in the individual streams, however, i.e., the multiplexing function does not handle any message protocol. This implies that agents (i.e., elements that can communicate by messages) must be connected to inlets and outlets of relaying functions in order to create an OSI-layer element. Consequently, a relaying function is an actor, and its inlets and outlets must therefore be defined as actor–agent interfaces.

Using OSI layers to identify where in a layer structure relaying and terminating elements exist has always been a definition problem. OSI layers normally make these functions parts of a "network layer," i.e., a layer that performs some type of connection control and is defined by one or several protocols. This principle does not work for connection-oriented networks, such as circuit-switching and virtual-circuit-switching networks, since layers that control switching are definitely separated from layers that perform switching in these networks. Therefore, in order to create a generally valid principle for modeling in AMLn, we define any layer element that contains a relaying/terminating actor and connects to agents that handle the protocols of this element as a "connectivity layer element", no matter how the relaying/terminating process is controlled, i.e., *control of relaying/terminating is always modeled as a separate actor or layer.*

If we exclude functions such as transcoding and regeneration, all relaying elements are **common actors**. Such an actor offers one or a few types of **specified actor interfaces** (SACs) to agents, and connects an often large number of identical or different types of agents. Note that a common actor defines the primitives of the actor–agent interface, contrary to the common agent case (where the agent defines these primitives). An example of a common actor (a *multiplexing actor*) is shown in Fig. 2.39. This is a partial model of the V5 multiplexing standard.

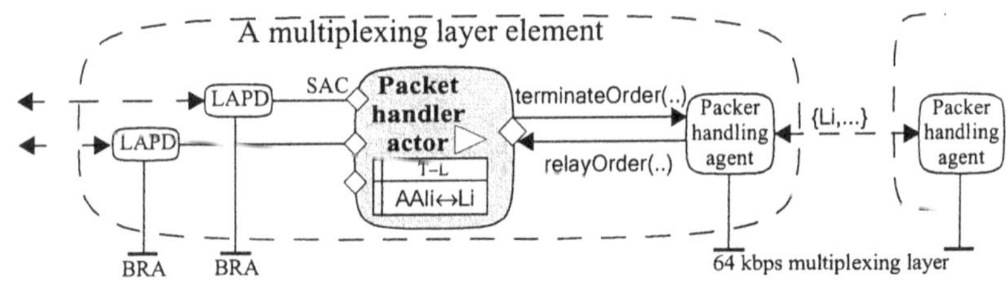

Fig. 2.39 Agents and a common actor in a multiplexing layer element. SAC=specified actor interface; Li=link identifier

This common actor connects many identical agents to the left and a single agent to the right. Its role is to multiplex LAPD frames that arrive on different BRA links (to the left) on a single 64-kbps link to the right (and the other way around). The protocol between multiplexing layer elements will therefore include a **link identifier** (Li) that identifies which of the BRA links is the source or the sink for user-data. The actor must be able to associate a link identifier (a parameter of the protocol between multiplexing layer elements) to a particular LAPD actor–agent interface. It therefore manages a table (T-L) that provides the translation between a link identifier and an actor–agent interface identifier (i.e., Li→AAli).

Since data that arrive at a common actor are not terminated by the actor, but are relayed by it to another actor–agent interface, we define the event of a piece of data arriving at a common actor as the primitive **relayOrder(...)**. The corresponding primitive from the common actor to an agent is called **terminateOrder(...)**, since the agent is supposed to terminate the relayed data somewhere. These primitives define the specified actor interface (SAC). The parameters of these primitives are identical: if the interface supports several streams in parallel, there are two parameters: **user data** (UD) and some type of **channel identifier** (CHi). If the interface supports just one stream, the user data is the only parameter.

The same modeling principle is applied to switching layers (see Fig. 2.40).

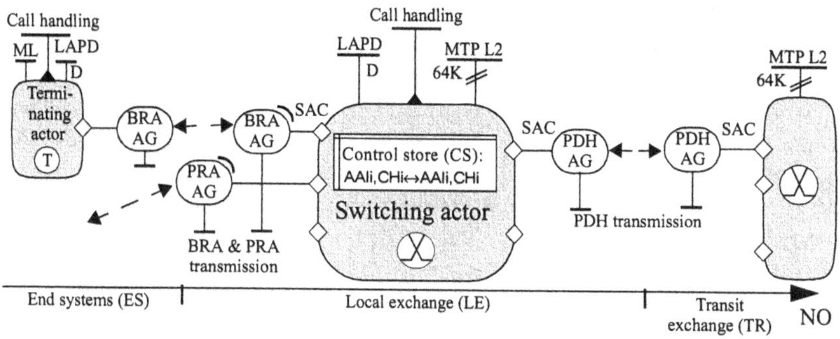

Fig. 2.40 Agents and actors in a circuit-switching layer. AAIi=actor–agent interface identifier; BRA=basic rate access; CHi=channel identifier; D=D-channel data; PRA=primary rate access; PDH=plesiochronous digital hierarchy; T=actor that terminates the switching layer; ML=media layer element

This model shows a part of the circuit-switching layer in an PSTN/ISDN. The layer consists of two types of actors: (common) switching actors and two types of terminating actors (for BRA and PRA). The actors are interconnected over agent layers of the BRA, PRA and PDH types (a number of other types of agent layers may be used between switching actors as well, e.g., an SDH layer). The switching actors in local exchanges (LE) and in transit exchanges (TR) relay 64 kbps synchronous bit streams between any of their actor–agent interfaces. Switching actors in LEs are designed so that they can connect a large number of agents on their "end-system side." These agents are either of the BRA or PRA types. LE switching actors also terminate signaling channels (called "D channels," used by the LAPD layer in DSS1) that are carried over BRAs and PRAs. On its "network side" an LE switching actor connects a smaller number of PDH agents, for example. It also terminates signaling channels (called "data signaling links," used by the MTP L2 layer in SS7) to other switching actors of LEs and TRs.

TR switching actors are basically identical, except that they do not connect BRAs and PRAs. A terminating actor (T) of the switching layer exists in an end-system and is connected to a BRA or a PRA. From the ISDN point of view, these are not switching elements. They just terminate ISDN user-data (at media-layer elements) and D channels (at MTP L2-layer elements). Note, however, that a terminating actor may be realized as a switching actor locally (i.e., as a private switch).

Circuit-switching actors do not themselves decide on how to switch streams. A separate call-handling layer (e.g., the one described in Sect. 2.3.2.4) decides that for every path through the switching layer, before the path can be used. Once the decision is taken, however, a call-handling-layer element stores its decision in the **control store** (Cs) of the actor as a mapping of an inlet to an outlet. Since a switching actor has no knowledge of which agent layer is actually connected to an AAI, it defines an inlet or an outlet as a particular channel in a particular actor–agent interface, identified by an AAIi and a CHi.

Since the specified actor interfaces (SAC) for circuit switching actors are synchronous, the relayOrder and terminateOrder primitives we defined in the previous multiplexing example, are implicit in this case.

Since connectivity layer elements are separated into actor and agents, we can create an LPM representation of such a layer. The model in Fig. 2.41 shows the LPM of an ISDN switching layer. We have assumed that both PDH and SDH agent layers are used in the core of this layer. The PDH layers are assumed to be 1.5 and 2 Mbps point-to-point links (T1 or E1), while the SDH layer is supposed to be a cross-connecting layer (note the different symbols used to denote circuit switching and cross connection respectively).

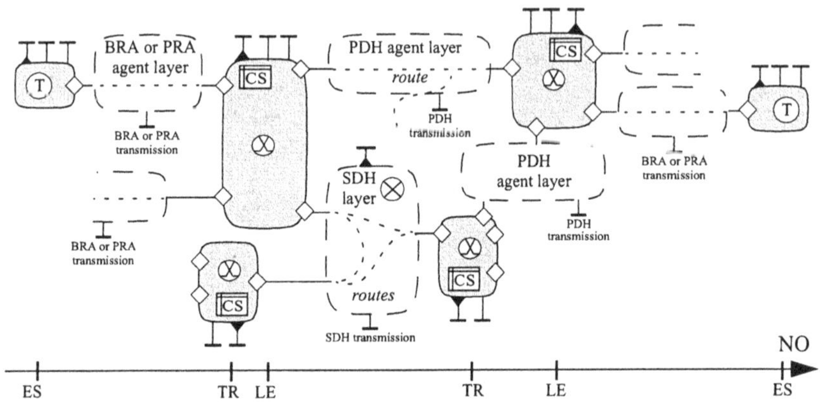

Fig. 2.41 The circuit-switching LPM in an ISDN. Cs=control store

From a structural point of view, this LPM model looks similar to the ISDN call-handling layer according to Fig. 2.36. There are, however, considerable differences supported by notations and annotations:

- The graphical symbol for actor–agent interfaces (the SAC symbol) is what formally defines this model as a model of a connectivity layer.
- Annotations by the circuit-switching symbol defines the layer as a circuit-switching layer.
- Actors support connectivity service points (actors in control layers support control points only).
- The fact that all actors support a resource control point (in control layers, only terminating layer elements do).

The most important difference between control layers and connectivity layers is, however, that there exists no layer state machine (LSM) representation for the latter. This is because *these actors do not interpret data, they only relay and/or terminate data*. Consequently, there are no states to be controlled, no abstract protocol to be handled, and therefore also no need for any meta-primitives (i.e., invokes).

We must therefore find another way to describe the relation between actors of such layers than by means of abstract protocol relation. The only relation between actors in a connectivity layer is that they are interconnected over some type of "connections." These are, in the circuit-switching case, realized by synchronous connectivity layers, and in the X.25 case by (asynchronous) LAPB layers.

In AMLn we use the generic term **route** for all types of connections (read more about this subject in Chap. 3). Suffice here to say that a "route" terminates in logical nodes of a logical network. By including an actor of a connectivity layer, all its agents, and all its users as functional parts in a logical node, we can let routes abstract how switching nodes are interconnected, and thereby create a simple logical network model. Figure 2.42 exemplifies this approach for the ISDN circuit-switching network.

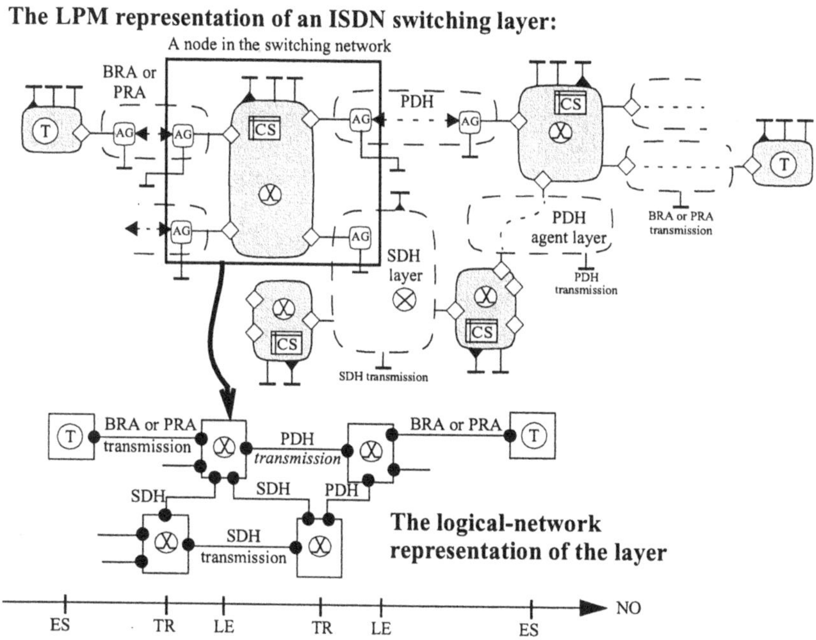

Fig. 2.42 The switching layer of an ISDN

Note that logical node symbols are chosen so that logical nodes cannot be confused with functional elements (such as layer elements), and route symbols are chosen so that routes cannot be confused with interfaces and protocols.

2.3.3.2 Submodels of Switching Actors

A common actor relays data either between AAIs, or between a layer interface and an AAI. To show how these two forms of relaying are controlled requires a common actor submodel. Such submodels also reveal the difference between circuit switching and packet-switching actors. For example, the circuit switching ISDN-actor submodel in an LE may look like Fig. 2.43. Note that this kind of model exists on a lower network level than the one of Fig. 2.42, and is therefore normally first defined by implementors.

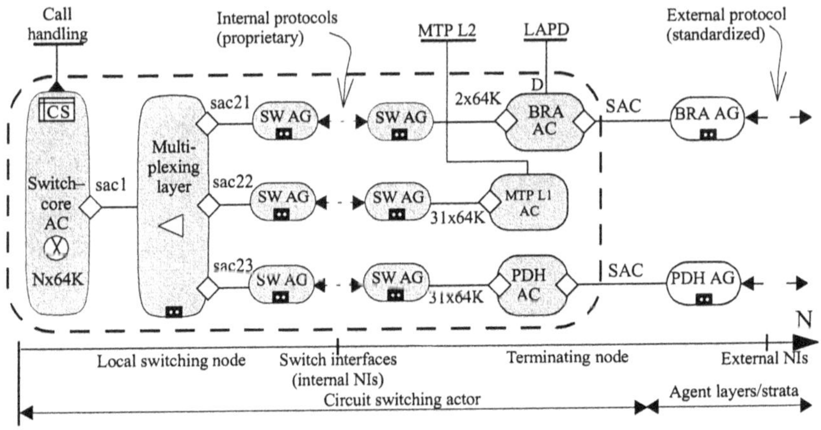

Fig. 2.43 A circuit switching actor in an ISDN local exchange. AC=actor; AG=agent; D=channel for signaling data; K=kbps; NI=network interface; SAC,sac...=specified actor interfaces

The model depicts a switching actor in a local exchange (LE) since it terminates a BRA agent layer (used for access; comprises two channels of 64 kbps) and a PDH agent layer (used for interconnecting switching nodes in the network core). In reality, an LE may connect a large number of BRA agent layers, many PRA agent layers (30 channels of 64 kbps each), and probably more than one core-network agent layer. All LE switching actors also connect thousands of PSTN access agent layers (one channel of 64 kbps each).

The circuit-switching actor may also be used for terminating 64-kbps channels. Figure 2.43 shows one example: the MTP L1 actor. It terminates channels that are used by MTP L2-layer elements of the of SS7 stratum. Similar actors (not shown here) provide layer interfaces to layer elements of the media stratum.

The element of the internal structure that actually switches anything is the *switch–core actor*. The basic switching rate in ISDN is 64 kbps, intended for on-demand traditional telecom services and for residential use of datacom applications. However, a number of 64-kbps channels can be switched in parallel, thereby providing higher bit rates (denoted as Nx64 kbps in the model). The standard defines N=2, 6, 24, and 30, i.e., 128, 384, 1536, and 1920 kbps switched channels. These channels provide unrestricted bit services and were intended primarily for video applications (e.g. video conferences) and bulk data transport between computers.

The internal structure is (in this case) designed so that the local switching node can be reused in all types of switching nodes, and so that it can connect several types of agent layers. The switch–core actor therefore defines a single type of (internal) specified actor interface, sac1, to which an (internal) multiplexing layer and network is connected. A sac1 can comprise up to several thousands of 64 kbps channels. Larger switching capacity can be achieved by adding more core switching actors. The multiplexing layer/network creates a small set of suitable actor-agent interface types (sac21, sac22, etc.), to which terminating actors and external agent layers can be connected. Note that all actor–agent interfaces and internal network interfaces are designed according to suppliers proprietary specifications, i.e., only external network interfaces are standardized.

The switch–core actor does not know if a stream goes out over an external agent layer or is terminated by the circuit-switching actor. The only element that knows that is the call-handling actor that orders the switch–core actor how to switch streams between channels in sac1. That is also why routing is performed in the call-handling actor, not in the switching layer. Once the call-handling actor has decided how to route a stream, it tells the switch–core actor (over the resource control point) which channels to connect. This information is stored in a control store (CS) in the switch–core actor.

The functions of the model in Fig. 2.43 are otherwise:

- The BRA agent layer carries two types of channels, B (there are two of them: B1 and B2, both 64 kbps) and D (16 kbps). The BRA agent (BRA AG) terminates the protocol and delivers all channels to the BRA actor (BRA AC). The BRA AC decides on how to handle the channels, depending on the application. When the BRA AC is a part of a circuit-switching actor, D-channel data are extracted and delivered over a layer interface to the LAPD layer element, while B channels are delivered to its local switching agent (SW AG).
- The same principle goes for the 2-Mbps PDH ("plesiochronous digital hierarchy" which means "almost synchronous") agent layer. The PDH AC does not do much (besides handling synchronization and management functions). It delivers its thirty one 64-kbps channels to its SW AG. Some of these channels are connected dynamically through the switching actor to BRA or PRA AGs, or to other PDH (and SDH, "synchronous digital hierarchy") AGs.
  Other channels are connected semipermanently to the local MTP L1 AC. These channels are used for signaling purposes, similar to D channels. The reason for not letting the PDH layer element extract those channels is that a more flexible SS7 network results from controlling these channels over the local switching network. Furthermore, it is not necessary to define a specific channel in the PDH frame that must always be the signalling channel (as in BRA and PRA). Signaling capacity between two network elements can also be increased simply by using more than one channel for signaling.
- The MTP L1 AC terminates many streams (called "signaling data links" in SS7) over a layer interface (to the MTP L2 layer).

Realizations of circuit-switching actors are obviously complicated systems. On the standardized ISDN network level we should however not be concerned by that. The separation of switching actors from agent layers and strata indicates what is of importance on the network level (see Fig. 2.44).

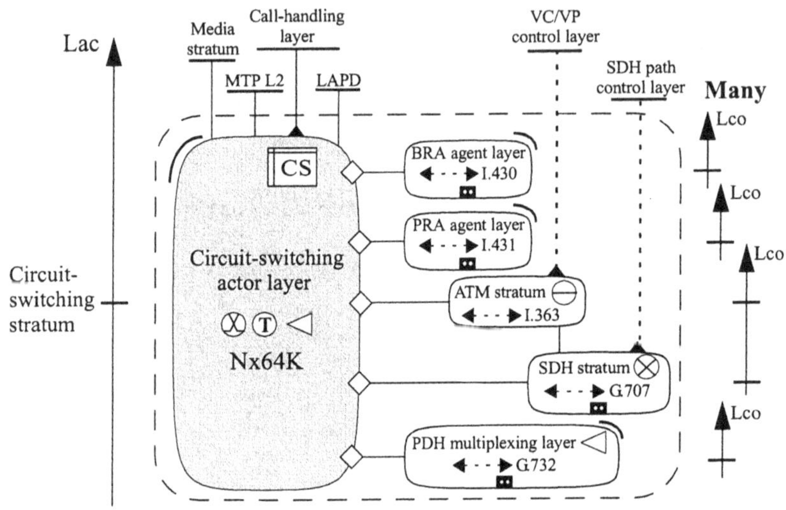

Fig. 2.44 The layer structure of the ISDN circuit-switching stratum. Some relevant ITU–T protocol standards are shown

The model shows all major agent layers that are used in an ISDN circuit-switching stratum (BRA and PRA to connect terminals to the ISDN network; ATM, SDH, and PDH to connect switching nodes within the ISDN). Agent layers and strata (except for ATM) interface physical media strata.[19] ATM and SDH strata are separately controlled: a "connection" through an SDH stratum is called "path;" a "connection" through an ATM stratum is called "virtual circuit" (VC). A "virtual path" (VP), is a number of VCs that are switched in parallel.

Submodels of packet switching actors (see Fig. 2.45), in particular for datagram switching, are considerably simpler than models for circuit switches. We choose

[19]A circuit-switching stratum is therefore often regarded as existing on OSI layer 1. Since the telecom world always tries to align their standards to the OSI RM, this was a problem when SDH became defined since this network system had a layered structure in itself that also included cross connectors (a kind of circuit switches). It was therefore difficult (to say the least) to regard SDH as just an OSI physical layer. This led to an effort to define a separate layered functional structure for transport strata (see ITU–T G.805). G.805 also defines how to describe stratification since an SDH network can carry data for PDH and ATM networks, where the latter is also a kind of transport network. However, not only is the modeling technique suggested by G.805 more complex than AMLn, its applicability is restricted to synchronous transport networks, such as SDH, PDH, and OTN ("optical transport network").

however to show the slightly more complex submodel for X.25 virtual-circuit (VC) switching in Fig. 2.45 since there are more similarities in this model to circuit switching.

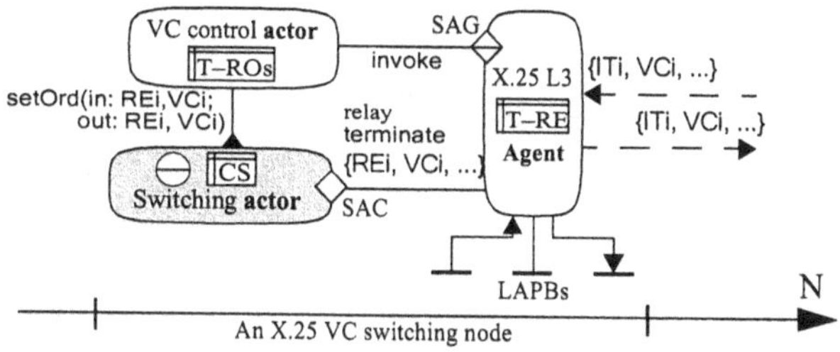

Fig. 2.45 An X.25 VC switching node. CS=control store; T–RE=route endpoint table; T–ROs=routing table

The model shows a switching node of an X.25 network. Such a node supports no users. It only relays data between LAPB links. Similar to circuit switching, there is a distinct difference between switching actors and control actors. Also similar is that the switching actor contains a kind of "control store" (CS), set by the VC-control actor. The data in this CS defines how to relay incoming data of a VC on a particular LAPB link to a VC on an outgoing LAPB link. In the actor–agent interface, LAPB layer interfaces are referred to by a neutral **route-endpoint identifier** (REi). The agent translates this identifier to a LAPB layer interface identifier through a special table called T-RE (more about these tables in Chap. 3). The purpose with REi is to isolate actors from knowing which connectivity layers are actually used by the agent.

Specific to X.25 L3 is that this protocol serves both the switching actor and the VC control actor. Thus, the single agent supports two types of actor–agent interfaces (SAG and SAC), which is why the protocol must include a parameter that identifies if a message contains data for the SAG or the SAC. We call a parameter that is used for discriminating between different interface types **interface type identifier** (ITi). The ITi in X.25 L3 is called "control bit" in the protocol specification. Note that X.25 L3 control elements communicate by an (implicit) abstract protocol since the X.25 protocol specification states explicitly all control messages.

As a comparison, ATM (see Sect. 6.4) solves control signaling differently: special VCs are used for carrying data for control layers, which is why the ATM protocol specification does not define control layer messages. The ATM switching layer is therefore insensitive to changes in the ATM control protocol. In reality, the ATM switching layer is (in line with the telecom tradition) defined as a "switching resource layer" that is controlled from a separate control layer (defined as part of the broadband-ISDN network system, B–ISDN).

Let's now return to X.25 and the model of the terminating X.25 nodes, where X.25 users access the VC service. Figure 2.46 shows an X.25 terminating node that serves a single user.

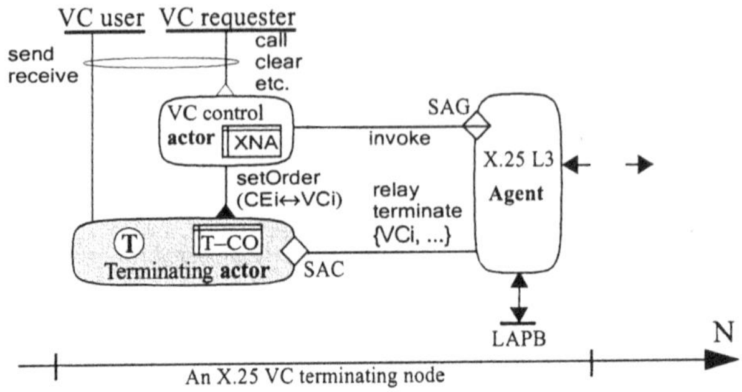

Fig. 2.46 An X.25 VC terminating node. T–CO=connection table; XNA=X.25 network address

The terminating node needs neither of the tables we defined for the switching node. It may, however, store its own network address (XNA), and the switching actor may contain a table (called T-CO) that records which VCs are in use by the user. This table may also translate between a local CEi and a corresponding Ci (VCi is the Ci in X.25).

A common layer interface is used both as a control point and as a connectivity service point (as for most packet-switching systems). Since X.25 can terminate a large number of VCs in a terminating node, the node may also be used for supporting many users concurrently, in which case the terminating actor must also perform discrimination between used-layer interfaces. This is exactly the "discrimination by CEi" case that characterizes the OSI RM–CON service, as was discussed in Sect. 2.2.2. Figure 2.47 shows how the CON service is defined in OSI RM and how it looks when we apply the actor–agent separation principle.

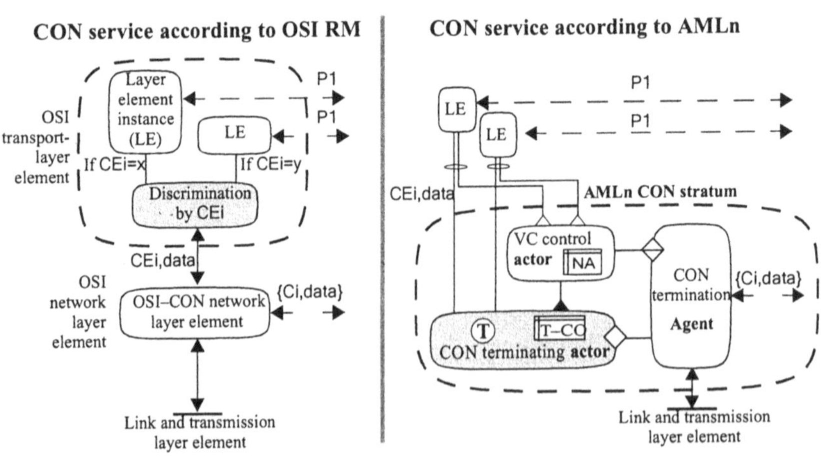

Fig. 2.47 The OSI–CON service modelled as an OSI layer element (to the *left*) and as an AMLn layer element (to the *right*)

We may recall that, in order to specify where and how discrimination is performed in the OSI RM definition of the CON service, we had to define a functional element in an OSI transport-layer element that did not deal with any message-handling aspect of this protocol. Such an element violates the OSI RM layer-interface concept (as well as the one in AMLn) since a functional element that connects to a lower layer (over layer interface) must handle a protocol (or a part of it).

The actor–agent separation principle solves this contradiction (see the model to the right in Fig. 2.47). The discrimination function is in reality performed by terminating actors of a CON switching layer. Multiple users are supported by adding a layer-interface identifier to the T-CO table. It is the responsibility of terminating actors to see to that connection endpoints for each layer interface are unique.

Note that the OSI RM model for CON services is identical to X.25 VC, except for how parameters are named (Ci and CEi instead of VCi, etc.). Thus, back in 1984, the OSI RM obviously thought of X.25 as the solution for packet-switching networks in general. However, we have seen many other CON technologies (FR, ATM, and TCP/IP) appearing since then, which cannot be fitted in the OSI RM CON model. By re-modeling this service using the actor–agent separation, we remove the ambiguity regarding the layer concept and arrive at a more precise and generally applicable definition of the layer interface that is offered by a CON switching layer, without changing anything in external interfaces of X.25 nodes (i.e., network interfaces and layer interfaces to X.25 users).

2.3.3.3 Discrimination In Connectivity Layers

We will end this chapter by discussing some other discrimination problems in connectivity layers. In an AMLn model, the actor is normally the element that handles users' layer interfaces (if any), while agents always handle used connectivity interfaces (CSPs) (see left of Fig. 2.48). However, in connectivity layers, users' layer interfaces may be distributed to both the actor and the agent, as we have seen in the ISDN model of Fig. 2.43. Figure 2.48 indicates this in the model to the right.

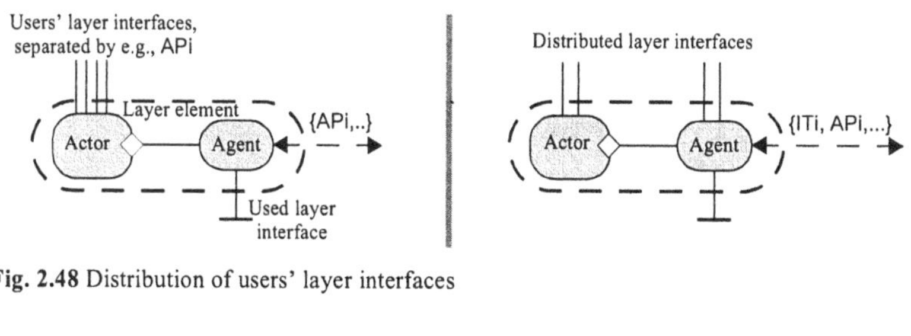

Fig. 2.48 Distribution of users' layer interfaces

This creates an additional discrimination problem: it requires an agent to be able to distinguish between layer interfaces and actor–agent interfaces, i.e., another application of the interface type identifier (ITi). The ITi parameter need only one bit and tells the receiving agent if the other discriminating parameter refers a layer interface (in which case it is an APi) or an actor–agent interface (in which case it might be an ALi, an ASi, or an Li).

In standards you will seldom find an explicitly-defined ITi since designers of network systems normally do not define any actor–agent separation. In case it happens, the ITi is probably hidden in a complex type of parameter that defines the APi for both agent and actor, as well as other forms of discrimination. Such an example is the PTI parameter in the ATM switching layer (see Sect. 6.4.3).

Sometimes no explicit discrimination parameters are needed because message parameters are bound to certain positions and of fixed length. Examples exist in the access part of the ISDN switching layer that was discussed before. Figure 2.49 shows how data in BRA frames are discriminated at termination points.

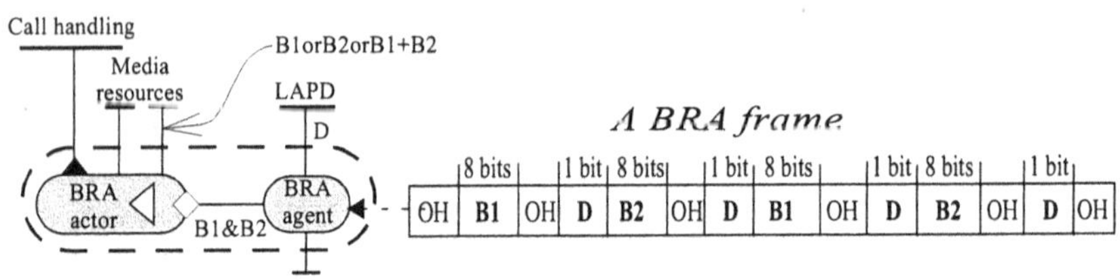

Fig. 2.49 Discrimination in the ISDN circuit-switching layer. B1,B2=B-channels; D=D-channel; OH=overhead bits

This model is also an example where the discrimination function in a layer element may be shared between its actor and agent. The agent discriminates between D and B channels and terminates the D channel, while the actor discriminates between media layer interfaces. It terminates (and multiplexes, if required) B channels.

2.3.4
Control Structures versus Connectivity Structures

The distinction we have made between control and connectivity layers reveals that interfaces between layers are not just interfaces that offer connectivity. Since the essence with a control layer is that it controls *resources* in a network system, some layer interfaces are used for control. A type of resource to control in ISDN are the actors of connectivity layers, but there are many other types as well. Let's therefore combine the ISDN call-handling-layer LPM from Fig. 2.36 and the switching-layer LPM from Fig. 2.41 in the same model (see Fig. 2.50). The model shows how the four types of interfaces (connectivity service point, control point, SAG, and SAC) build up the functional structure of an ISDN (in the L–N plane).

In this rather complex model, the different symbols for layer interfaces and actor–agent interfaces help us understand how the network works. Control points are also drawn with dotted lines in order to make the model more intelligible. Added functionality in this model are a number of strata[20] (LAPD and the SS7 connectivity strata) that are needed in order to provide the call-handling layer with asynchronous connectivity (since the switching layer is not a suitable datacom network). In addition, media layers that use service points of switching layer actors are indicated. Another possible layer that relies on service points is a packet-handling layer (not shown in Fig. 2.50, however).

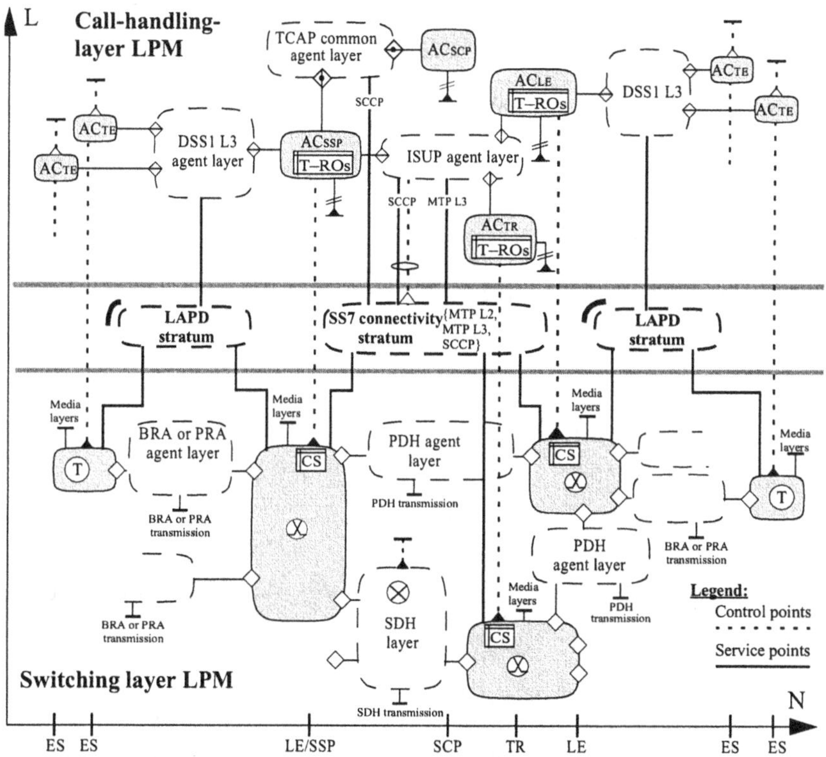

Fig. 2.50 The layer structure of an ISDN. L=layer dimension; N=node dimension

We can reduce the complexity of this model by collapsing the N dimension and showing control structures and connectivity structures separately, interconnected over types of actor–agent interfaces, as in Fig. 2.51. This implies that the L dimension dissolves into an Lac dimension for the control structure and four Lco dimensions for connectivity structures. Except for their mutual dependence on actor–agent interfaces and on the switching layer, these structures are independent, also as to how many layers they comprise. Since the N dimension is collapsed, this model is valid for every ISDN network.

[20] In AMLn, a stratum is a structure or an aggregate of layers.

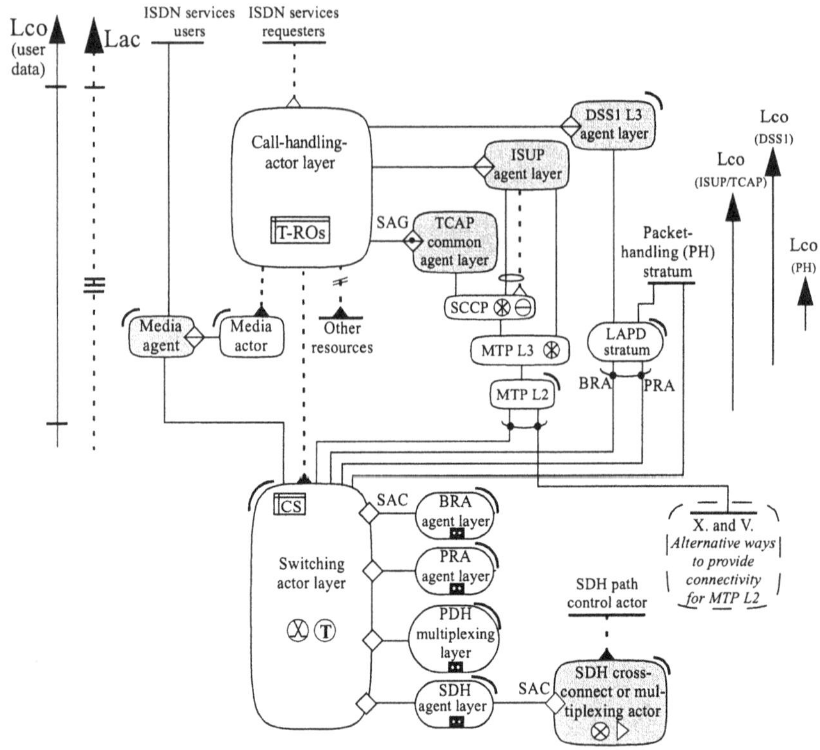

Fig. 2.51 The functional structure of the ISDN network system, separated into its control structure and its connectivity structures. SAC=specified actor interface; SAG=specified agent interface

In an AMLn model of the ISDN network system, a layer interface realizes either a control point or a service point, and the structures behind these types of interfaces are separated (corresponding to the "user-plane/control-plane" separation, defined in the ISDN standard). This is a major architectural difference between circuit-switching and packet-switching network systems (such as the SS7 or TCP/IP). In the latter case, there is no such distinction between control and service points. For datagram systems (such as IP and MTP), layer interfaces realize service points only, while in connection-oriented systems (such as TCP and SCCP) a single layer interface realizes both a control point and a service point.

Note also that the separation of an OSI layer structure into control and connectivity structures can be recursively applied. For example, SDH is separated into a control structure, consisting of cross-connecting and path-controlling actors (that setup and manage 2 Mbps SDH paths between the ISDN switching actors), and SDH agent layers. The SCCP layer also indicates that it can be separated into separate control and layer structures, since it exhibits a service control point (that is used by the ISUP agent layer when requesting connection-oriented SCCP services).

Media layers consist of resources that create media connections (or "calls") to their users, e.g., ordinary telephony connections (transports 3.1 kHz voice signals). A media-agent layer element manipulates media signals that are transported through the switching layer. Example of media-agent-layer elements are:

- Transcoders for translating between different types PCM coding, or between compressed and uncompressed formats.
- Conference bridges and mixers for multimedia calls.
- Echo suppressors and echo cancellers.
- Different kinds of interworking units (IWU) for circuit-switched data between ISDN and PSTN (analog) terminals.
- Since the same switching layer is used both for analog and digital accesses, analog-digital converters are also parts of media layers.

The call-handling actor layer also makes use of media layers as a means of communication with end-users. To that regard it uses a number of media actors, such as digit receptors, voice recognition and answering machines, announcement machines, etc. It uses the control points for sending messages (that are stored in the actor) over a telephony connection, and for receiving information on messages that users send over such a connection. In the media layer, such messages may be transported as tone signals ("dual-tone multi-frequency" (DTMF)) or as voice signals that are interpreted by a voice recognition media actor.

Control structures may have any number of layers. For example, a possible improvement of the ISDN network systems is to separate the call-handling actor layer into a connection-control actor layer and a service-control actor layer[21], the former controlling all resources for setting up and handling on-demand synchronous connections in the switching stratum, the latter for controlling user's requests for services and other resources, e.g., the use of media layers. Figure 2.52 shows how we can model this separation by defining additional resource control points (rCP). In this solution, the service-control-actor layer will now take care of service analysis, media handling, charging etc., while the connection-control-actor layer will take care of connection handling by performing routing and by controlling switching actors.

The effect of this separation is that service-control actors no longer need to control the switching actors. The ISUP protocol can therefore be reduced to service handling only (called ISUP/.S in the model). The same goes for DSS1 L3 that we reduce to DSS1 L3/S. On the other hand, the connection-control-actor layer needs connectivity as well. In this example we choose to use TCAP to support the abstract protocol of this layer in the core network (which is a simple but not neces-

[21]Also referred to as the "call-connection separation," frequently discussed over the years but first realized in "next generation networks" (NGN), that is now under development.

sarily ideal solution), and a protocol called DSS1 L3/C to support the abstract protocol between core actors and actors in terminals. The DSS1 L3 agent layers will still use LAPD for their connectivity needs. The problem of distinguishing between the two DSS1 L3 agent layers is solved by using different LAPD service–access points (there are still many unused).

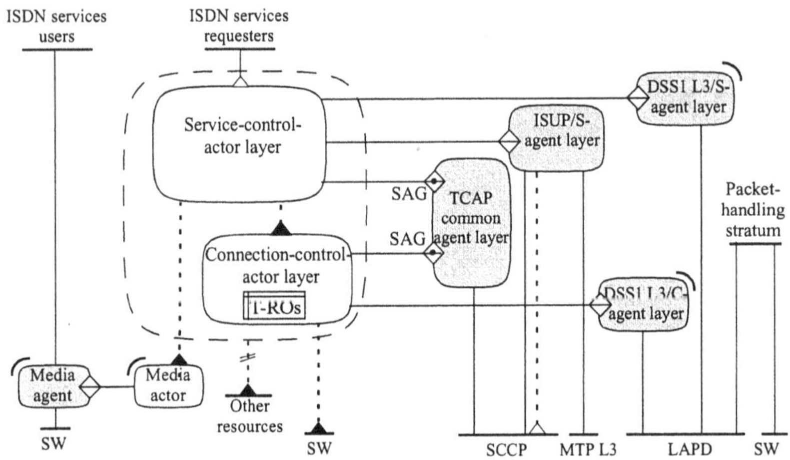

Fig. 2.52 A solution for separating call handling into service control and connection control. SW=switching stratum

The ISDN structure according to Figs. 2.51 and 2.52 still exhibits a rather complex network system, especially when compared to most packet-switching network systems. The corresponding X.25 model looks much simpler (see Fig. 2.53).

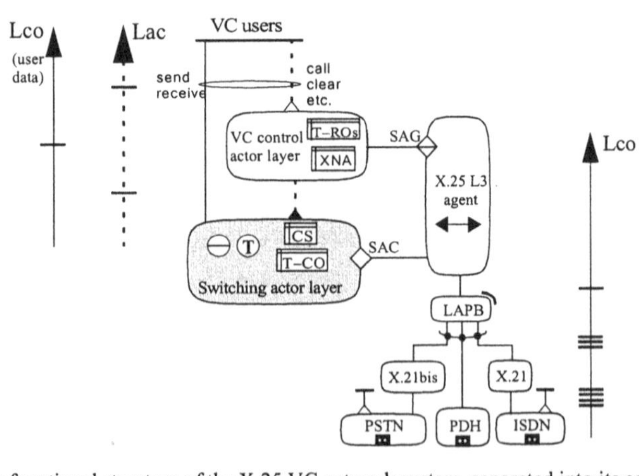

Fig. 2.53 The functional structure of the X.25 VC network system, separated into its control structure (LAC) and connectivity structures (LCO)

Note, however, that packet-switched network systems may have very complex functional structures as well, e.g., as the broadband ISDN which is based on ATM (we describe that structure in Sect. 6.4).

The reason for the simplicity of X.25 is that it does not offer anything other than a raw packet service. The complex ISDN structure, on the other hand, was designed to become an evolution from the PSTN with the same (or even improved) requirements on isochronous services (telephony, etc.) as in PSTN. Consequently it was based on the same circuit-switching stratum as the PSTN. Since signaling within the call-handling layer requires asynchronous data transfer, however, one had to make all the layers of the DSS1 and SS7 strata part of ISDN as well. At the same time, the growing need for data communication over ISDN accesses was met by adding additional complexity in terms of packet-handling functions and inter-working needs with X.25 networks. As Fig. 2.51 shows, packet services are offered to ISDN users both over LAPD links (SAPI= 16) and over switched channels.

Let's now summarize our findings on control and connectivity structures:

- By applying the actor–agent separation approach, any OSI layer structure dissolves into a single control structure (over at least two levels of control) and a number of mutually independent connectivity structure, which may run on a switching stratum. Conceptually we say that the L dimension dissolves into an Lac dimension and one or several Lco dimensions.
- The relations between control and connectivity structures are fully defined by SAGs (in ISDN only provided for TCAP) and SACs.
- Control structures are strictly hierarchical, while connectivity structures normally are not (as the seven-layer structure of OSI RM makes us believe).

As mentioned before, the ISDN standard describes ISDN only as an OSI layer structure. Actor functions are only indicated by including actor description fragments in message protocol specifications. This standard therefore (more or less) hides the distinction between the control structure the connectivity structures, which certainly does not facilitate the understanding of ISDN.

2.4 Stratum Levels

We have seen that functional structures of real network systems have very little resemblance with the OSI RM (seven layers in a hierarchical layer structure). Their functional structures are not strictly hierarchical, and consist often of more than seven layers (see, e.g., the ISDN layer structure depicted in Fig. 2.51). Parts of a layer structure are also frequently grouped or aggregated into larger functional elements for different reasons. For example the SS7 connectivity system (MTP L2, MTP L3 and SCCP), which is defined as a part of the PSTN/ISDN layer structure, could be treated as such a group in ITU–T standards.

When a group of layers is specified in a way that makes it possible to handle it as a black box in a network, we call it **stratum**. The network system, or network of which the stratum is a part, is then said to be *stratified*. To be able to treat such groupings as black boxes, the layer interfaces on which users of the stratum rely must be thoroughly specified (e.g., by using AMLs, see Appendix D). Note that by "user" we do not necessarily refer to end-users, but any other layer or stratum that relies on the stratum. Once these layer interfaces are specified, the internal layer structure of the stratum becomes invisible to using layers and strata.

There can be many reasons for defining parts of a layer structure as a stratum:

- A group of layers plays a certain role in a network. The SS7 connectivity stratum, being the nervous system in the core of PSTN and ISDN networks, plays such a role. The set of adaptation layers in ATM also plays a specific role in broadband ISDN systems, which is to adapt the raw cell transport performed by the ATM switching stratum to connectivity services that different types of media service are used to (this kind of adaptation is sometimes called "emulation").
- Parts of a layer structure that can be foreseen to be affected by technology changes are candidates for being hidden in strata. This is a very important reason, considering that networks have been and are in constant evolution, both as regards low level functions, such as physical media, transmission and transport technologies, and high-level service-oriented functions.
  This was also the reason for the only effort to define a stratum concept that ITU–T has published: the "generic functional architecture of transport networks" (which we call GTA), applied to create a functional model of the SDH stratum.
- In Sect. 2.1.1 we discussed the importance (for network system management in a context of continuous evolution) of keeping layer definitions simple. A layer should offer a single service so that the use and design of a layer and its protocols becomes easy to understand. One aspect of this is that in AMLn, layer management functions are modeled as separate layers (in particular when connectivity for layer management is provided by an APi of the very layer). One of many examples of this is LAPD that we discussed in Sect. 2.1.1. This means that many network functions that are defined as a single layer/protocol in standards, should dissolve into layers that belong to different AMLn planes (the management plane and the traffic plane respectively).

In the present chapter we will only take a look at how strata can be identified in layer structures. The ISDN network system, for example, is not known as being stratified. Nevertheless we can model it as such. For example, the model of Fig. 2.51 indicates a possibility to define its layer structure into a media-services stratum and three connectivity-service strata (a circuit-switching stratum, a packet-handling stratum and diverse X. and V. strata), as depicted in Fig. 2.54.

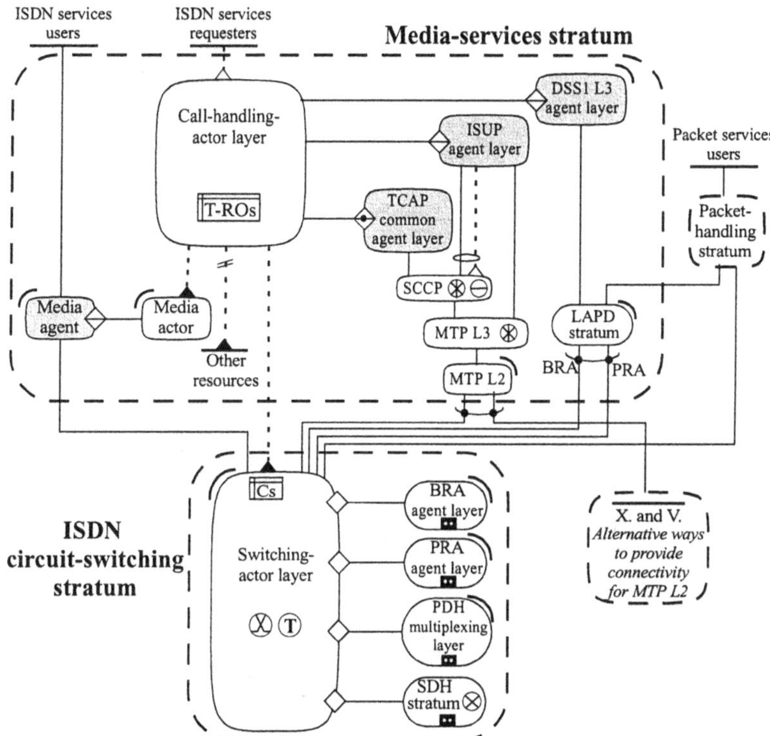

Fig. 2.54 A basic stratification of the ISDN network system

The model now indicates that the ISDN circuit-switching stratum may be replaced with another stratum (e.g., a packet-switching stratum), the ISDN still being able to offer the same (and possibly improved) end-user services. Obviously, however, if we would pursue such a replacement, we would have to adapt to other types of layer interfaces. Such adaptations can be very costly, unless existing layer interfaces between strata are preserved. This implies that, for a stratum to be open to replacement of internal layers, it must include adaptation layers (such as the ATM adaptation layers).

To qualify for being a real stratum, all layer interfaces that a stratum provides must be standardized and defined in details. In the ISDN case, connectivity service points of the switching stratum are standardized as to their properties, but the control points over which switching and media actors are controlled are not. Every supplier implements its own solutions to these interfaces. This is also part of the explanation to why ISDN is not characterized as "stratified."

The ISDN structure is also an example of how strata can be defined recursively. Figure 2.55 depicts four internal strata that exist in the ISDN media-services stratum. All substrata depend on the same switching stratum. In addition the SS7 connectivity stratum can run on other strata as well, such as V.- and X. networks. Each stratum can be specified and designed separately, i.e., defines its own internal layer structure.

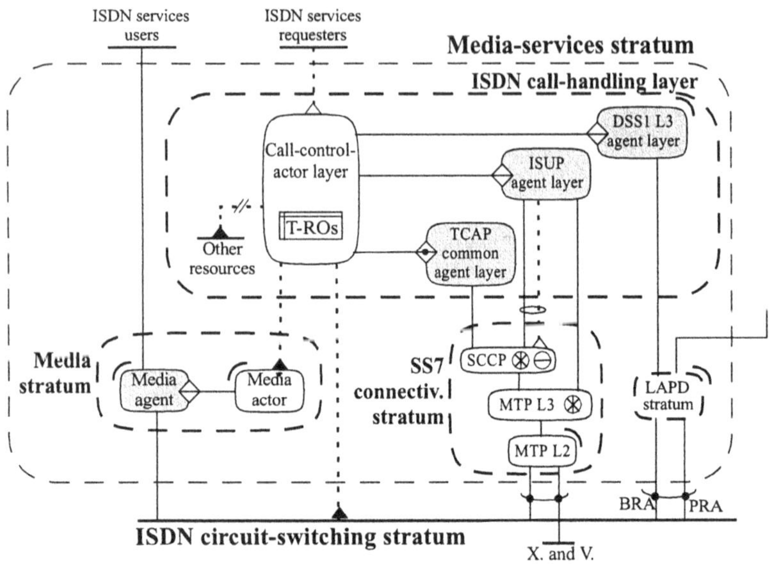

Fig. 2.55 Internal strata in the ISDN media-services stratum

This structure reveals a stratum substructure of four types of strata:

1. A *media stratum* consisting of diverse media agents and actors.
2. At the top, the *ISDN call-handling layer*, consisting of call-handling actors and three types of agent layers, relying on two other internal strata in the ISDN media-services stratum.
3. The *SS7 connectivity stratum*, consisting of SCCP, MTP L3 and MTP L2 layers.
4. A number of *LAPD strata*.

The SS7 connectivity stratum relies on services (called "signaling data links") of primarily the ISDN switching stratum, which is part of the explanation of why it is not regarded as a *common* packet-switching stratum (such as TCP/IP or ATM). However, as all strata, the internal structure of the SS7 connectivity stratum can be modified, provided that layer interfaces used by its users are kept intact. Part of the evolution of PSTN/ISDN is also to replace this stratum with IP- and ATM- ("asynchronous transfer mode") based connectivity strata.

Thus, layer structures can be modeled and viewed on several **stratum levels**, as soon as you introduce stratum definitions in the layer structure of a network system. Fig. 2.56 shows the top level (called stratum level 0) of ISDN strata we have discussed (compare with Fig. 2.54).

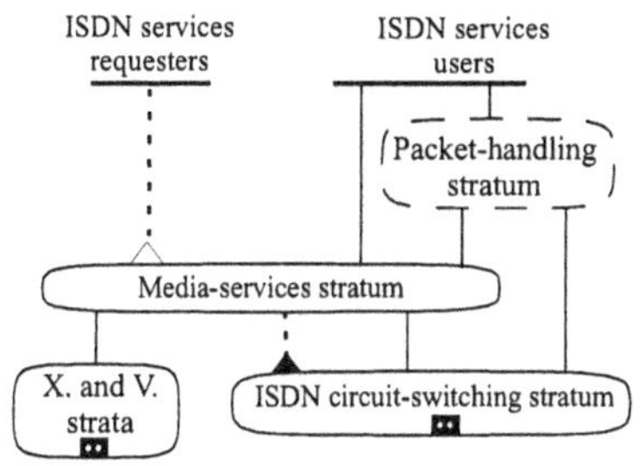

Fig. 2.56 ISDN stratum level 0

Each of these strata can now be assigned an internal structure of strata and layers. In the previous models we have used an intuitive framing notation for this. In a correct AMLn model we use the **substructure** symbol to identify stratum levels. Fig. 2.57 shows the internal structure of the media-services stratum (on stratum-level 1).

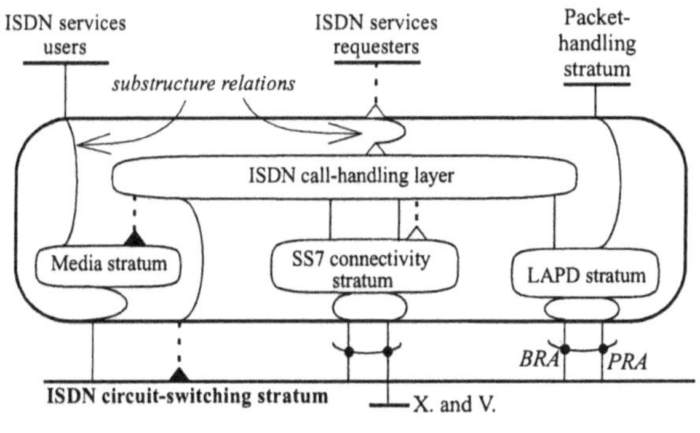

Fig. 2.57 The ISDN media-services stratum is modeled on stratum level 1

This means that the ISDN call-handling layer will be modeled on stratum-level 2 (see Fig. 2.58).

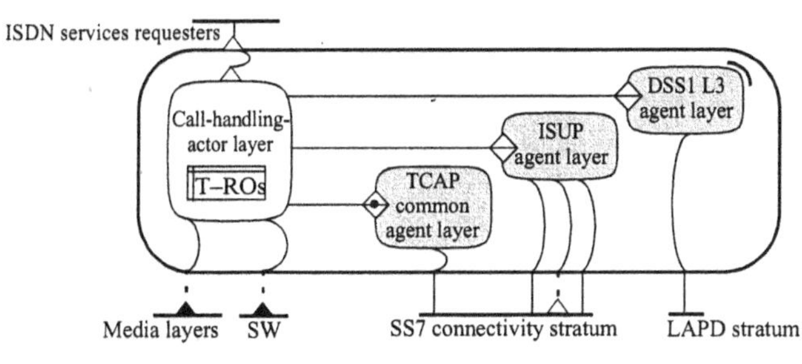

Fig. 2.58 The ISDN call-handling layer is modeled on stratum level 2.

3 Node Structures

3.1 Introduction

Node structures represent a completely different view on network systems. While layer structures describe how layers are related over functional interfaces called "layer interfaces," a node structure shows some kind of "nodes," interconnected by some kind of "connections." The intuitive understanding of how nodes and layer elements are related is that a node is some kind of "package" for layer elements.

Packaging layer elements of a given layer structure (thereby producing models of **derived nodes**) can be made in many ways, however. Furthermore, there are not only many types of connections around, we also know that one type of connection can be a bearer for other types of connections, such as a 64-kbps bit stream connection that carries a voice connection, or a route in a switching network that defines a number of channels. Some principles for packaging on the network level are therefore needed to support modeling of network structures.

If we look in network standards as well as inside implementors solutions, we will find that almost all nodes are defined by applying one or both of two packing principles, commonly called **vertical partitioning** and **horizontal partitioning**. We can use the OSI RM model for defining the difference between these principles (see Fig. 3.1). Note, however, that OSI RM did not define any of these concepts.

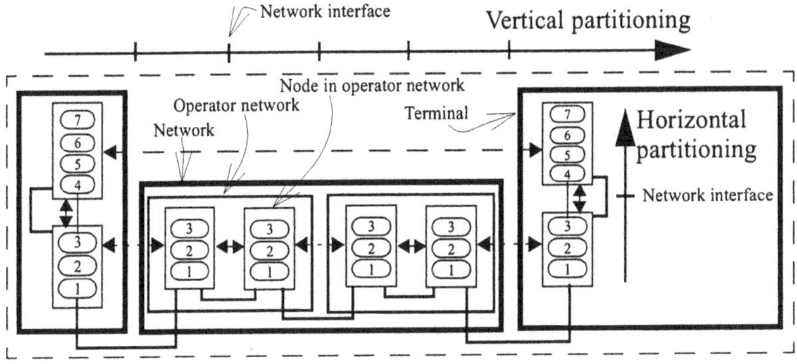

Fig. 3.1 Partitioning strategies exemplified for the OSI RM

The difference between these approaches lie in how the functionality of the overall network is partitioned.

- **Vertical partitioning** implies that a given interlayer structure is allocated to nodes so that all layer interfaces become local interfaces inside the nodes. As a consequence, a layer becomes partitioned into layer elements that communicate by means of **intralayer protocols** between the nodes. Interfaces between nodes are commonly called **network interface** (NI).[1] As the model indicates, a network interface can be a very complex entity defined by a number of protocols and possibly a physical medium.
  Figure 3.1 shows how this approach can be recursively applied: the overall network is separated into nodes that are *terminals* and a node that represent the *network* between all terminals; this network is separated into *operator networks*, each with its own internal structure of nodes.
  Vertical partitioning has until now been the only principle for node definitions in standards, primarily because it defines how terminals interface the network, how operator networks interwork, and how different types of nodes inside operator networks interwork (which is needed so that operators can buy their nodes from different suppliers). It was also regarded as a competitive advantage for suppliers that standards did not require any particular implementation of layer interfaces. A disadvantage that has become very obvious over time is, however, that nodes defined in this way become very complex. As a result, introducing new services and managing functionality in "vertical networks" is slow, tedious, and error prone.
- **Horizontal partitioning** implies that a particular type of layer interface is specified as a network interface[2] as well, exemplified in Fig. 3.1 for the layer interface between OSI layer 3 and 4. As a consequence, layer elements inside a node that are defined through vertical partitioning can be allocated to different internal nodes in all nodes that are concerned. Note that horizontal partitioning, contrary to vertical partitioning, does not cause any partitioning of the existing layer elements themselves, i.e., only layer interfaces are concerned by this approach.

The benefit with horizontal partitioning is that we can describe and implement a complex network as a structure of less complex **logical networks**, which, however, requires that we define how layer interfaces are transformed into network interfaces, and how horizontally partitioned nodes are interconnected. Let's make an example of the previous OSI RM model, by defining a logical network that comprises OSI layers 4 through 7 only (see Fig. 3.2).

---

1 When vertical partitioning is applied in the context of the OSI RM, network interfaces appear as vertical, which is why this method is regarded as "vertical partitioning."

2 As Fig. 3.1 shows, these network interfaces would appear as horizontal interfaces in the OSI RM, which is why this form of partitioning is regarded as "horizontal partitioning."

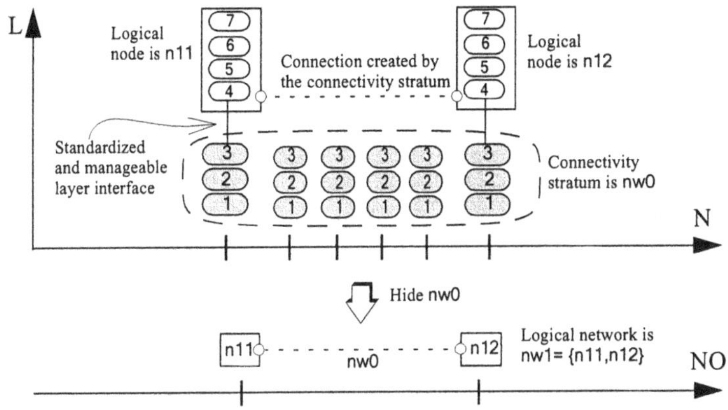

Fig. 3.2 A possible logical network in the OSI RM. L=layer dimension; N=node dimension; NO=logical-node dimension; n...=(logical) node name; nw...=(logical) network identifier

The two **logical nodes** n11 and n12 are created by horizontal partitioning of the model in Fig. 3.1. We then used the fact that all layers below exist only to create some kind of logical connections for layer 4, on which layers 5 through 7 also depend. A logical connection is the equivalence to a connection over a physical media in the OSI layer model. If we can specify how this logical connection is created and used, we can also disregard all layer and network structures below the layer 3/4 interface, which we regard as just a logical network (nw0) that offers a defined type of connections. As a result we can define the logical network nw1 as depicted in Fig. 3.2. For this logical network to make sense in network management, the layer 3/4 interface must be standardized as some kind of network interface. Note also that each logical network must have its own N dimension, since it exists only in some of the nodes of a network. This is why we introduce the **logical-node dimension** (NO), i.e., we let each logical network define its own structure of logical nodes.

A complex layer structure (e.g., the OSI layer structure) may also be horizontally partitioned on several levels by defining more than one layer interface as network interfaces, resulting in a structure of nested logical networks (see Fig. 3.3). Note that, at the extreme, each layer in a layer structure can define a logical network.

The connection between logical nodes on a higher level depends on connections on a lower level (this is the essence of the OSI RM layer concept). We interpret this as that an upper logical node is a **hosted** node and the lower one is a **hosting** node. The relation between these node types is a **hosts** relation. The hosts relations define a structure of logical networks. We therefore say that these relations exist in the **logical-network dimension** (NW).

By using this terminology, the whole network is modeled as a **logical-network structure**, i.e., a structure of logical networks that are related by hosts relations between logical nodes. Thus, we actually dissolve the N dimension into an NO and an NW dimension.

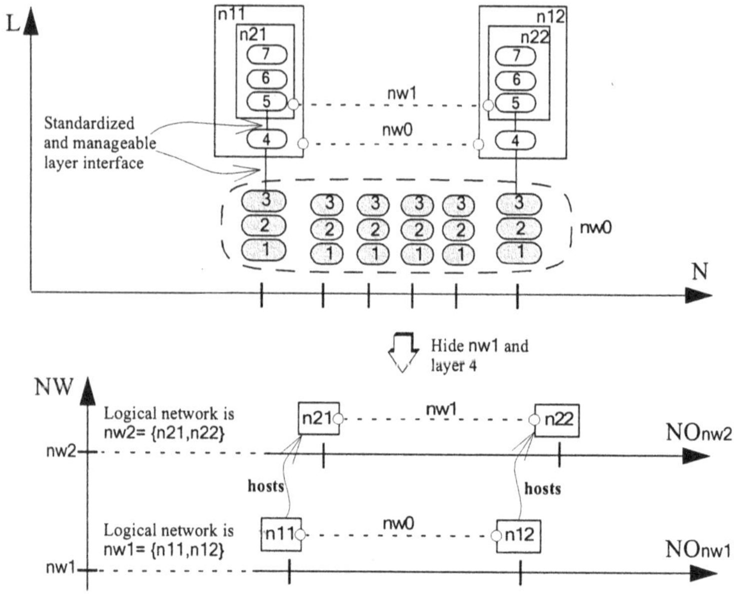

Fig. 3.3 A possible logical-network structure in the OSI RM. NW=logical-network dimension

Both vertical and horizontal partitioning is recursively applied in a number of refinement steps when a network system is developed. Each step refines the model to a lower **network level**. It is very important that each network level is modeled separately from the previous one, since protocols and layers on different network levels are independent of each other. We therefore show network levels in models by the *substructure* relation. For example, the complete node structure for the model in Fig. 3.1 would be modeled graphically as in Fig. 3.4. Note that there may be different numbers of lower network levels inside nodes of the same logical network.

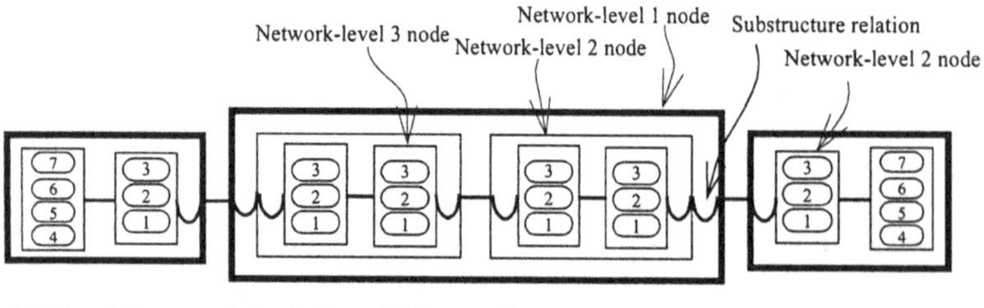

Fig. 3.4 Possible network levels in an OSI network

In this case we know that the internal nodes in terminals have been separated by vertical partitioning, since we reveal information about which layer elements are included. In general, however, a node structure model does not, and must not reveal such information. Therefore, a node-structure model over several network levels will not tell whether two nodes have been separated by vertical or horizontal partitioning. To find out which it is, the layer structure model and the mapping relations (called **allocates**) between a node structure and that model must be consulted.

An obvious distinction between the layer structure inside a logical node (i.e., its functionality) and the logical network model is that the latter holds for many other layer structure as well. Thus, a logical network model shows a network-wide package structure that is highly insensitive to changes of the functions it performs. This is also why we normally do not define protocol relations in logical network models, since protocols imply a particular functional structure inside logical nodes. In case we want to show this functionality in the context of a logical network model, we define a network interface (NI) as depicted in Fig. 3.5.

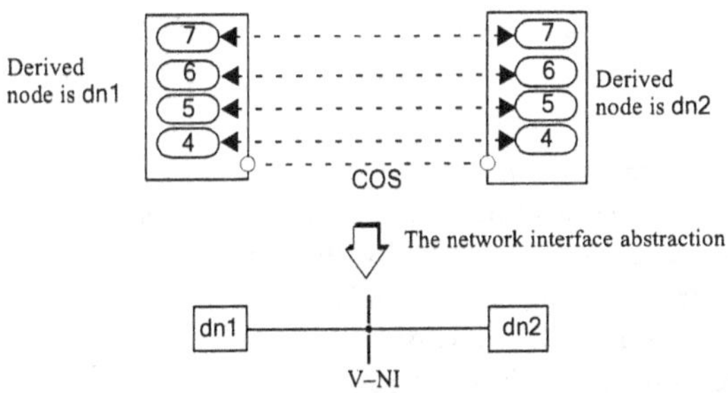

Fig. 3.5 The network interface abstraction. dn...=derived node; V–NI=network interface created by vertical partitioning

A network interface in AMLn is confined to a particular logical network. It comprises both the layer structure of the logical network and the connectivity relations on which it relies. The nodes that are related over a network interface are *derived* by mapping a logical-network structure on a layer structure, which results in node types with completely defined functional contents (which the "logical node" is not). We therefore call such nodes **derived nodes**.

Derived nodes and network interfaces are very volatile entities since they are affected by every possible modification of basic AMLn structures (layer structures and node structures). Obviously, network structures based on derived nodes and network interfaces need not be explicitly modeled in AMLn since they can be derived from the basic structures.

Derived nodes and networks act as requirements on suppliers' solutions to network elements. Such models should therefore define if the separation of two derived nodes is created by vertical or horizontal partitioning since logical network models cannot. To demonstrate that we show in Fig. 3.6 the effect of partitioning a derived OSI node that implements OSI layers 4–7 in both ways. In the vertical partitioning approach we separate the layer 7 element in two interworking nodes. In the horizontal partitioning approach we separate layer elements 4–6 from 7 in separate nodes.

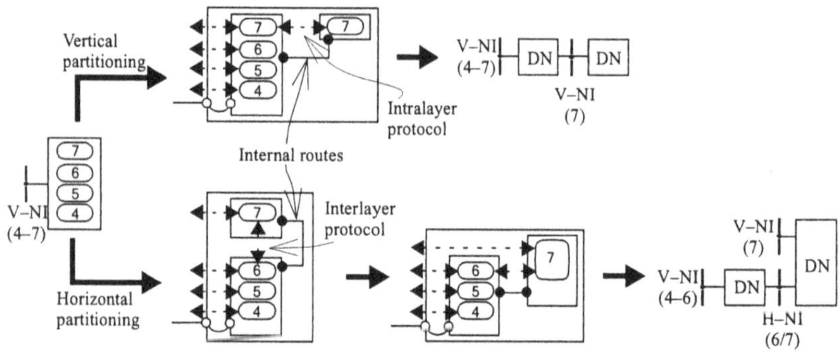

Fig. 3.6 Vertical and horizontal partitioning of the same derived node. DN=derived node

Even if we examine an NI specification, it is not likely for us to realize that the protocol between the derived nodes (DN) is an intralayer protocol in the vertical case and an **interlayer protocol** in the horizontal case. It is therefore very important that network interfaces are identified with regard to whether they represent a vertical or horizontal partition, and which layers or layer interfaces they implement. Figure 3.6 shows a possible way of identifying NIs.

Network models that appear in standards always describe some kind of derived network. Nodes in such models are called either "logical node" or "network element" (the concept of "derived node" is based on AMLn; the hosting versus hosted node definition has not been observed in any standard so far). The concept of "network interface" is commonly used in standards, in a similar sense as in AMLn. The way an NI should be identified (see Fig. 3.6), in order to be able to relate it to a layer structure in an unambiguous way, is not specified, however. In an attempt to produce something similar, network interfaces are commonly called either UNI (user-–network interface) or NNI (node–node, node–network or network–network interface). Both UNIs and NNIs normally denote interfaces in vertical partitions (compare with Fig. 3.1). Thus, the distinction between vertical and horizontal partitions (as suggested by Fig. 3.6) is never reflected in standards by NI names or specifications.

3.2 Logical Networks

3.2.1 Definitions, Parameters, and Tables

The concept of "logical network" has been used from time to time in network standards, as well as in implementor's solution models. There exists, however, no formal definition of this concept (it is also not included in the OSI RM). For example, why is it "logical" and not physical? What then is a "physical" network? Why do we need to define logical networks? We spend a considerably amount of pages in this book to shed some light on these issues.

When we discuss or describe a network system, we often refer to only a part of all functions that exist in an operator network. For example, the network system ATM is defined by the ATM switching layer and a number of ATM adaptation layers (AAL). An operator who operates an ATM network must also implement a number of layers that together can serve the ATM part with transmission functions. Such parts could be a number of SDH layers (another network system) and, in addition, different types of transmission systems for access transmission. Since a particular transmission technology can be used over different kinds of physical media, the system of physical media on which the operator network relies constitutes another "layer of functions," that is not necessarily an instantiation of any of the other network systems.

Another well-known example is the Internet: if we regard all specifications that are published as Internet standards, we have the standardized version of the Internet network system. The Internet consists of much more than that, however. For example, without all the networks that are used for connecting IP nodes to each other (PSTN, ISDN, FR, ATM, SDH, IEEE LANs, DSL technologies, etc.) and used as transmission systems, there would not be an Internet. All these supporting networks are instantiations of other network systems. Furthermore, what users normally perceive as Internet functions are a very large number of services which are not at all part of the standardized Internet. The functions that produce such services exist as application layers on top of standardized Internet layers.

Altogether, these scenarios reveal that real networks are built by at number of "groups of layers" that each is specified in, and an instantiation of, a particular network system. It is therefore essential that layer elements belonging to such a "group of layers" can be isolated in some kind of logical nodes that can be interconnected into a logical network. In doing so, one creates a model of a real network that consists of a number of logical networks that rely on each other, that are configurated on top of each other, and that may be managed separately. The main purpose of this is to provide maintainable relationships between what is standardized in a particular network system, and one or several interdependent logical networks in a real network. By these relations, any changes in functional requirements that are specified in a network system standard can be traced to a particular logical network, and from there to components in operator networks. Logical networks are also excellent targets for network configuration, derived networks for function testing and for fault tracing, provided that they are adequately defined, and that layer interfaces between logical networks are identifiable and manageable in system solutions.

Furthermore, most networks are in constant evolution, which implies that parts may be replaced by other parts that are instantiations of newer network systems (that offer more attractive technologies). For example, ATM networks that rely on SDH networks have been built so far, but the advances in optical switching technologies means that, in the near future, it will be possible to replace existing SDH networks with optical transmission networks.

How well logical-network definitions serve these purposes is, however, a matter of how forward-thinking network system designers have been. It is also not necessarily wise to put all functions of a particular network system in a logical network of its own. This all depends on the extent of the network system definition. For example, to allocate all functions of the PSTN/ISDN-network system in a single logical network would not do any good, considering the enormous evolution these network systems have already been exposed to for many years.

Basically, without an understanding of how *services* in a network rely on each other, it is difficult to define adequate logical networks. The industry therefore tries to define a service hierarchy for networks, so far without any essential result. Nevertheless, since the days of the OSI RM, the industry seems to agree at least on one basic paradigm for separating functionality in networks, which is in a structure of connectivity layers and connectivity strata ("**connectivity structures**" in AMLn). The OSI RM layer structure relies completely on this paradigm, as we showed in Chap. 2, although it defines neither physical nor logical network concepts

Let's now look more thoroughly at how logical networks and logical-network structures are modeled. Figure 3.7 shows the logical network nw1 in its graphical representation and in tabular form, as well as in the context of its hosting networks nw01, nw02 and nw03 (the tables are used by a management system when configurating the logical network as a part of another network).

A logical network model shows a number of logical nodes interconnected over some types of connectivity relations. Due to the large variations in structure and properties of connectivity relations, we use the generic name **route** for all types of such relations (route specializations are discussed in Sect. 3.3).

All logical networks rely on one or several hosting networks for transporting messages between logical nodes of the same logical network. The **(logical) network identifier** (NWi) for the hosting network that realizes a route is annotated on the route. The logical-network structure that is implied by these hosting networks is indicated in Fig. 3.7. Note that a hosted node (such as n11) may be hosted by more than one hosting node (n11 is hosted by both n011 and n021).

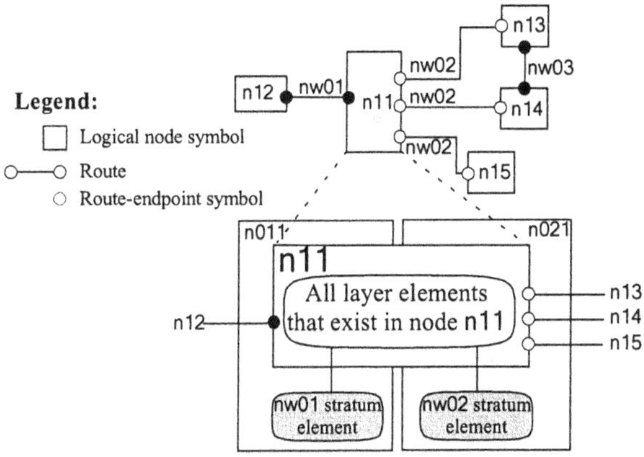

Logical network nw1:

Route	Hosting network
{n11,n12}	nw01
{n11,n13}	nw02
{n11,n14}	nw02
{n11,n15}	nw02
{n13,n14}	nw03

Logical-network structure:

Hosting node		Hosted nodes	
NN	NWi	NN	NWi
n011	nw01	n11	nw1
n021	nw02	n11	nw1
n012	nw01	n12	nw1
n022	nw02	n13	nw1
Etc.			

Fig. 3.7 A logical network. n...,NN=node name; nw...,NWi=logical network identifier

To be able to handle a logical network separately in configuration activities, all nodes are assigned **node names**. These names belong to a known or implied **numbering system** that may be unique to the particular logical network.

A layer element in a node must identify a remote node with **address parameters**, (AddrPar) that are defined by the hosting network, see Fig. 3.8.

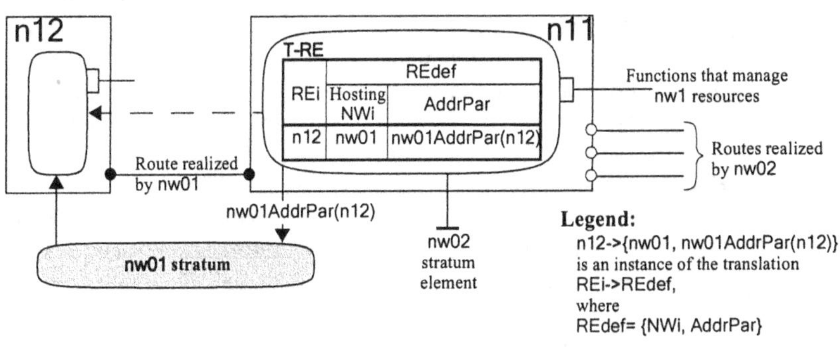

Fig. 3.8 Route identification and addressing in logical networks

The nodes n11 and n12 are logical nodes of the logical network nw1. In this scenario, n11 wants to send something to n12. n11 therefore must refer the route to n12 by the endpoint on n11. We say that n12 is the **route-endpoint identifier**, REi, in the scenario. However, in this scenario, the hosting network nw01 is assumed to know nothing about how nodes are identified in nw1. It is therefore necessary (in this case) to store information in n11 about how to refer its route endpoints in communication with hosting nodes. We call this information **route-endpoint definition** (REdef).

Thus, in general, a route endpoint in a logical network is defined by an REi (which is equivalent to a node name in the very logical network) and an REi→RE-def translation. Translations for all routes that terminate in a logical node are stored in a **route-endpoint table** (T-RE) in the logical node. These tables are normally set and updated from a central management system although decentralized updating (by special protocols between the logical nodes themselves) exists as well.

The model in Fig. 3.8 exemplifies that the routes on which a logical node relies can be realized by several hosting networks. These may be of different types, which implies that they define different types of addressing. In general, therefore, the REdef for a route must comprise two parts: a hosting-network identifier (NWi) and a number of address parameters (AddrPar) for that network.

Each network system may define its own address parameters. This is because they operate differently. For example, a network that offers switched on-demand services requires a destination network address (destination NA). Some networks require an access-point identifier (APi), others don't. There are also networks that accept names of hosted nodes as address parameter. These networks commonly refer to this address parameter as a **global title** (GT).

We must live with such differences in addressing. However, hosting networks that operate in a similar way (e.g., MTP L3, IP and X.25 datagram), and therefore offer the same type of address parameters to hosted networks, use (with no exception) different parameter names. This is a very unfortunate state of affairs for everybody who tries to understand network system other than those already known.

Figure 3.9 gives an overview of address parameters and names used in a number of well-known network systems. The symbols shown in the right-most column may be used as annotations for functional elements and logical nodes in models to denote the type of relaying (or "transfer mode"). Since there are only three types of address parameters that are needed for routes (GT, NA, APi), the table exemplifies how one can complicate things by not using the same name for the same thing. For example, instead of calling the TCP access point TCP–SAPI (which was indirectly suggested already by the OSI RM), one chose "port." Furthermore, some systems use a single name for a set of addressing parameters that belong to different layers, such as the "socket" in TCP/UDP and the parameter PTI in the ATM protocol (the PTI is discussed in Sect. 6.4.3).

The logical nodes of network nw1 in Fig. 3.8 depend on the services and addressing methods of the hosting networks nw01 and nw02. This dependence may become a problem when designing the functions and protocols of the logical network nw1, and for managing this network. For example, a hosting network may offer an unreliable service. Some hosting networks may offer non-switched services only where a switched service may be more useful. Furthermore, if we look at the layer element depicted in Fig. 3.7, it is obvious that the routes of the logical network nw1 depend on where in a hosting network the hosted nodes are allocated. Any reconfiguration of nodes of nw1 will therefore cause problems, since a number of T-RE tables must be updated.

Network system	Addressing parameters			Comments	Relaying symbols
	GT	NA*)	APi		
PSTN/ ISDN	UPT num.	Address E.164 E.168	—	Point-to-point E.168 is the UPT application of E.164	⊗ Circuit
GSM	MSISDN	Address E.164 E.213	—	Point-to-point E.213 is the PLMN application of E.164	⊗
X.25	—	Data number X.121	—		⊖ Virtual circuit ⊛ Datagram
ATM VP/VC	—	Address I.330	—	I.330 is an application of E.164.	⊖
IP	—	IP-addr. RFC1166	Protocol	"Protocol" refers TCP, UDP etc.	⊛
TCP/UDP	***)	IP-addr.	Port	TCP is CON, UDP is CNL	
MTP	—	Point code	Service indicator	"Service indicator" refers SCCP, ISUP, etc.	⊛
SCCP	GT	—	—	Used GT refers some numbering system, e.g. E.164	⊛
	—	Point code	Subsystem number		
Ethernet (LANs)	—	MAC address	—	Point-to-point. Point-to-multipoint Broadcast	⊛
LAPD	—	—	Service access point identifier	Point-to-point Broadcast	
PPP	—	—	Protocol	Point-to-point. "Protocol" refers e.g. IP and other networks	
SDH	—	—	**)	Point-to-point, several channel rates	⊗ Cross-connect
PDH	—	—	**)	Point-to-point, several channel rates	⊗
BRA/PRA	—	—	**)	Point-to-multipoint of D and B channels	

*) Networks that use NA are switching networks

**) Discrimination is by position

***) Domain names (a kind of GT) are supported by DNS

Fig. 3.9 Addressing parameters defined by different network systems

These are all good reasons to modify nw1 to a logical-network structure instead. In Fig. 3.10 we have re-defined nw1 as a *common* hosting logical network. Two hosted networks are shown, nw2 and nw3. The nw1 model tells us that it is hosted in three logical networks, nw01, nw02, and nw03, which are not shown here, however.

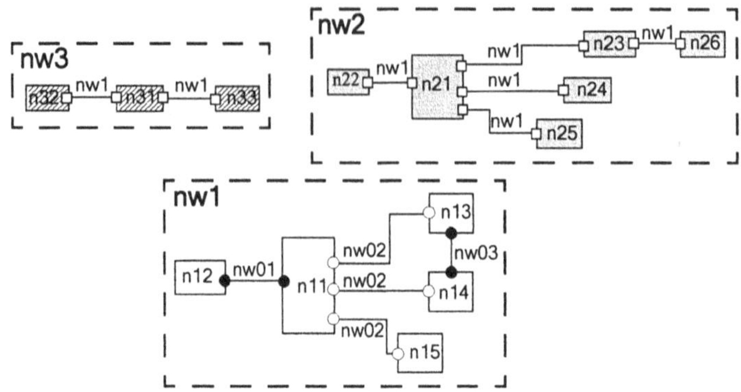

Fig. 3.10 nw1 is a hosting network for nw2 and nw3

The information given by these models tells, through different route symbols (read more about this in Sect. 3.3), which type of connectivity service a particular logical node must rely on. The annotations on these symbols identify the hosting logical networks (e.g., the route between n13 and n14 is created through the logical network nw03). This information is the part of the logical network dimension (NW) that is not dependent on the exact configuration of hosted nodes in hosting nodes. It is therefore a relatively stable form of information of the model. We can refine these models to define the configuration of logical nodes in a particular network by adding hosts relations between nodes of different logical networks, which is the more volatile part of the NW dimension. Figure 3.11 shows several ways in which this information can be specified.

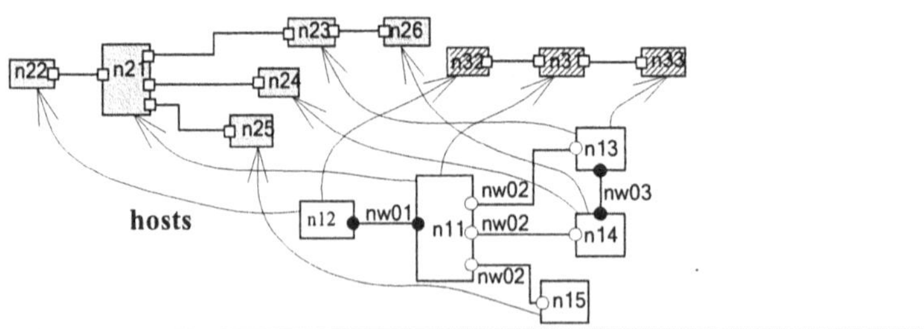

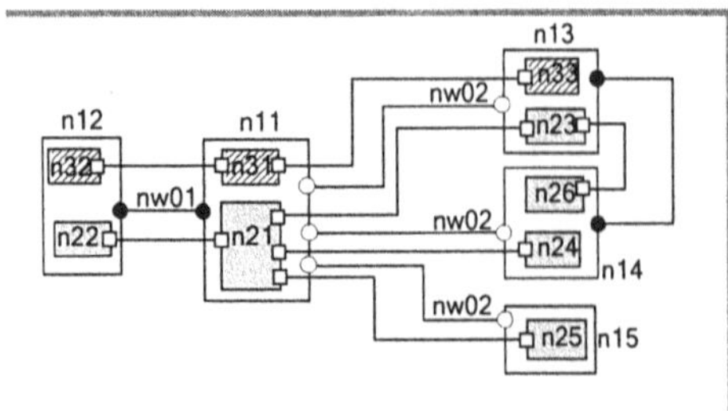

Logical-network structure table:

Hosting node		Hosted nodes	
NN	NWi	NN	NWi
n11	nw1	n21 n31	nw2 nw3
n12	nw1	n22 n32	nw2 nw3
n13	nw1	n23 n33	nw2 nw3
n14	nw1	n24 n26	nw2 nw2
n15	nw1	n25	nw2

Fig. 3.11 Different methods to define hosts relations, i.e., logical network structures

At the top we have refined network nw1 with graphical hosts relations. Note that a logical node (e.g. n11) can be hosted by nodes of several other networks. At the lower left we used the method to draw a hosted node inside its hosting node to show the same thing. This method may seem attractive in this case. However, in more realistic cases, it can create models that are so cluttered with route symbols and annotations that the model is not intelligible. The lower right corner of Fig. 3.11 shows a table representation.

Translating a logical network (here nw1) to a common hosting network requires us to extract such elements out of the original layers of nw1 that can be useful for many different functions, and put them in a *common layer* that interfaces all hosting networks and has a standardized and manageable layer interface to other functions/layers in nw1, as shown in Fig. 3.12. The model shows one of the nw1 nodes, n11, in the L–N plane. The internal structure can be directly derived from the logical-network structure of Fig. 3.11.

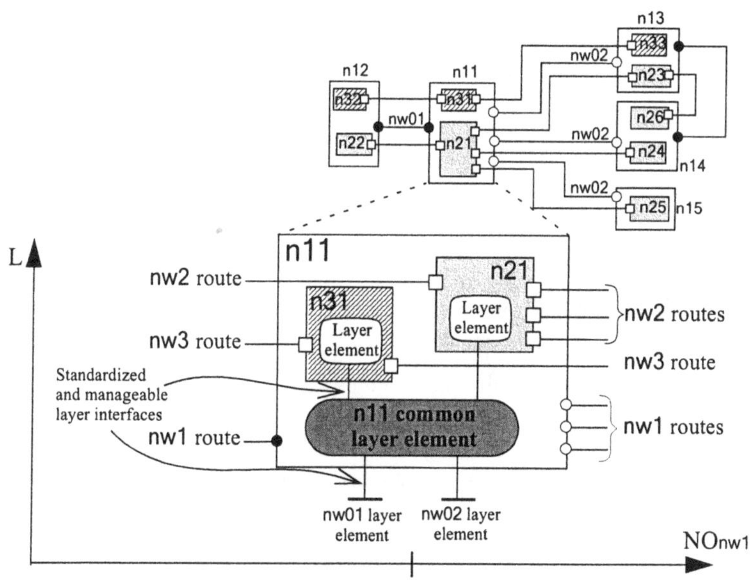

Fig. 3.12 A hosting node of nw1

The actual properties and design of such a common layer varies a lot in existing network systems. For nw1 one can imagine any or all of these functions of a common layer:

- A common layer may allow nw1 to act as a hosting network for a number of hosted networks that operate concurrently. In that case the common layer must perform some form of discrimination, as was discussed in Sect. 2.2.

- It may allow networks that are hosted in nw1 to use their own numbering system for addressing, i.e., making them independent of whatever addressing parameters the hosting network uses.
- It may protect hosted networks in nw1 from any changes of layer-interface primitives in case any of the hosting networks nw01 or nw02 are replaced or modified. This requires a modification of the common layer, but not of layer interfaces that are used by hosted networks in nw1.
- It may protect hosted networks in nw1 from using an unreliable hosting-network service. In such cases the common layer will handle a service-enhancing protocol.
- nw01 or nw02 may be synchronous, and the common layer may create an asynchronous service for its hosted networks. This requires the common layer to handle a framing protocol (such as in MTP L2). It may also be the other way around (as in ATM/AAL1 circuit-emulation).
- It may protect hosted networks in nw1 from changes of connectivity type. For example, in the case where nw01 or nw02 is a connection-oriented network, the common layer in nw1 may create a connection-less service for its hosted networks. This also requires the common layer to handle a protocol.

If we transform nw1 to a hosting network, we have to define such a common layer, as Fig. 3.12 shows. The common layer element in node n11 is part of the common layer in nw1. Note that nodes of hosted networks are volatile entities in the model in relation to the nw1 itself. nw2 and nw3 nodes may appear, disappear, or be reconfigurated in the nw1 network at any time.

The way configuration tables in nw1 look depend on how nw1 supports discrimination and address translation. Figure 3.13 gives an example. This set of tables is based on the following requirements:

1. We assume that nw1 uses discrimination by the layer APi method, according to Sect. 2.2.2 (which implies that layer elements of the same layer all are associated to the same APi value).
   This problem is solved by realizing a table called T-AP in the common element that relates an APi value to a hosted logical-network identifier (NWi).
2. We assume that the common element allows hosted nodes to use their own numbering system for addressing, i.e., node names of networks nw2 and nw3, respectively. When these names are used in the layer interfaces, the common element regards them a global titles (GT).
   This problem is solved by adding another table, the GT translating table, T-GT. This table defines the translation of a global title to a route-endpoint identifier (REi). The REi identifies the route between n11 and the nw1 node that hosts the node that is referred to by the GT. Note that the output of the T-GT table is an input to the T-RE table.[3]

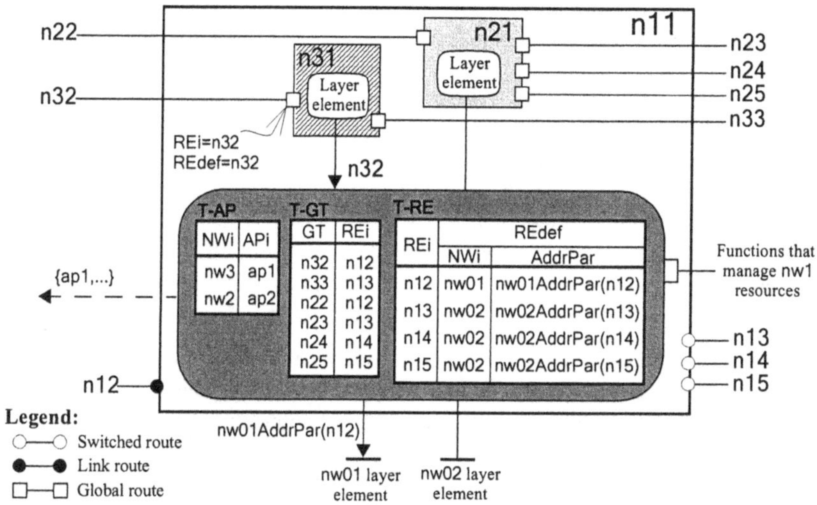

Fig. 3.13 Tables in a hosting node that supports address transparency

The model also shows a scenario where node n31 sends something to n32:

- n31 addresses this node by the global title n32.
- By using the T-AP table, the common element creates a message that includes the APi value for n32 (which in this case is equivalent to the APi for n31).
- The common element must then find the nw1 route that shall be used, which is given by the T-GT table (as n12).
- The common element then consults the T-RE table to find out which hosting network to used (here nw01) and how n12 is addressed over nw01.
- The data is then sent over nw01.

Through the realization of these tables in common layer elements, a high degree of *addressing transparency* is achieved. We regard this as the most important property to be shown in graphical models of logical networks. Route symbols are therefore defined primarily for different addressing methods (see Sect. 3.3).

3.2.2 Simulation and Realization of Logical-Network Structures

Layer interfaces are the entities in logical-network structures that create the functional structure of a network. However, the model in Fig. 3.13 identifies (deliberately) layer interfaces only indirectly by referring to identifiers (NWi) of hosted and

[3] The T-GT table translates a GT to an REi in this case, since n11 has direct routes to all other nw1 nodes. If nw1 had been a switching network, the T-GT would have translated a GT to an nw1 network address (NA) instead, to be used as an input to a routing table, T-ROs.

hosting logical networks. Hosting networks are identified in T-RE tables by their NWi. Hosted networks are identified in T-AP tables by their NWi. This is because we want the model to be valid regardless of how a network system will be realized.

Layer interfaces, as we have seen in Chap. 2, are defined only with respect to which operations and abstract service primitives (ASPs) they support. What is missing, therefore, is the connectivity support that may be required to allow two layer elements to address each other and communicate over a layer interface. For vertically-partitioned networks, this is actually not an AMLn problem, but something to be solved by implementors. For horizontally-partitioned networks, and when one wants to simulate the behavior of a model, communication over layer interfaces of concern must be specified. In principle, it can be done in two ways:

1. If one wants the two layer elements to be remote to each other, one must define an interlayer protocol that can carry the events (i.e., instances of ASPs) of the interface between the elements. An additional connectivity stratum is then also needed to support that protocol.
   This is actually a refinement of the original model (since it creates additional protocols and layers), and may therefore be defined as a part of the model itself, but on a lower network level. We describe this kind of refinement in details in Chap. 4.
2. In simulation, and on the lowest network level in solution models, layer elements are no longer remote, but run as applications on the same processor or on a common processing platform, which are the elements that provide the connectivity for layer interfaces.

There are many different design languages and platforms available for this kind of connectivity. However, no matter which method is used to configurate layer elements correctly in that context, identification of layer interfaces will be required. We call such an identifier **layer-interface identifier** (LIi). A layer structure can be refined to a model that can be simulated or executed by defining the LIi for each layer interface, and then define the platform application address (PA) of the layer elements that connect to the endpoints of a layer interface. The translation between NWi and LIi is then defined in a T-LI table, and the definitions of platform addresses are defined in a T-PA table for the particular platform.

Figure 3.14 shows the corresponding tables for the layer structure of Fig. 3.13. All elements of this model are assumed to run on the same platform (which may be a simulator). The platform addresses of these elements are defined for each LIi in the T-PA table as hostingPA and hostedPA respectively. Logically, the depicted scenario goes as follows:

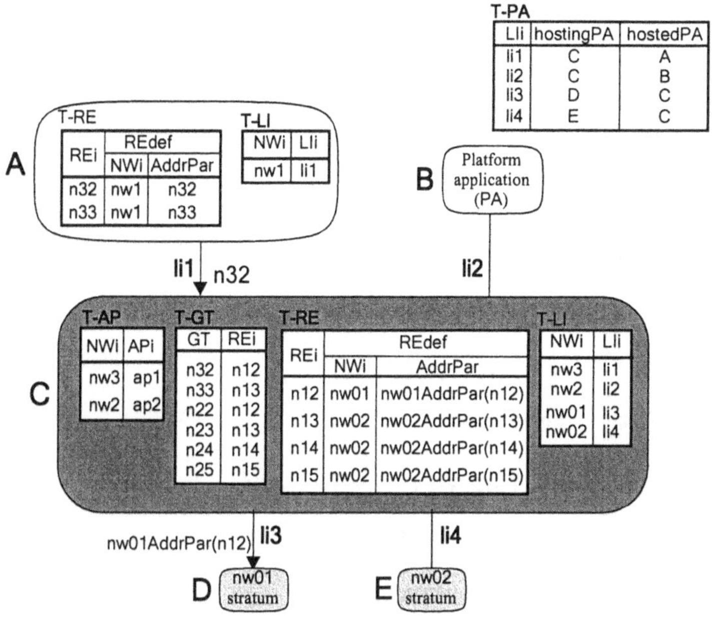

Fig. 3.14 An executable layer structure. All layer elements are also platform applications. LIi=layer-interface identifier; PA=platform address

- Element A (in node n31) wants to send something to n32. Its T-RE table shows that n32 can be used as address parameter and the hosting network is nw1.
- A gets the platform address (C) to nw1 through its T-LI and the T-PA tables. The address parameter n32 together with message data are sent over the platform, referring C.
- C (in node n11) reads the address parameter n32 and finds via its T-GT and T-RE tables that the data are to be sent over nw01.
- C gets the platform address (D) to nw01 through its T-LI and the T-PA table, and sends the data over the platform, referring D.

Note that the only table that depends on the actual platform is the T-PA. Thus, the same model can be run on different platforms just by associating it with a relevant T-PA. Note also that in real simulation and execution, the T-LI and the T-PA tables are not run-time tables, but data that will be used for compilation of the model.

3.3 Route Properties and Symbols

The properties of routes must be specified in network models. To produce expressive models of logical networks, we also need symbols for routes. The large variation of route types makes it necessary to select a few criteria to affect the route symbol, and add all other properties of routes as attributes in logical-network models. We normally put the name of the hosting network that realizes the route as an annotation, as exemplified in Fig. 3.15. This annotation also servers as a reference to the transfer properties (rate, quality of service, etc.) of the route.

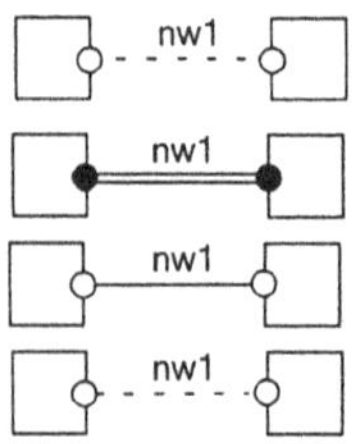

If nw1 type is ISDN → a route in, e.g., the ISDN media service logical network, realized by the ISDN switching stratum.

If nw1 type is LAPD → a route in, e.g., the ISDN call-handling logical network (in the access part).

If nw1 type is IP → a route in, e.g., the TCP logical network.

If nw1 type is TCP → a route in an application logical network of an IP-based network.

Fig. 3.15 Defining route properties in logical networks

A number of other criteria may influence the design of route symbols:

1. One criterion may be the *position* in the layer structure of the layer that realizes the route.
2. Another is how the route is *addressed* (as mentioned previously).
3. Some routes are realized *on-demand*, others are realized for *permanent* or *semi-permanent* use.
4. Another property of interest is if the data flow over the route is a single flow, or separated into multiple flows (connections or channels).
5. A property that must be represented in a route symbol is its *structure*, i.e., whether it is point-to-point, point-to-multipoint or broadcast.
6. A property of interest might be if the data flow is *synchronous* (as in PSTN/ISDN/PDH/SDH) or *asynchronous*, as in all routes realized by packet-switching systems (adaptation layers such as AAL1 in ATM are disregarded). We do, however, not use this criterion in the route symbols that are defined here.
7. Some routes are *duplex*, some are *simplex*. Duplex routes may be *symmetric* (the same transfer rate in both directions) or *asymmetric*. These criteria are also not used in the route symbols that are defined here.

Criterion 1: If we had modeled an OSI system, it would have been natural to define a specific route type for every one of the seven layers. However, we are not pursuing such a model (the author regards the fixed number of layers and the position of layers in this layer structure as a severe mistake of OSI RM). As we have seen, real networks exhibit very complex interlayer structures and no network system is comparable as far as their layer structures are concerned. There is actually no upper limit to how many layers may exist.

A lower boundary exists in every network, however, and that is the interface towards a **physical-media stratum**. We therefore introduce only one special route symbol within this criterion the **physical route**, which is a route that interconnects logical nodes over such strata. Since nodes in such a network model are special compared to every other logical node type, we say that physical routes interconnect **physical nodes** in **physical networks**. This "physical" picture of a network is important for several reasons:

- Through the depicted physical connections, the model can be used as a basis for describing all details of how to connect nodes of the operator network by physical media.
- Physical media and the electromechanical design of the nodes constitute a major part of the costs for implementing and managing an operator network. The model of a physical network is therefore important for analyzing the cost aspects of an operator network, and for developing a plan for installation and management of physical elements.

Note, however, that physical routes are abstractions, as any other type of route. A physical route abstracts a path between two or more physical interfaces to a physical-media stratum. Such a stratum is built by electromechanical devices, such as physical connectors, antennas, cables, cable concentrators, and manual cross-connecting devices, that in wide-area networks constitute a major part of the total cost for building the network. In this book, however, we do not deal with how to model and design physical-media strata. Instead we hide such strata behind physical routes.

The symbol for physical route has a black square at its endpoints. It is also the only route symbol that reveals something about its properties (wireline or wireless) through the symbol, as depicted in Fig. 3.16.

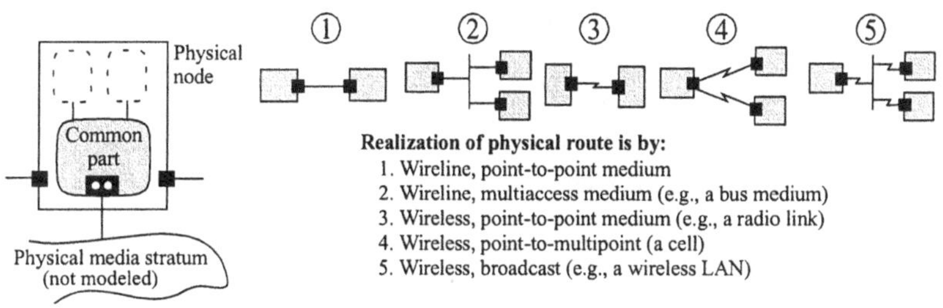

Fig. 3.16 Symbols for physical routes. Such a route is realized by a physical-media stratum

Also shown to the left in this figure is the special layer-interface symbol that is recommended for denoting interfaces to physical media in layer structures. Note that a physical node is the lowest hosting level that can exist in a logical-network structure.

Criterion 2: There are many ways a route may be addressed, as well as many different addressing systems. We therefore base route-endpoint symbols only on whether a T-RE table is required, and, if so, if a T-RO table is needed. This results in five types of route endpoints which, by their symbols, indicate how the route is addressed (see Fig. 3.17).

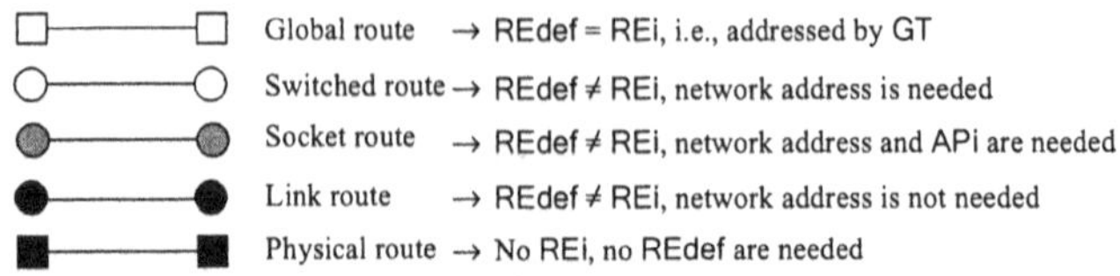

Fig. 3.17 Route-endpoint symbols distinguish addressing methods

1. **Global route**: this is a route that requires nothing but a global title for addressing (implies that the route endpoint need no REdef specification in the hosted node).
2. **Switched route**: switched-route endpoints require an REdef specification that contains only some kind of hosting-network address (CEi may be required). The switching network may support an APi, but then used according to the layer APi method
3. **Socket route**: This type of route requires both a network address and an APi. The switching network supports the layer-element APi method.
4. **Link route**: both global, switched and socket routes imply some kind of routing in the common element. Routes may, however, be realized just as point-to-point hosting networks, in which case no hosting network address is needed (other parameters may, however, be required in an REdef). Such routes are called link route.
5. **Physical route**: a physical route belongs to a logical network that interfaces a physical-media stratum. It requires neither an REi nor an REdef.

Criterion 3: routes that are established on-demand are drawn with dotted lines, all other routes with solid lines. Note that this distinction may be used both for global routes, socket routes and switched routes.

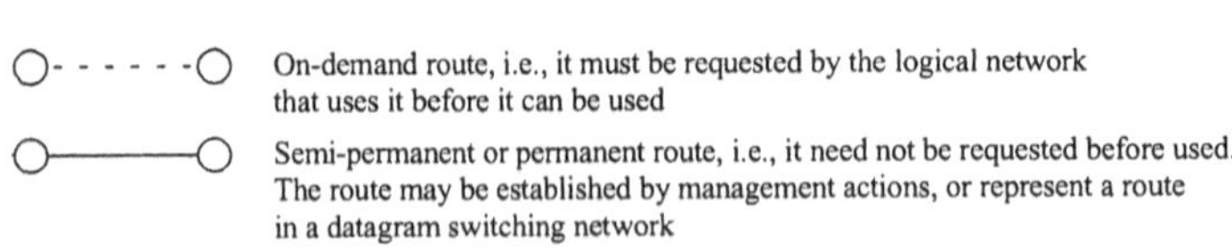

Fig. 3.18 Route symbols that identify the permanency of a route

Criterion 4: routes that comprise more than one channel are drawn with double lines, as the example in Fig. 3.19.

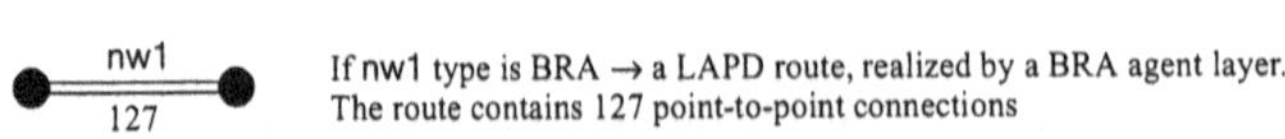

Fig. 3.19 Route symbol for routes with multiple connections/channels

The symbol can be annotated with the number of connections or channels in the route. Normally, all channels in a multiple-channel route are identical. When such a route is used, data to be sent or received must be identified as to which channel it belongs. Methods that are used differ between synchronous and asynchronous routes:

- In *synchronous* routes, channels are identified by their position in the frames of the hosting network. Such channels are normally called *timeslot* (such as the 32 timeslots in an E1 transmission system, or the eight timeslots of an GSM carrier). Since such channels are available to a hosting node in precise, recurrent time intervals, no identifier other than a synchronization signal is needed in sending and receiving.
- In *asynchronous* routes, channels must be identified by some type of **channel identifier** (CHi), both by the using hosted network and in the protocol that the hosting network uses. The OSI RM (which calls channels "connections") suggests that different identifiers should be used at the endpoints of a channel (**connection-endpoint identifier**, CEi) and in the protocol (**connection identifier**, Ci) in order to allow local references to be used inside hosting nodes. As usual, however, each standard tends to use its own terminology: the connection-oriented X.25 network calls its channels *channel*; ATM calls its channels *virtual circuit* and *virtual path*; LAPD calls its channels *connection*. Some standards do separate CEi from Ci, others don't. Identifiers are normally named in accordance with how channels are named (e.g., VCi for "virtual circuit identifier" in ATM), with the exception of LAPD that calls its Ci *terminal identifier* (TEI).

Criterion 5: The route structure is modeled as shown in Fig. 3.20.

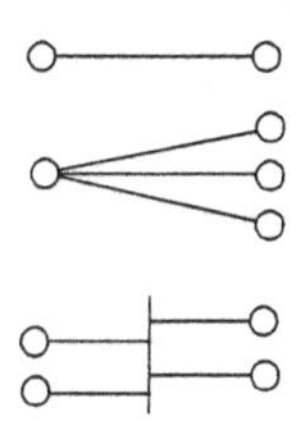

Fig. 3.20 Route symbols for route structures

It is obvious that a much greater number of route symbols could be created by combining these criteria with other properties of connectivity services. This would make network models more expressive but less intelligible. In general, therefore, AMLn recommends only the basic route symbols that were discussed. However, an AMLn user must be allowed to define route symbols that can express the specific characteristics of the network system being modelled. The only requirement from the AMLn point of view is therefore that the user creates a symbol with identifiable route endpoints, to indicate the boundary of a logical node.

3.4 Route Type Examples

3.4.1 Physical Routes

Since physical routes abstract paths through physical-media strata, they terminate in nodes that are hosting nodes only. Such nodes frequently terminate more than one physical route, possibly of different types. Figure 3.21 shows an example taken from GSM: the model shows part of the functionality of a GSM base station ("base transceiver station" (BTS)) in the L–N plane.

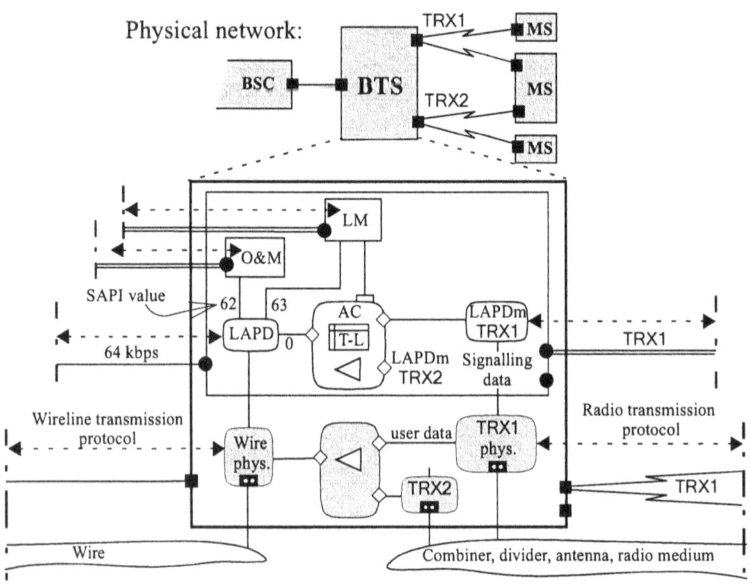

Fig. 3.21 Part of a GSM base station (BTS). AC=actor; BSC=base station controller; LM=layer management; MS=mobile station; O&M=operation and maintenance; TRX=transceiver

A BTS performs multiplexing of two types of streams between many mobile stations (MS, to the right) and a "base station controller" (BSC, to the left). These functions are performed by two logical nodes in a BTS: a physical node that multiplexes user data and that is also a hosting node for a logical node that multiplexes signaling data. The latter also acts as a hosting node for several logical nodes (e.g., O&M and LM) that deal with management.

The physical layer on the radio network side is built by a number of agent layers, with one per carrier. The standard defines an agent as a part of a TRX module (that also includes a LAPDm agent, see page 135 for definition of LAPDm). A single agent layer on the fixed network side handles all physical layer communication between BSC and the BTS. The physical node includes a multiplexing actor that connects all agent layers and multiplexes user-data channels.

A physical route on the radio network side is a wireless point-to-multipoint route. The radio transmission protocol uses the capacity of this route to create eight timeslots (a call occupies one timeslot). A BTS may terminate one or several routes of this type (one per TRX or "carrier"), depending on the expected traffic intensity in the cell it serves. All carrier routes are realized over a single physical medium layer that consists of antennas for reception and sending, filters ("combiner" and "divider" that allow the same antennas to be used for several carriers), and the radio medium.

The single physical route on the fixed network side is realized by a wireline physical medium (as depicted) or a point-to-point wireless medium (a "radio link"). The transmission protocol uses the capacity of the physical route to transport several 64-kbps channels (normally 24 or 32), most of which are used for user-data streams, and at least one for signaling streams. Since a voice call is compressed to 13 kbps in this physical network, a single channel on the network side can carry four calls.

No configuration tables are needed in the physical layer multiplexer, since user-data channels have fixed positions in physical routes on both sides.

The physical node also acts as a hosting node for a hosted node that performs multiplexing of signaling data. The physical node performs its hosting node role by separating signaling channels from user-data channels and deliver signaling data to LAPD and LAPDm hosted nodes. The discrimination method is "by position" on both sides:

- The physical layer agent on the radio network side extracts a number of different types of signaling channels from the frames of the radio transmission protocol. There are special channels for paging (PCH), access (RACH), call setup SDCCH), etc. The physical layer agent delivers data of these channels to the LAPDm agent over the layer interface, also defining to which signaling channel the data belong (as a kind of connection-endpoint identifier). Thus, the physical layer realizes a multichannel link route for the hosted node, with one route per carrier.
- The physical layer agent on the fixed network side extracts the data on the single 64-kbps signaling channel from the frames of the wireline protocol and delivers them to the LAPD agent. Thus, the physical layer realizes a single channel link route for the hosted node on this side.

Let's also briefly look at how the hosted network operates for signaling data received from the radio network side:

- The LAPDm agent performs a typical link protocol function on every received LAPDm frame (error detection, etc.). The protocol is based on LAPD. The possibilities that LAPD gives to define connections (through SAPI and TEI identifiers) are not needed, however, since the hosting network already has identified each type of signaling channel. When LAPDm receives a correct frame, it takes its user data and adds a signaling channel identifier (PCH, RACH, etc.) and sends the resulting datum as user data to the multiplexer over the actor–agent interface.

- The multiplexer must handle a T-L table that relates each actor–agent interface (i.e., each TRX) on the radio network side to a a link identifier (Li) on the fixed network side. The multiplexer delivers the user data with a link identifier value for the actual TRX to the LAPD agent.
- The LAPD agent associates everything coming from the multiplexer as data belonging to access-point identifier 0 (i.e. SAPI=0). It also uses the TEI parameter to identify links (i.e., TRXs). The agent therefore puts the user data in a LAPD frame and sets SAPI=0 and TEI=Li. The data of the frame is then sent over the layer interface to the physical layer agent on the fixed network side.
- This agent puts the data into the single 64-kbps channel over the physical route to the BSC.

As figure 3.21 shows, the LAPD agent layer is also used for making this layer a hosting network for other networks. Other SAPI values are used for that. For example, all received LAPD frames with SAPI=63 are delivered to a logical network for layer management (LM), e.g., for maintaining the T-L table in the multiplexer. The BSC is responsible for the operation of the BTSs it connects. Frames referring SAPI=62 are used for operation and maintenance of the BTS.

3.4.2 Link Routes

We will use LAPD ("link access procedure over the D channel")[4] as an example. LAPD was originally designed for realizing multichannel link routes for many concurrent hosting networks in ISDN access networks. LAPD therefore defines an access-point identifier (called SAPI, "service access point identifier") and a connection identifier (called TEI, "terminal identifier").

The LAPD logical network is an asynchronous, duplex, symmetric, non-switching hosting network that realizes link routes with multiple connections. It is used as a general hosting network in, e.g., the access part of an ISDN network. Hosted nodes belong to diverse control networks, including nodes of the LAPD management logical network. Discrimination between hosted nodes is done by SAPI according to the layer APi method. Figure 3.22 shows the link routes that LAPD realizes[5] and the tables that are used by LAPD layer elements.

---

[4] "D channel" is an ISDN concept. Since LAPD is used also within other network systems that define no D channels (e.g., in GSM), the name LAPD sounds inadequate today.

[5] In this model we have assumed a LAPD application that uses a point-to-point route to connect a LAPD node in a terminal to a LAPD node in a network. When LAPD is used in ISDN, a point-to-multipoint route called the D channel is used instead to connect a number of terminals to the network.

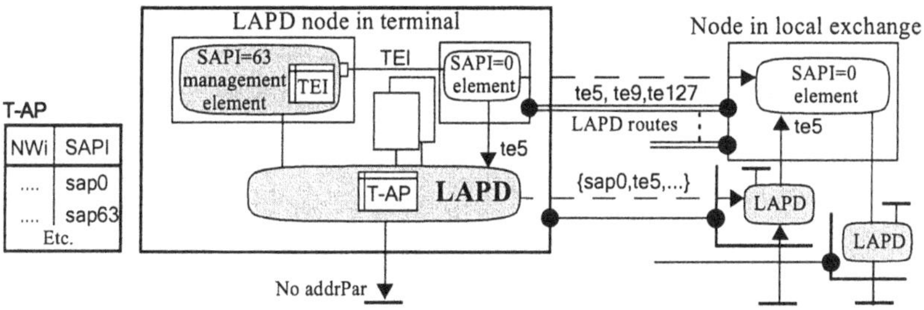

Fig. 3.22 Link routes realized by LAPD. TEI table=list of connections that the terminal can use

A single link route interconnects two LAPD nodes. It is assumed to be a single-channelled, permanent route (in ISDN called "D channel"), i.e., it has no REi and no REdef, which is why no T-RE table is needed in LAPD layer elements of terminals.

A LAPD hosting network realizes multichannelled, permanent point-to-point routes for a number of hosted networks. Layer elements of a particular hosted network are identified by SAPIs. When LAPD is used in the context of another network system, the assignment of a SAPI to a particular hosted network is defined in that context. For example, when LAPD is used in ISDN, SAPI=0 identifies the ISDN call-handling layer, SAPI=63 identifies the LAPD layer-management layer. When LAPD is used in GSM, assignment of SAPI values is different (see the discussion at the end of this chapter).

Hosted nodes in terminals terminate a single LAPD route. Since the hosting network (i.e., LAPD) is non-switching, and since the layer APi method is used for discrimination, hosted nodes in terminals need no REi or REdef for this route, i.e., no T-RE table.

However, since the route is multichannelled, a hosting node must deliver a connection identifier (Ci) to the LAPD layer element at sending. LAPD uses the name TEI ("terminal identifier") for Ci. The range for the TEI is 0 through 127. These connections are not created on-demand, but by a special LAPD management network (that has hosted nodes in every terminal and in the network) when a terminal initially connects to the network. This network is (normally) hosted on SAPI=63. The connection identifier TEI=127, a broadcast connection that exist for every SAPI, is the only connection that need not be assigned by the management network.

Since LAPD was initially defined for the ISDN application, the range 0 through 126 of TEI values is shared between a number of terminals on a terminal site. This is why a particular hosting node has access only to a subset of the total number of TEIs. For example, the model in Fig. 3.22, shows that the hosted node on SAPI=0 uses a LAPD route with two connections, denoted as te5 and te9 (besides the broadcast connection te127). The indicated scenario shows the hosted SAPI=0 node in the terminal sending on te5. Note that the LAPD network does not know how a hosting network uses connections. It just delivers the TEI value to the receiving hosted node.

Let's also take a look at how hosted nodes on the network side make use of LAPD. Figure 3.23 shows the hosted node on SAPI=0, which in ISDN is part of the call-handling logical network. This node depends on a very large number of LAPD hosting networks (one for each subscriber line), of which we only indicate two, here denoted nw1={n11,n12} and nw2={n21,n22}. The call-handling network is called nw3={n31,n32,...}.

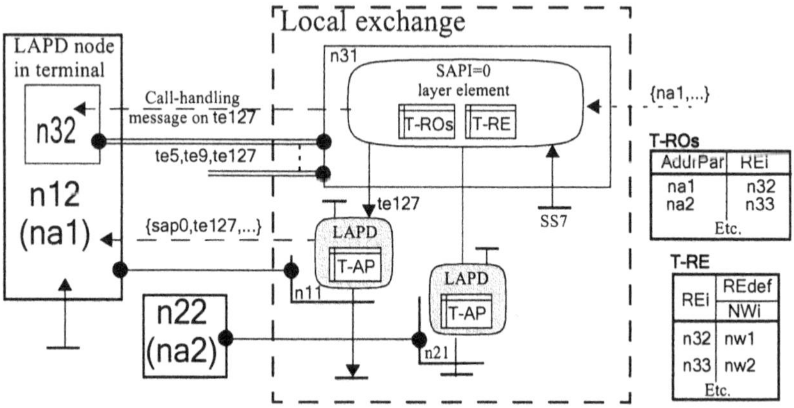

Fig. 3.23 LAPD layer elements in a local exchange

A scenario is depicted:

- The call-handling layer element in the local exchange receives an (ISUP-) message over SS7 that requests a call to a subscriber terminal, identified by the ISDN network address na1 (actually called "subscriber number" in ISDN).
- The layer element analyzes this message and then creates a call-handling (DSS1 L3-) message to be sent to the actual subscriber terminal.
- It must then find the LAPD route to that terminal. Since there are many such routes, this requires a routing analysis and therefore a routing table, T-ROs. The routing table gives the REi=n32, which is the call handling node in n12.
- The call-handling layer element in the local exchange then uses a T-RE table to find the LAPD hosting network to use. As was shown previously, LAPD does not require any address parameters, which is why only the NWi part of an REdef is needed in this table.
- The data of the DSS1L3 message is then sent over the broadcast connection te127. The reason is that (in ISDN) a terminal site may actually comprise many internal physical terminals, each having an n32 node. All of them listen to te127, but the node that actually takes the call must, when answering, use a TEI that is assigned to it, which in this scenario is either te5 or te9. The chosen TEI will then be used in the dialogue that follows between this terminal node and n31.

This is the method used to support multiple terminals on an ISDN user site that all belong to the same ISDN network address.

The GSM application of SAPI and TEI in LAPD and LAPDm (see Fig. 3.21) is very different to ISDN since the connectivity structures to support are different:

- LAPDm is a variant of LAPD used in the GSM radio network. LAPDm uses the SAPI parameter for defining priorities between different types of signals, which is not in line with the intended use of SAPI. The TEI is (most likely) not used at all.
- The T-L table defines the Li→AAli relations. The LAPD agent must deliver the Li to the actor. It uses its TEI parameter for that, which is also not in line with the intended use of TEI.
  The model in Fig. 3.21 shows that LAPD must perform discrimination between data to be relayed in this network and data to be terminated in the BTS. According to AMLn, this requires an interface type identifier (ITi) in the LAPD protocol. LAPD uses the SAPI parameter for that, where SAPI=0 defines data to be relayed and all other SAPI values define data to be terminated. Using SAPI as an ITi is not in line with the intended use of SAPI.

There is a lesson to be learned from this: the author initially learnt about LAPD from the LAPD standard (and therefore, basically, its use in the ISDN context). When later encounting GSM, he naturally expected that TEI and SAPI were used as described in the LAPD standard. It took him many frustrating hours to understand the LAPD applications in GSM. After finally understanding the actual needs in the BTS, and translating those needs to AMLn defined tables, it became clear how and why SAPI and TEI were used in GSM the way they are. If AMLn had been available when GSM was standardized, other more relevant names for SAPI and TEI would have been used in describing LAPDm (probably also another name would have been chosen for the protocol). It had saved a lot of time for all engineers who over the years have tried to understand what is going on in the BTS.

3.4.3 Switched Routes

3.4.3.1 Introduction

The variation of switching techniques and route types in this area ranges from circuit-switched (CS) to packet-switched (PS) routes, from connection-oriented (CON) to connection-less (CNL) routes, from routes that are addressed by network addresses (NA) only, over routes that are identified by both an NA and an access point-identifier (APi), and from switching layers/strata that realize a single level of switching to layers/strata that realize several levels of switching.

- A main difference between circuit and packet switching (as route identification is concerned) is that, in the CS case, the user of the route need not identify the route in any way. In the case of PS, some type of identifier is always needed, either an NA or some type of virtual circuit identifier (VCi). An NA is used also in the CS case, but only by the control layer in the set up phase.

- CON switching is one mode of operation of X.25. Other CON-switching networks are FR and ATM. CNL switching (or "datagram switching," as it is more often called) is another mode of X.25, and is also the mode in which IP and MTP L3 work.
- In CS networks and in X.25, FR and ATM, NA is the only means to identify a route. In MTP L3 and IP, routes are identified both by an NA and an APi. The user of the route is not bothered by the APi, however, since the layer APi method is used by these layers, which implies that the APi is the same for all layer elements of a using layer, and is set at installation time by the management system.
- All PS network systems except ATM include a single level of switching. The ATM/AAL stratum includes three levels of switching. At the basic level, virtual paths (VP) are switched. This is a level that can be used as a cross connecting level for a large set of virtual circuits (VC). VCs can also be switched individually, which is the second level of switching. The third level is realized by the layer AAL2 that can switch parts of user data of cells ("cell" is the packet type that is switched on VP/VC levels).

By definition, when an NA is the only part of the address parameters of an REdef, a **switched route** is referred. A destination network address refers a remote node in the hosting network that hosts a node of the hosted network. The hosting network is a switching logical network that contains a switching layer or stratum. Figure 3.24 shows the principle for realizing and using switched routes in a simple datagram-switching network (more or less similar to datagram switching in X.25).

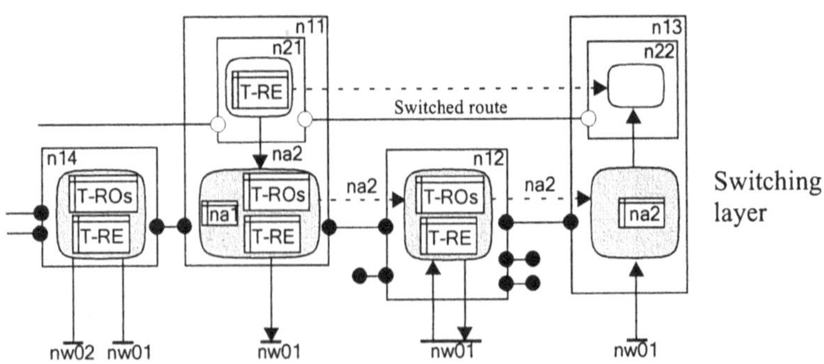

This scenario:

In T-RE(n21): n22→na2
In T-ROs(n11): na2→n12
In T-RE(n11): n12→nw01addrPar
In T-ROs(n12): na2→n13
In T-RE(n12): n13→nw01addrPar

Fig. 3.24 Switched routes in a datagram network

The difference to link routes is that the node (here n11) that hosts a calling node (n21 here) does not necessarily have a route to the called hosting node (n13 here). The hosting node may have to select a hosting network route that connects to another hosting node that can relay/switch data. In general, both the hosting node at the calling site and intermediate relaying hosting nodes may have several routes that may eventually lead to the called hosting node. This is what requires the hosting nodes to handle a table that can translate a network address to a route-endpoint identifier (i.e., NA→REi) of the switching network. We denote such tables T-ROs.

In Fig. 3.24, the switching network nw1={n11,n12,n13,n14,...} hosts the network nw2={n21,n22,n23,...}. nw1 relies on link routes that are realized by two hosting networks, denoted nw01, nw02. The switching network consists of different types of nodes: some nodes both route and terminate (e.g., n11), some terminate only (e.g., n13), and some route only (e.g., n12 and n14). Every node that routes data must include a T-ROs and a T-RE. Nodes that terminate only must know their own network addresses (e.g., na2 for n13) but need no tables. Such nodes are, however, still nodes of the switching network, since the role of a switched logical network is to route, originate, and terminate.

In AMLn, every layer terminates data in terminating nodes. *A switching logical network definition therefore always includes the logical nodes where data are terminated*, even when such nodes do not perform any routing function. In implementers' descriptions, however, terminating nodes are often regarded as existing externally to the "network," i.e., they are not regarded as parts of the switching layer or network. This is because implementers tend to use a closed, vertical partitioning view on networks. From an operational point of view this makes sense: terminating logical nodes exist in terminal equipments that are normally not operated by operators. Operator network parts on the one hand and terminal equipment on the other may also be supplied by different suppliers. These aspects are irrelevant, however, when modeling switching layers and logical nodes of network systems, where an open view on switching networks should be applied.

Let's now look at the network address (NA) that is used for switched routes: the model in Fig. 3.24 indicates that every node in the switching network that hosts nodes must also have and know its own network address (na1 for node n11 and na2 for node n13). However, the depicted scenario also indicates that it would work equally well if the hosting node name n13 had been used instead of the network address na2. So, why do we use two different parameters for the same thing? There are at least three good reasons:

1. Similar to the network in Fig. 3.24, most public switching networks contain many switching, multiplexing, and relaying nodes that do not host any nodes, but need to be identified by their node names in the T-ROs and a T-RE tables of the switching network. Thus, network addresses only refer the subset of hosting nodes that must be known by hosted nodes. This subset and how to identify it is defined by the numbering system of the switching network.

2. If a hosted node is moved to another hosting node, it would prefer that other hosted node could still reach it by the same network address. If this is supported in networks, network addresses and node names must be kept separate.
3. In traditional public networks, a terminating node can never become a routing node, which assumes that there is always a single operator who operates the routing part of network (and thereby all T-ROs tables). In modern datagram switching networks (such as TCP/IP and MTP L3), the boundary of the routing part of the network is open ended. It allows the switching network to grow without central control, and many switching network operators to exist in the same switching network.
   For example, we can easily imagine node n13 in Fig. 3.24 upgraded from just being a terminating node to a gateway, i.e., a routing node for a large number of hosted nodes (each with its own network address). This would imply that the hosted node n22 could be moved to some hosting node in the new network, the node name of which would not be known by any of the existing switching nodes. The gateway node and the switching network it serves would be operated separately. This is the principle behind how the IP stratum in the Internet is operated, for example, supported by IP addressing methods and separate routing protocols for intra- and inter-network routing.

3.4.3.2 Routes in Circuit-Switching Network Systems

There are two categories: systems that switch end-user data (e.g. PSTN and ISDN) and systems that switch data on behalf of such systems. Examples of the latter are PDH and SDH. The former are systems that offer single channel routes with rather low transfer rates (less than 2 Mbps). The latter offer high transfer rate (up to several Gbps), multichannel routes. PSTN and ISDN realize (primarily) routes on demand by end-users, while PDH and SDH routes are created by control from a management system. In the industry, the term "circuit-switching" is actually reserved for on-demand systems such as PSTN and ISDN, while PDH and SDH are called cross-connecting systems. However, when we classify systems after the type of switched routes they offer, PSTN, ISDN, PDH, and SDH all belong to the same class of systems (i.e., circuit switching).

We will study the structure of route abstractions in one of these systems here: ISDN. The ISDN switching layer operates synchronously, i.e., it realizes synchronous connections. These connections are used by other layers in ISDN for creating asynchronous routes (such as LAPD routes) and media routes between ISDN users (e.g. voice connections). Figure 3.25 shows an ISDN in three views.

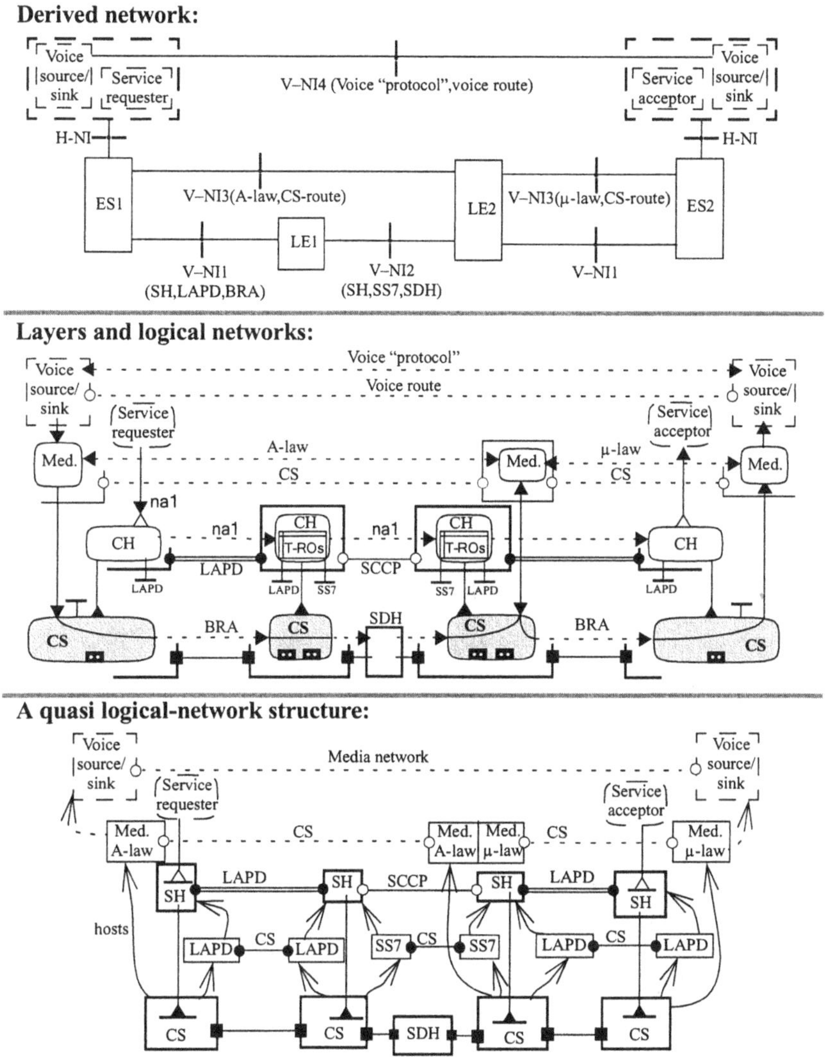

Fig. 3.25 An ISDN modelled in three views. CS=circuit-switching; ES=an ISDN end-system; LE=an ISDN local exchange; Med.=media encoding function; CH=call-handling layer

The top portion shows a network model based on **derived nodes**, while the middle section shows the same model expanded in the layer dimension with possible logical networks indicated. The model at the bottom indicates a possible logical-network structure, extracted from the model in the middle. We included only two terminating nodes of the switching network, both of the BRA type. In reality, the network would support PRA-type terminals as well, and have thousands of terminating switching nodes.

Some comments on these structures are needed:

- In the **derived-network model** we apply the AMLn definition of a network interface, comprising protocols and the routes on which the protocols run. All network interfaces, except those that define how users connect to the ISDN, represent vertical partitions of the ISDN (the only interfaces that are standardized in ISDN). The network interface between users and the ISDN represents horizontal partitioning. This interface is, however, not standardized, i.e., each supplier of terminals designs it in its own way.
  ISDN users use primarily media connectivity services (in this model we have assumed that a voice service is used). A user may be a person using a telephone or a machine (e.g., a voice-message-storing or voice-answering machine), i.e., the horizontal network interface (H–NI) is a man–machine or a machine–machine interface.
  This H–NI comprises two logical interfaces: one is used by the user in his requester/acceptor role, one in his service using role. In man–machine interfaces the person using the interface takes both roles. In machine-machine interfaces, there may exist two interfaces, one for requesting the service, another for using it.
- The *layers and logical networks* model in the middle reveal the distinct separation in ISDN between control and signaling layers on the one hand, and layers that transport user data on the other (in the ISDN standard referred to as the "control plane" and the "user plane").
  This structure is characterized by the requirement that, before a user can use a media route, it must first be established by the call-handling layer (CH) that controls all switching layer elements and all media resources. A destination network address (in the scenario called na1) is only used by service requesters in communication with CH elements in terminals. The CH layer handles no APi and CEi, i.e., all media routes are single-channelled.
  The CH layer establishes a media route in two phases (sometimes referred to as "call handling" and "connection handling"): First it decides on which kind of media resources are needed in terminals (if a choice is possible), and if any transcoding resources are needed in between (as the A/μ-law transcoding element in the scenario). As to the latter, the CH layer must also find out where such a resource is available in the network (in this scenario in LE2). In a second phase it creates circuit-switched routes between media resources. This scenario shows clearly the distinction between a "circuit" and a media connection, since two circuits were needed to create the media connection in this case.
  The effect of this structure is that when media routes and circuit-switched routes are used, there is no need for identifying any of these routes by an REi, or deliver any addressing parameters. Both types of routes appear as single-channel link routes, although they are created on-demand.
  ISDN is also an example of a network system that (with the exception for PABX

systems) does not allow terminals to act as gateways, i.e., routing elements. The model reveals that in two ways: the destination network address is not delivered to terminating CH nodes; these nodes do not store their own network addresses or any T-RO. Thus, the network defines (through its routing tables) which node is a routing node and which is a terminating node.

- As has been said before: *logical network structures* can only be defined if the realization of layer interfaces are standardized (which implies the specification of H–NIs). Since there are no H–NI interfaces defined in ISDN, the whole ISDN is a single logical network, in relation to logical networks defined by user elements.

  In the model at the bottom of Fig. 3.25 we have, however, indicated a possible path to an internal logical-network structure for ISDN. To turn this model into a correct logical-network structure, one must define H–NIs for all types of control points, in which case all control point will vanish from this model. Layers that are defined by control point protocols (i.e., interlayer protocols), as well as some new layers for supporting these protocols, will appear instead. We will discuss this subject in details in Sect. 4.3.

  Such a modification of the existing ISDN would lead to a new architecture, exhibiting a horizontally-partitioned network system. An evolution in that direction of public networks is already on-going, sometimes denoted the "next generation network" (NGN) and already applied, e.g., in the UMTS network system ("3G").

3.4.3.3 Switched Routes in SS7

In SS7, the switched routes are realized by MTP L3 which, as route identification is concerned, is an X.25 network enhanced by the layer-APi method (see Fig. 3.26).

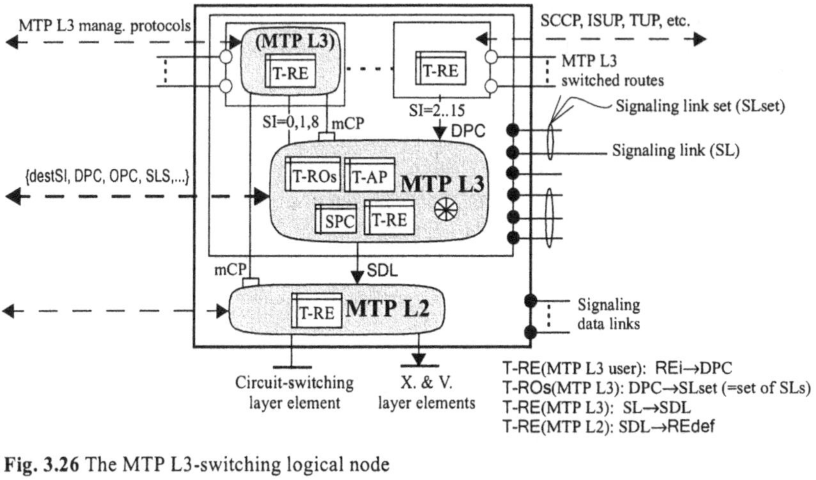

Fig. 3.26 The MTP L3-switching logical node

Hosted nodes in MTP L3 are of two classes: nodes which perform "signaling network management" (a control layer) and nodes of logical networks (primarily the PSTN/ISDN call-handling layer) that use the MTP L3 datagram service. MTP L3 offers a single type of switched routes: datagram routes, which are addressed by hosted nodes by MTP L3 network addresses, called *signalling point code* (SPC). Thus, hosted nodes handle a T-RE where the REdef for a route comprises a single *destination point code* (DPC). Each MTP L3 node knows its own SPC. When an MTP L3 message is sent by a hosted user, the MTP L3 layer elements adds its SPC as the *originating point code* (OPC). When a message is just relayed through the node, nothing is changed in the message.

MTP L3 applies the layer-APi method. It must therefore consult a T-AP table (NWi↔APi translation) both when sending and receiving. The APi in MTP L3 is called *service indicator* (SI), and has a range of 16 values. MTP L3 is used both in national and international networks. How to assign SI values to hosted nodes is defined only for international use as follows: SI values 0,1 and 8 are reserved for layer-management networks and SI=3 for SCCP. The remaining twelve SI values are intended for other hosted networks, which (with the exception of SCCP) all perform some type of call-handling functions in PSTN/ISDN networks. Still, almost 20 years after the standard was published, only seven of these values are assigned.

MTP L2 nodes are interconnected by link routes that are called *signaling data links* (SDL). These rely primarily on 64-kbps circuit-switched connections through the switching network (often referred to as MTP L1 routes). SDLs are semi-permanent links, established by MTP L2 management.

An MTP L3 node must handle a routing table (T-ROs), since every node in an MTP L3 network can both route and terminate data. To allow for some flexibility regarding failures on the MTP L1 level, as well as for higher transfer rates than 64 kbps, several SDLs may connect the same two MTP L3 nodes. A specific layer that runs on SI=0 defines and manages such groups of SDLs.[6] It calls such a group *signaling link set* (SLset), and the individual channels *signaling link* (SL). T-ROs and T-RE tables in MTP L3 therefore rely on SLset, SL, and SDL, as shown in Fig. 3.26.

Note that an MTP L3 message also includes a parameter called *signaling link selection* (SLS). This parameter has several applications, depending on the hosted network that initiated the message. One is to identify the signaling link so that data can be transferred in sequence in the MTP L3 network, similar to the purpose with the virtual circuit identifier in CON networks (as e.g., in ATM). This feature actually makes MTP L3 behave like a CON network without having to establish routes on-demand.

---

[6] Note that this layer is not described as a separate layer in the standard, but just as a function, well intertwined with other functions in MTP L3. This could not happen if AMLn had been used for modeling. A similar set of management functions are used in IP (see Sect. 3.4.3.4), but they are described as separate layers (as they should be).

Requirements on reliability of MTP L3 services is very high, since its intended use is for call handling in public networks. This is achieved by reliable signaling data links, by error handling functions in MTP L2, and by the signaling link set handling in MTP L3 that includes redundant links and link failure handling.

3.4.3.4
Switched Routes in the Internet

In IP-based networks, switched routes are realized by the IP layer. The fact that route types offered to hosted nodes, and the way APi is applied, are identical to MTP L3 means that the same type of tables must be part of an IP node (see Fig. 3.27).

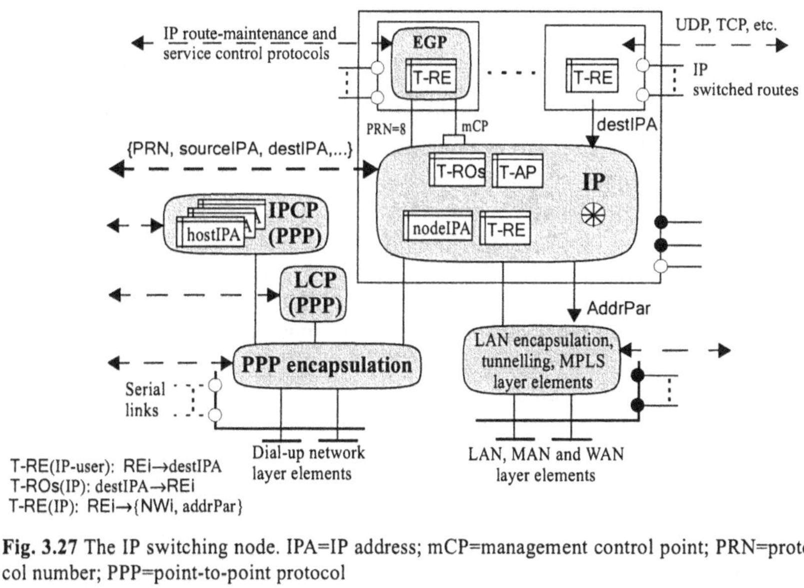

Fig. 3.27 The IP switching node. IPA=IP address; mCP=management control point; PRN=protocol number; PPP=point-to-point protocol

As could be expected, names of configuration parameters are different from MTP L3: the APi is called *protocol number* (PRN), the NA is called *IP address* (IPA) and, in messages, the sender's address is referred to as a *source address* (instead of "origin" as in MTP L3)

As in MTP L3, hosted nodes are of two classes: nodes that contain management functions (several types of route maintenance and service control protocols) and nodes of logical networks that use the IP datagram service. Different from SS7 is that management protocols are defined as separate layers, which considerably facilitates the understanding of IP management.

The mode of operation in the IP layer is equal to MTP L3. A difference though is the large number of nodes that only terminate IP messages and therefore need no T-ROs. These are called *hosts*. In the Internet application of IP, many of these hosts connect only temporarily to the Internet over dial-up networks. For such hosts, a route between the host and an IP node must first be established (through some dial-up network), and a temporary *hostIPA* must be assigned. This is achieved by, e.g., the "point-to-point protocol" (PPP) stratum element of the model (see Fig. 3.27.) The PPP protocol specification does not describe a single layer/protocol, but a stratum of three layers: an encapsulating link layer (similar to HDLC); a link control layer (LCP) that defines and controls properties of the link, and a layer that assigns IPAs to hosts (IPCP).[7]

APi values (i.e., "protocol numbers" in IP) are registered permanently to different types of layers by the Internet Assigned Numbers Authority (IANA). The protocol number range in IP is 256, i.e., much larger than for MTP L3, but only a few values are still unassigned.

Although the mode of operation of the IP and MTP L3 layers are similar, the routes they realize do not have equal properties. IP services are not guaranteed to be reliable since the IP layer can run on any type of supporting LAN, MAN, and WAN strata, and there is no correspondence to a link layer (similar to MTP L2) that checks user data for errors. Instead the Internet architecture defines a number of encapsulation and tunnelling layers, with the main purpose of framing IP-message data for transfer over different types of strata.

The correspondence to signaling link supervision and maintenance in the MTP L3 layer exists, however, in terms of diverse routing layers and maintenance layers that run on specific protocol numbers (one of them, the "exterior gateway protocol" (EGP) runs on protocol number 8, as shown in the model). These are all layers that control the T-ROs and T-RE tables in the IP layer (as indicated by the management control point, mCP, in the model).

3.4.4 Socket Routes

In general, switched routes in packet-switching networks do not offer an ideal service for application layers. There are at least four problems to consider:

1. The number of possible application layers may far exceed the range of the APi defined for the switching layer (16 in MTP L3, 256 in IP). For example, the Internet must be dimensioned for a very large number of applications.
2. In general, datagram networks do not guarantee in-sequence delivery of data.
3. To make the switching layer efficient, the size of message that can be sent is to be restricted, or even fixed (as in ATM).
4. The service may not check the correctness of user data (IP does not).

[7] PPP is intended for many types of dial-up access applications (e.g,. Appletalk, Xerox, DECnet, Novell IPX). The encapsulation layer and the LCP are common for all PPP applications, while the IPCP is specific for IP networks.

Some of these aspects were already considered when the OSI RM defined its transport layer. Here we will only comment on the first bullet in the list, starting with how SS7 deals with it.

From the outset, only a few MTP L3 call-handling applications were foreseen (only six are defined, e.g., TUP and ISUP). As time went by, it became clear that more suitable services would be needed and that more applications would appear than could be accommodated by the APi. Since the range of the MTP L3-APi could not be increased, MTP L3 was instead *extended* with another layer (the SCCP) that could support a larger range of APi. This resulted in an SCCP/MTP L3 switching stratum, that consists of two layers, each defining its own APi, as shown in Fig. 3.28.

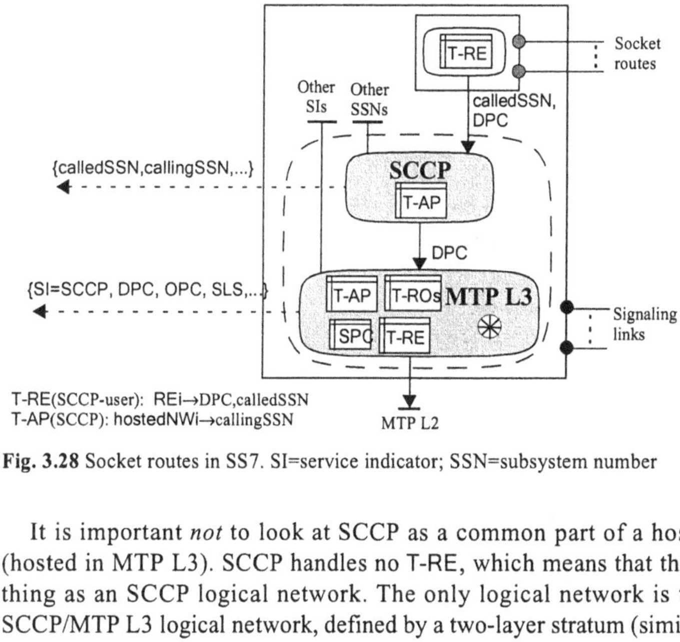

Fig. 3.28 Socket routes in SS7. SI=service indicator; SSN=subsystem number

It is important *not* to look at SCCP as a common part of a hosting network (hosted in MTP L3). SCCP handles no T-RE, which means that there is no such thing as an SCCP logical network. The only logical network is the combined SCCP/MTP L3 logical network, defined by a two-layer stratum (similar as we view the TCP/IP stratum in IP-based networks).

Again we have to learn some new terms for the same things:

- The APi of SCCP is called *subsystem number* (SSN), the sending side in SCCP is a *calling party*, the receiving side a *called party*. SCCP applies the layer-element APi method, which implies that routes might be used between hosted nodes that are assigned different SSNs.
- Let's also repeat the MTP L3 terminology: the APi of MTP L3 is called *service indicator* (SI). Network addresses are *signalling point codes* (SPC). A destination network address is therefore a *destination point code* (DPC) and an originating network address is an *originating point code* (OPC). MTP L3 applies the

layer APi method which implies that routes between hosted nodes are assigned the same SI.

Figure 3.28 demonstrates how the distinction between layer APi (applied by MTP L3) and layer-element APi (applied by SCCP) affects operations:

- The SCCP user must deliver as address parameters both the MTP L3 network address (DPC) and a called SSN.
- SCCP gets the calling SSN from its T-AP table. SCCP must include both the called and the calling SSN in its messages.
- MTP L3 requires only the DPC as address parameter, since the called SI is the same as the calling SI. The actual SI for SCCP is stored in the MTP L3 T-AP table.

Thus, by adding SCCP with its subsystem numbers, one introduces two levels of APi. The address parameters that the SCCP user uses for referring a particular route is a destination network address and an APi. We call such routes **socket routes**[8] (using a terminology introduced by TCP/IP). The model in Fig. 3.28 shows that these parameters belong to different layers, which actually contradicts the OSI RM definition of layers. We do not mind, however, since the SCCP user is not bothered by that.

Since SS7 was never intended as a packet-switching network for public use, the creators of SCCP most likely found an SSN range of 256 (eight bits) to be more than enough for future needs. They seem to have been right, since today only 10 numbers are assigned. As a matter of fact, most of these are not even used as APi, but as a substitute to the absence of **actor-layer identifiers** (ALi) in TCAP (a common agent layer that runs on SCCP), as is discussed in Sect. 6.3.

The definition and use of an APi parameter in SCCP cannot be compared to how the corresponding APi, the *port*, is used in the IP stratum. The purpose with IP was never to support applications that run directly on IP, but to support different kinds of management layers, as well as layers that could adapt the raw, unreliable datagram service to connectivity services that were better suited to application networks, especially TCP (protocol number 6) and UDP (protocol number 17). TCP and UDP are therefore integrated parts of all IP node realizations (as SCCP is with MTP L3 in most SS7 networks). We should therefore look at TCP, UDP, and IP as the resident layers of a {TCP UDP IP} logical network, as depicted in Fig. 3.29. This logical network realizes socket routes and is therefore, from a structural point of view, quite similar to the SCCP/MTP L3 model (in Fig. 3.28). The terms used for APi, NA, etc. are different from any other network system (what else could be expected?).

[8] Note that this term is not used by the SCCP standard, which defines this way of addressing as one of several addressing methods in SCCP (e.g., global title addressing, see next section).

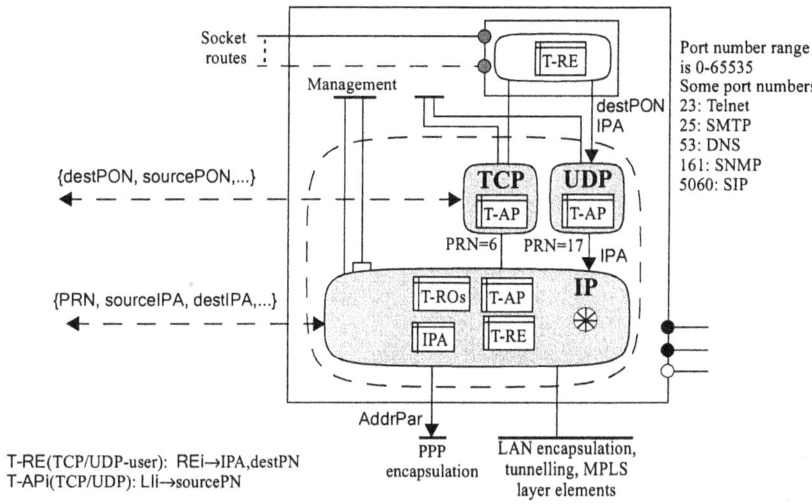

Fig. 3.29 Socket routes in IP networks. PON=port number; PRN=protocol number

There are, however, functional differences as well:

- All applications are assigned to port numbers (in SS7, MTP L3 also supports applications). The range of port numbers is very large (0-65535). To the knowledge of the author, all assigned port numbers are used by "real applications," i.e., there exist no such thing as TCAP that listens to a whole set of port numbers.
- TCP offers a CON service, UDP offers a CNL service. They are standardized separately, each defining its own protocol. Port numbers are normally common for both TCP and UDP (SCCP also defines both CON and CNL services, but within the context of a single and very complex protocol specification).
- A minor difference to CON services in SCCP is that socket routes are always referred to by {destIPA, destPortNumber} in TCP messages (in SCCP an OSI approach is used in data phase, identifying SCCP connections by a kind of connection identifier (Ci), called *destination local reference)*.
- Although not shown by the model, there exists management types of layers as well that run on TCP or UDP, e.g., BGP (an internetwork route management protocol), DNS (the domain name system, see Sect. 3.4.5.3) and SNMP (the Internet management system).

3.4.5
Global Routes

3.4.5.1
Introduction

Switched and socket routes have the disadvantage that hosted nodes must refer a route with a network address of a hosting node, which they cannot control and which cannot be easily changed, since it is used for routing in the hosting network. Hosted nodes may be reallocated to other switching nodes for many reasons, and must then normally be addressed with a new network address. Such changes will affect other nodes of the same hosted network, by requiring modification of their T-RE tables. An alternative method of handling this problem would be to modify the T-ROs tables of switching nodes, so that hosted node can use the same network address even when reallocated to another hosting node. This is normally not realistic for large switching networks.

This problem is more serious in open networks such as the Internet, but similar needs have been felt in SS7, as well as in public networks such as PSTN/ISDN. The solution is to allow hosted networks to define and use their own numbering system, separated from the numbering system of the hosting network. The principle is depicted in Fig. 3.30.

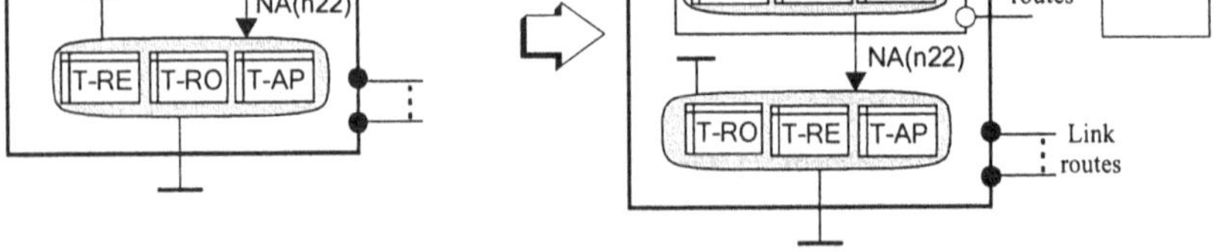

Fig. 3.30 Realizing global routes. GT=global title; T–GT=GT→{REi,APi}

- To the left is a switching logical node n11 that applies the layer APi method. n11 hosts a number of logical nodes, e.g., n21, that rely on switched routes that are addressed by NAs. These NAs are defined by the numbering system of the nw1 switching network.
- To the right we have transformed the logical network nw2 to a hosting network that applies the layer-element APi layer method. nw2 allows hosted networks (e.g., nw3) to use its own numbering system for addressing. We say that nw3 uses **global routes** and that nw2 views these addresses as **global titles** (GT). The

term "global title" is used in SS7 and in PSTN/ISDN. It corresponds to the *domain name* in IP-based systems.

Obviously, before a routing analysis in nw1 can take place, an address translation must be done. For this purpose, a new type of table is needed, the T-GT table, to be handled by nodes of nw2. This table translates a GT (part of the numbering system of a hosted network) to an REi of a switched route, and a destination APi of nw2. Its T-RE table is then used for translating the REi to an nw1 network address.

3.4.5.2
Global Routes in SS7

Let's now look at how global titles are supported in some network systems, starting with SS7. In this system, SCCP has been assigned the task to support global routes (see Fig. 3.31).

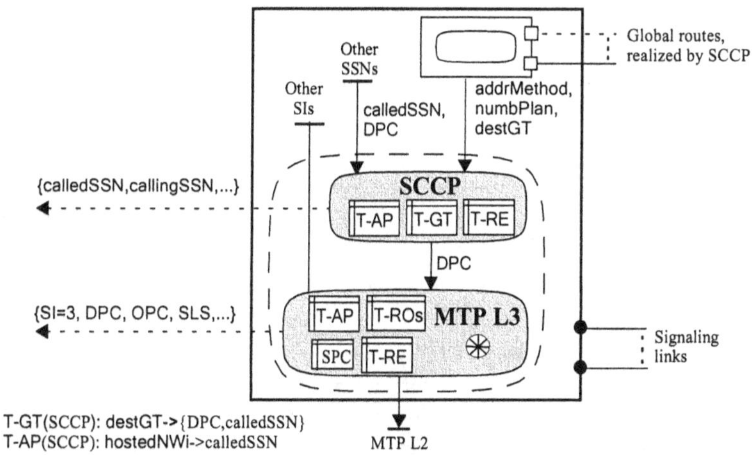

Fig. 3.31 Realization of global routes in SS7

Note that global title addressing in SCCP does not substitute socket addressing. It is just an alternative addressing method, i.e., the hosted node is identified by the same SSN, whether it uses socket or global routes. Global routes do, however, complicate the SCCP layer interface, since a hosted node must, for every invocation, inform SCCP about which *addressing method* is used and which *numbering system* (of the hosted network) is referred (every hosted network may have its own numbering system).

The SCCP protocol is not affected by the introduction of global title handling in SCCP, as long as every SCCP node has a relevant T-GT. However, SCCP does not stop there: in order to facilitate T-GT table management, SCCP also supports centralized handling of T-GT tables, i.e. all T-GT translations for a particular hosted network may be handled by a single SCCP element in an MTP L3 node somewhere in the SS7 network. This implies that there is no longer a direct MTP L3-switched route between the nodes that are hosted by SCCP. For every invocation, SCCP must analyse where in the MTP L3 network the T-GT handling node for the referred numbering system exist, and then send its messages to that node for address translation. This results in a considerable increase of SCCP complexity, since it turns SCCP into a switching network, and requires extensive changes to the original SCCP protocol, as indicated by the model in Fig. 3.32.

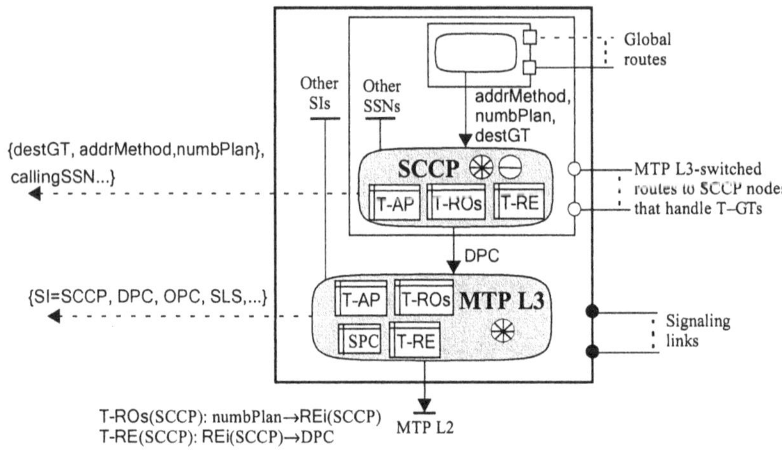

Fig. 3.32 Handling centralized T–GT tables in SCCP turns SCCP into a switching network

Note that a routing table (T-ROs) is now needed in an SCCP node. Normally a network address is the search key for such a table, but in the SCCP case it is more likely that the identity of a numbering system is used as the key.

SCCP supports both CNL and (several forms of) CON services, user-data error handling, segmentation of user data, and, in addition, three methods of address handling. No wonder that SCCP has become one of the most difficult ITU–T standards to understand. SCCP is an excellent example of how one can complicate a network architecture by squeezing too much functionality into a single layer.

3.4.5.3
Global Routes in IP networks

If we look at IP-based networks instead, several of these pitfalls have been avoided:

- CON and CNL services are handled by different layers (UDP and TCP respectively).
- Only one mode of global route handling (centralized) is supported. Translation tables are handled by a separate logical network — the *domain name system*

(DNS) — not by TCP and UDP. A *domain name* is the equivalence to a global title.

Figure 3.33 depicts the GT problem and how DNS solves it.

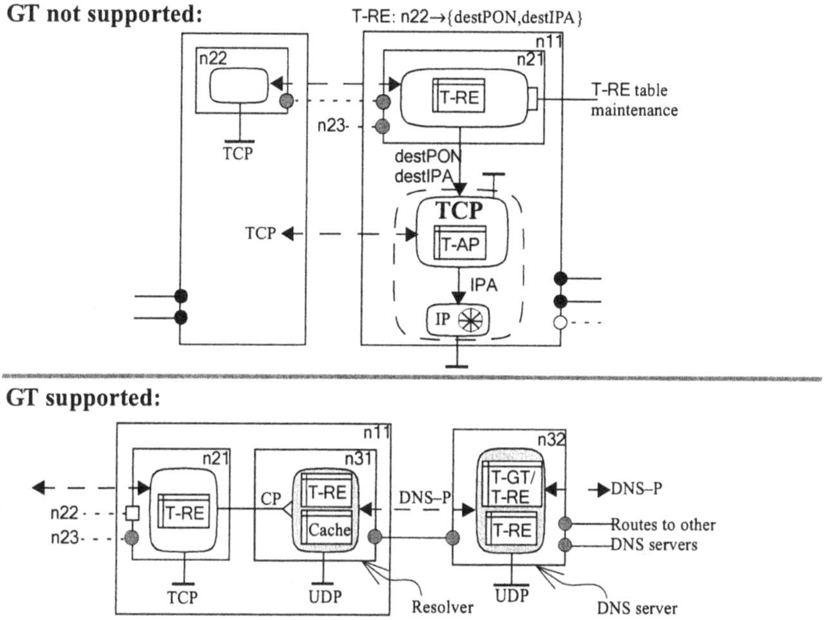

Fig. 3.33 Introducing DNS in IP-based networks. CP=DNS access control point; DNS=domain-name system; DNS-P=DNS protocol for access and T–GT/T–RE maintenance

Every layer element has some (explicit or implicit) means of translating an implied logical node name of its own network to an REdef that defines address parameters of a hosting node. Application logical networks in IP-based networks can be relieved from such translations and maintenance by allocating GT→REdef translations for all application networks in a logical network of its own, the DNS network (a GT→REdef translation is an entry in a combined T-GT and T-RE table).

This is an example of vertical partitioning of functionality, i.e., a function (GT→REdef translation) that may be needed by a great number of logical networks is extracted and allocated in a common logical network. Application networks access the common network over a DNS-defined access control point (CP). Logically, such a control point exists in every switching node, e.g., in n11 in this model. Note, however, that this control point is not standardized.

The element that provides this control point is called *resolver.* Although the resolver is normally just a library procedure that is compiled into a software realization of an application-layer element, it is logically a part of the DNS logical network. The resolver does not handle translation tables, but a DNS access protocol that is used for retrieving GT→REdef translations from remote DNS servers. The DNS protocol is also used for maintaining GT→REdef translations. Figure 3.34 shows the scenario for retrieving a GT→REdef translation from the DNS network.

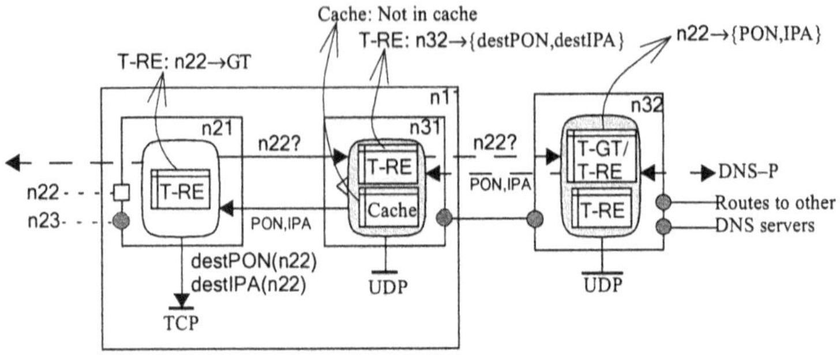

Fig. 3.34 The scenario for retrieving an REdef from DNS. IPA=IP address; PON=port number

- The layer element in n21 wants to send a message to n22. Its T-RE shows that the route to n22 is a global route, which is why n21 has to contact the resolver.
- The resolver (n31) will first look into its cache memory to see if the wanted translation is stored there (as a result of previous request from n21). If not, it uses its own T-RE table to contact the nearest DNS node (n32 here), that may have the translation.
- n32 may be able to provide socket information for n22. If not, n32 contacts other DNS nodes, using its own T-RE, to get the answer.
- When the answer is finally delivered to n21, n31 stores the GT→REdef translation for n22 in its cache memory for future use. Next time n21 asks for the same domain name, n31 may use this information (if not too old) instead of asking a remote DNS node.
- In some IP-based networks, e.g., the Internet, there is no central authority that guarantees that the information stored in DNS is correct, i.e., that it accurately reflects the configuration of applications in the network. Application nodes, such as n21, may be reallocated in the hosting network nw1 or vanish from the network. A discipline must therefore be used by resolvers to check regularly against DNS servers that the information stored in its cache is still valid.

The DNS logical network has a hierarchical structure of nodes that store GT→REdef translations, according to a hierarchical *domain-name structure* (see the logical-network structure in Fig. 3.35).

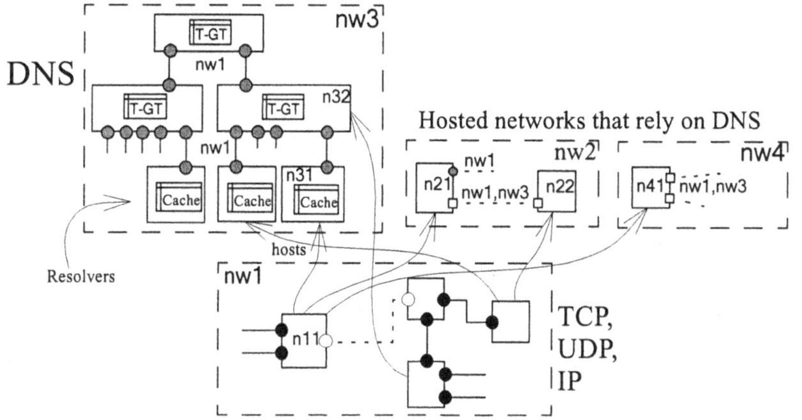

Fig. 3.35 The logical-network structure of an IP based network. All routes of nw2 through nw4 rely on nw1

This exercise is an example where three logical networks (nw2, nw3 and nw4) may be related by relations other than routes. In this case, nw3 resolver nodes offer access control points to nw2 and nw4 nodes (see Figs. 3.33. and 3.34). Since routes say nothing about control points, the only way this is reflected in this logical-network structure is by the fact that, e.g., n21 and n31 are both hosted in the same node (n11) of another network (nw1 here).

Thus, to reveal if there exist relationships, and what type of relationships such collocated nodes have, one has to consult the corresponding layer structure. Figure 3.36 shows the effect on a layer structure of extracting the GT→REdef functionality of two layers (L2 and L4) and allocate it into a common layer (DNS), thus creating three layers instead (L2', L4', and DNS).

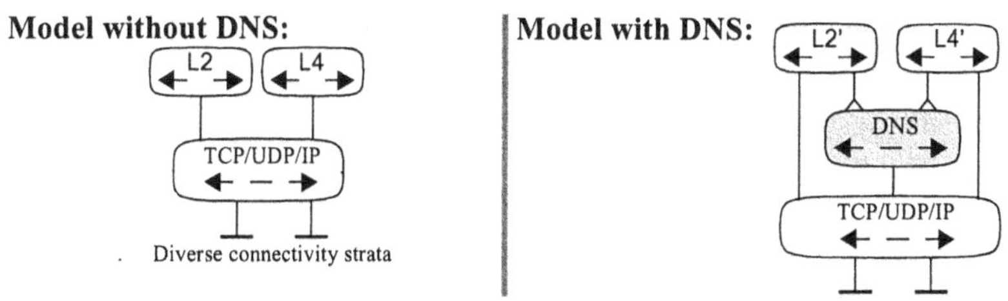

Fig. 3.36 The effect of DNS on a given layer structure

3.4.5.4 Global Routes for Mobile Services

Since global routes are insensitive to where in a switching network the logical node resides, they imply that some kind of global route handling may also be performed for networks where terminating nodes and/or users of terminating nodes roam around. Such networks are cellular networks and the fixed access, wide-area networks PSTN and ISDN. The two levels of mobility that are supported by these networks are normally called *terminal mobility* and *personal mobility*.

Terminal mobility. This must be supported by all cellular networks. In these networks, terminals are roaming around. The identification of the terminal is stored in the terminal, and is checked by the network both when a call is made from the terminal and when the terminal is called. Since a terminal can be accessible from any fixed access point (i.e., base station) of the network when the call is made, this identifier corresponds to a global title. Note, however, that it identifies the terminal and not the terminal user. Thus, a global route is used by a terminating node (i.e., mobile phone) of the mobile switching network.

The distinction between a global title and a network address (NA) is upheld by different names and structures of identifiers for this kind of networks. For example, in GSM, the correspondence to a GT is called *mobile station ISDN number* (MSISDN), while an NA (for routing purposes) is assigned temporarily to a terminal (when involved in a call) as a *mobile station roaming number* (MSRN). Both identifiers are however part of the same general numbering system for telephony systems (according to ITU–T E.164). Thus, the independence between numbering system for GTs and NAs that we assumed for packet-switching network systems is not entirely implemented.

Personal mobility. In fixed access networks (PSTN/ISDN), terminals are fixed, but *users* are roaming around. Thus, a global route (if the network supports something like that) exists between users of network terminals (normally persons).

Persons are inherently mobile. Traditionally, however, PSTN/ISDN operators are not concerned with who uses a terminal (i.e., a telephone). They only see the person who subscribes to the fixed access, which is the one who pays the bill, no matter who used the telephone. In that sense, PSTN/ISDN has always supported mobility without using any GT (for persons). This way of providing services has, however, some awkward consequences as well:

- Since the network only handles NAs (i.e., telephone numbers), it implies that if I move my telephone to another access, I must normally accept another NA, since otherwise the routing strategy (reflected by the number structure) would break down. It may take some time before all my friends find me again.
- A user who is not close to the access subscribed to cannot be reached. Fixed access telephony networks try to overcome this problem by diverse redirect and voice mail box services.
- Services are bound to accesses, i.e., to the person who subscribes to the access. Within the network, every access is associated to a particular *service profile*, which defines which services are subscribed to over that access. Thus, if I use an access other than my own, I am constrained to the services that are available over that access. This was not a problem as long as ordinary voice calls were almost the only service used. However, with the advent of IN services in public

networks, it became of interest to be able to use one's own service profile, even from other accesses.

- When I use an access other than my own, I cannot direct the charges to myself (except from pay-phones of course).

All these problems arise when the network does not distinguish between the *terminal user* on the one hand, and the *terminal* and the *access* on the other (or, in other words, between GTs and NAs). To be able to do that requires technologies for address translation, authentication, and advanced charging methods that could not be implemented before digital accesses and the IN technology were introduced in networks some ten years ago. The ITU–T then quickly defined the network service for *universal personal telecommunication* (UPT), described in ITU–T F.851. In this standard, UPT is defined as

> "the ability of a user to access telecommunication services at any terminal on the basis of a personal identifier (i.e. the *UPT number*), and the capability of the network to provide those services delineated in the user's service profile. Personal mobility involves the network capability to locate the terminal associated with the user for the purpose of addressing, routing and charging of the UPT user's calls."

Thus, the UPT number corresponds to a global route between service users, of which at least one is an UPT user. Similar to the case in GSM, however, UPT numbers are taken from the same numbering system as all other telephony numbers (i.e., ITU–T E.164). Since the network needs to know if a called or calling number is an UPT number, special rules for selecting UPT numbers are defined (see ITU–T E.168). This standard suggests three methods in which E.164 numbers may be used to support UPT numbers distinguishing between using the service: only within a single operator network; within a country; international. The last method implies assigning a special country code ("878," defining a "world wide UPT community") to UPT users who want to roam world wide.

The network accepting a call to an UPT number must translate this number to the terminal identity, called *UPT global title* (UPTGT), where the UPT user will take the call. The choice of the term "global title" may seem odd when the terminal is connected to a fixed access, because, in that case, it is used by the network as an NA. However, UPT also works over mobile terminals, in which case the UPTGT is actually a global title.

The fact that the UPT user can access the network from a large number of accesses, as if using one's own fixed access, requires certain new procedures, to be used by UPT users and to be realized in the network. The basic procedures are:

- To be reachable, the UPT user's location must always be registered in the network. If the user wants to take and make his call from a fixed access, he must exercise a registration procedure, telling his UPT operator the network address of the access he will use. If the user wants to take and make his call from, e.g., a GSM mobile, he must give the MSISDN of that terminal. These procedures are equivalent to the location update procedures in mobile networks, only that they are not made automatically.

- Whether the UPT user receives a call, makes a call, want to register his whereabouts or change his service profile, he must exercise an authentication procedure to avoid frauds, i.e., somebody else using his UPT number for getting services from the network.
- Charging an UPT user who makes a call to a non-UPT user is as any other call. However, when somebody makes a call to an UPT user, the same problem as in mobile networks arises: if I call somebody in my own cellular network I do not expect to have to pay extra for the fact that he might take the call on the other side of the earth. At least I want to be made aware of that fact, before I decide to go through with the call. Thus, UPT requires more advanced charging method than regular telephone calls.
- One of the basic ideas behind UPT is that an UPT user will, regardless of location, have access to the same services that a fixed access subscriber has over his or her own access. The only restriction is that services that require special capabilities that the actual terminal does not have cannot be provided. This also implies that the UPT user (as well as the network) must be allowed to modify the user service profile from any suitable terminal.

Realization of UPT services in WANs affect only the call-handling layer of the network. Handling the additional translation tables, registration, charging, and authentication procedures requires centralized solutions, such as home location register nodes (HLR) and service-control point nodes (SCP).

4 Modeling Vertical and Horizontal Partitions

4.1 Introduction

Particular layer structures and network structures are the result of how a given functionality is *partitioned* and how *distribution* is to be supported. Both activities result in refinements of layer and network structures. In existing network systems, the dominating approach to functional partitioning has been *vertical partitions*. By introducing the logical-network structure concept, we have indicated how a network can be viewed as a system of *horizontal partitions*. In this chapter we will take a closer look at how these two partitioning approaches are made visible by different types of relations in layer and node structure models.

We will find that applying vertical partitioning to a given layer structure implies:

- Logical network structures are not defined (in other words, all functions exist in a single logical network).
- Layer interfaces are not refined. They should always be defined by abstract service primitives (ASP) and, if relevant, by a layer interface behavior description in order to facilitate the understanding of how the layer service must be used. Such a specification also provides an opportunity for later horizontal partitioning. Standards normally focus on the **intralayer protocols** that are the results of vertical partitioning. Sometimes they provide ASPs, sometimes also behavior descriptions of layer interfaces, but mostly nothing of the kind.
- Some layers will be refined to strata.
- Vertical partitioning turns layer elements into intralayer structures and adds extra layers for supporting such structures.

Applying horizontal partitioning to a given layer structure implies:

- Group of layers are allocated to different logical networks. Nodes of different networks thereby become related by **hosts** relations.
- Layer interfaces are standardized and may be refined to **interlayer protocols**.
- Horizontal partitioning does not affect existing layer structures. It may, however, lead to the appearance of new layers to support layer interface communication.

When refining a layer structure, both vertical and horizontal partitioning may be applied recursively. Major refinements of network structures must be distinguished as models on different **network levels**. Adjacent network levels are identified by sub(-structure) relations.

4.2 Vertical Partitioning

Vertical partitioning (VP) of network systems follow the OSI RM paradigm, which places layer interfaces as local (non-standardized) interfaces in nodes. No logical-network structure is defined, i.e. the concept of **hosting** network and **hosted** networks, and thereby the hosts relation, is neither defined nor modeled. As a result, VP is only applicable to a single logical network.

Normally, a model created by VP only shows how distribution of layer elements is supported by connectivity layers or strata. Figure 4.1 shows a simple example, where the layer L1 in the network nw1 is partitioned into two layers, L11 and L12. L1 offers some type of control-point service and is also assumed to interface a physical-media stratum. Note, however, that the reasoning is the same for a layer that offers connectivity services, or is based on some other type of connectivity stratum.

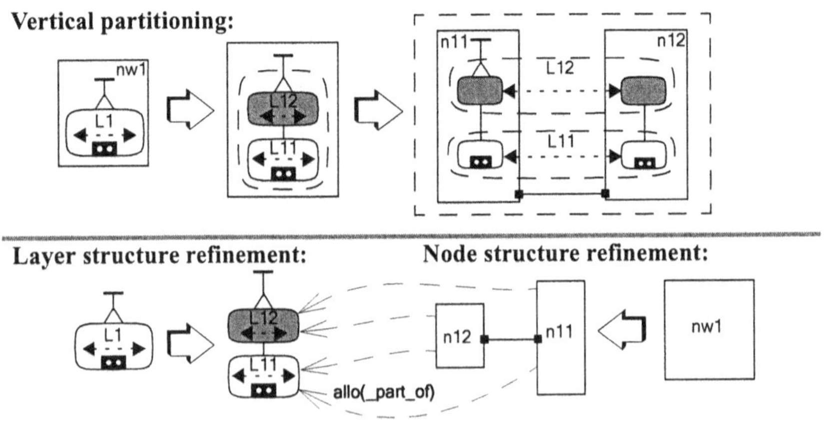

Fig. 4.1 Separating connectivity functions from other functions

Figure 4.1 shows three VP-refinement steps of L1 and nw1:

- L1 is first vertically refined into two layers, in order to separate functions that provide network connectivity (layer L11) from functions (L12) that handle the actual service that is offered over the control point.
- A particular distribution solution is then applied to define (in this example) nw1 as consisting of two nodes interconnected by a physical route.
- When we separate the layer structure from the node structure, the fact that the nodes will allocate parts of the layers is modeled as **allo(-cates)** (in this case allocates_part_of) relations, either as graphical pointers as in this figure, or as a

table. Note that allo relations must be defined by the node structure model and not by the layer structure, since a given layer structure holds for an infinite number of node structures.

This approach can then be repeated recursively inside an nw1 node. In Fig. 4.2 we have partitioned the L12 layer element inside n11 into two local layers in order to support vertical partitioning of this layer element. The distribution solution defines an internal network (nw2) in n11 that consists of two nodes, n21 and n22, interconnected by an internal physical route.

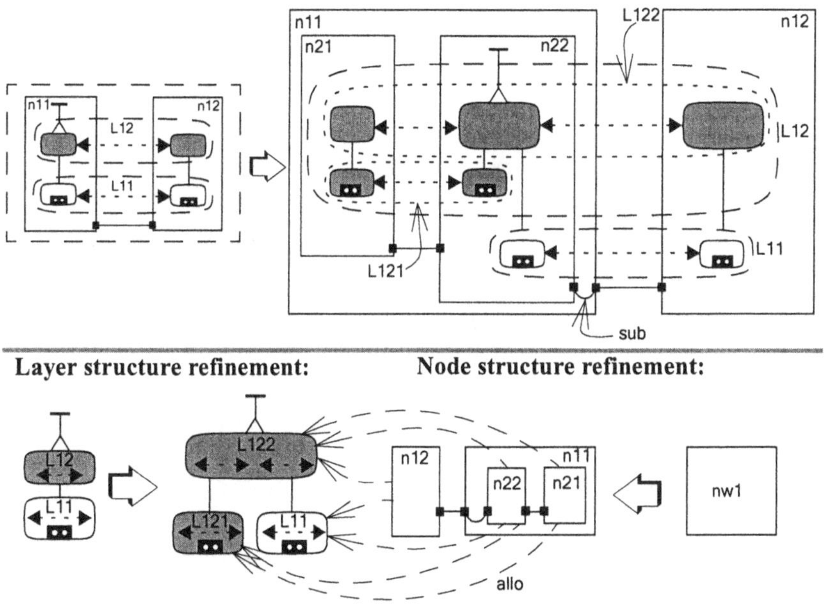

Fig. 4.2 Separating connectivity functions over several network levels

Since the internal node structure of n11 is invisible from n12, and completely independent of any particular internal node structure in n12, nw2 must be defined as a node structure on a lower network level than nw1. Considering possible later refinements of the modeled network system, it is important that the independence between nw1 and nw2 is shown in the model, which is done by the sub relation.

These examples have both focused on **connectivity structures**, i.e., to separate connectivity functions from other functions (possibly in several steps) in order to support distribution of the latter. Another approach to vertical partitioning is to separate a given layer according to its internal **control structure**, provided that resources that exist on a lower level of control must be distributed in a network. We encountered such an example when discussing DNS in Sect. 3.4.5.3. In that analysis we saw that DNS is a common distributed resource that offers control points which are used by other layers for address translation (GT→REdef). Thus, layer interfaces between the DNS layer and other layers are interfaces in a control structure, as depicted in Fig. 4.3.

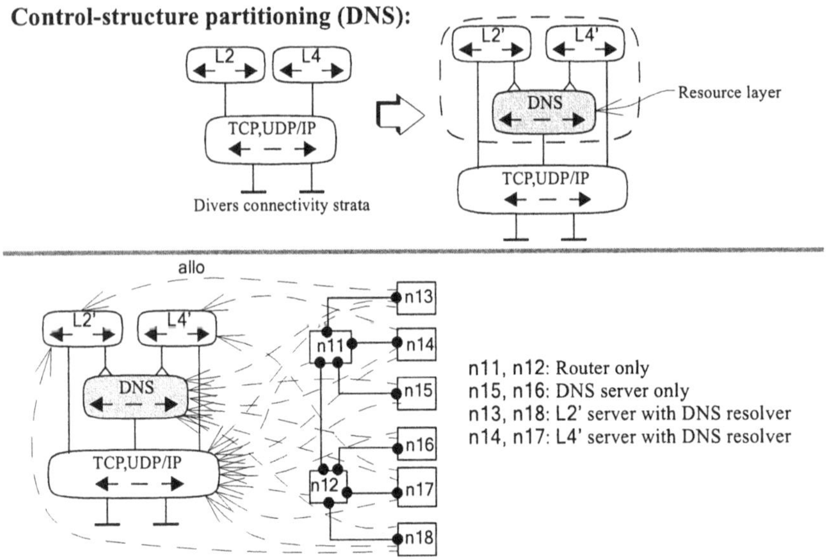

Fig. 4.3 Separating functions of a control structure

As long as we do not define separate logical networks for different layers (or group of layers), there is only a single, flat node structure that tends to be more and more complex as we continue to apply VP. This complexity consists not so much in the connectivity structure of the network, but in an increasing complex structure of allo relations, as Fig. 4.3 indicates. These relations are important since they tell which functional contents a node has or may have.

Note, however, that most standards that describe VP structures do not explicitly define allo relations. Instead, complex nodes of a **derived network** are given specific names that are supposed to associate to their functional contents, e.g., as depicted in Fig. 4.4. The amount of information (indicated by the allo relations of the model in Fig. 4.3) that is implied but not defined when such a simplified model is create is obviously considerable. In general, only the creators of the model in Fig. 4.4 can be expected to understand which assumptions are made by it. Furthermore, this approach also tends to generate a large name space of node names, since each name represent a particular combination of layer elements from different layers.

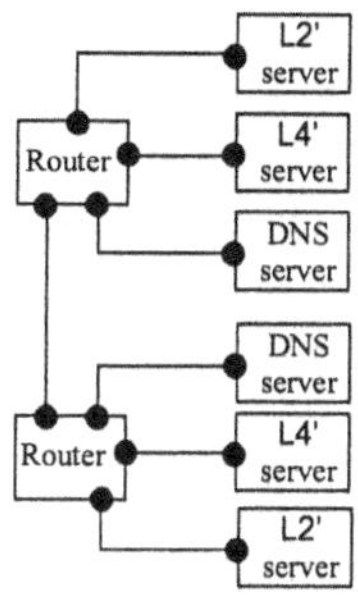

Fig. 4.4 Network structures in standards are normally based on (functionally) very complex derived nodes

This effect unnecessarily contributes to the impenetrable mass of terms that characterizes telecom standards, as well as suppliers solutions of standardized network systems.

4.3 Horizontal Partitioning

4.3.1 Introduction

As mentioned before, there are three aspects on creating a horizontal networks (in this order):

1. Definition of a logical-network structure.
2. Specification of layer interfaces.
3. Refinement of layer interfaces to network interfaces.

We will use the model of Fig. 4.3 to exemplify how these refinements affect the model of a network system. Let's first define separate logical networks for the {IP,TCP,UDP} stratum, as well as for each of the layers DNS, L2' and L4'. This simplifies the model considerably, since most of the allo relations in the original model now turn to hosts relations, as depicted in Fig. 4.5. For every logical network in this structure, the protocol handling aspect is now defined by allo relations, while the layer interface handling aspect is defined by hosts relations. Note that this separation of concerns in the model makes it an excellent tool for requirement tracing.

- Modifying a protocol of a layer is specified in the layer structure. The allo relations show which logical networks are concerned. If the modification does not affect the provided layer interface, this is all that matters.
- Modification of a type of layer interface is also specified in the layer structure. The allo relations show which logical networks are concerned, and hosts relations which logical nodes are involved.

- Configuration changes do only concern logical networks, and can be specified per logical network. The hosts relations show which nodes and tables are affected. For example, if we would move the DNS server n36 from the IP node n16 to n19, we would move the hosts relation from n16 to n19. The model will then tell us that the T-RE tables of the DNS resolvers n37 and n38, hosted in n17 and n18, must be modified.

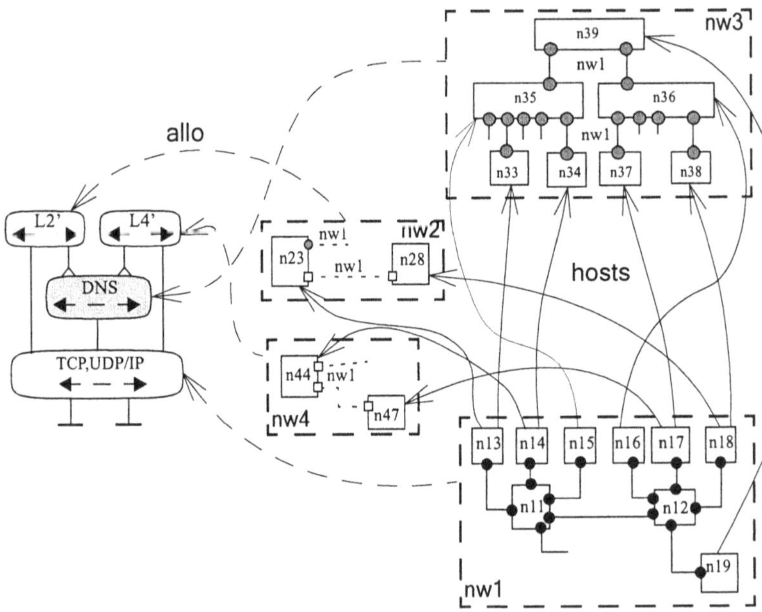

Fig. 4.5 The model of Fig. 4.3, modified by logical network definitions

If we want to refine a layer interface, we must first have access to a specification of that interface. If a layer structure has been produced (assumable as a result of vertical partitioning), each layer interface type that is defined by that structure should also be associated to a **layer interface specification** (in ITU–T standards sometimes called "service definition"). However, since we cannot rely on that (as regards existing standards), we may have to update the layer structure with such specifications, before we can approach the refinement activity of that interface type. Figure 4.6 shows how the TCP and UDP layer interface can be specified. The AMLs formalism for defining ASPs has been used here (other formalism may do as well). The AMLs variant of finite state machine (FSM) is used to describe the behavior in the interfaces.

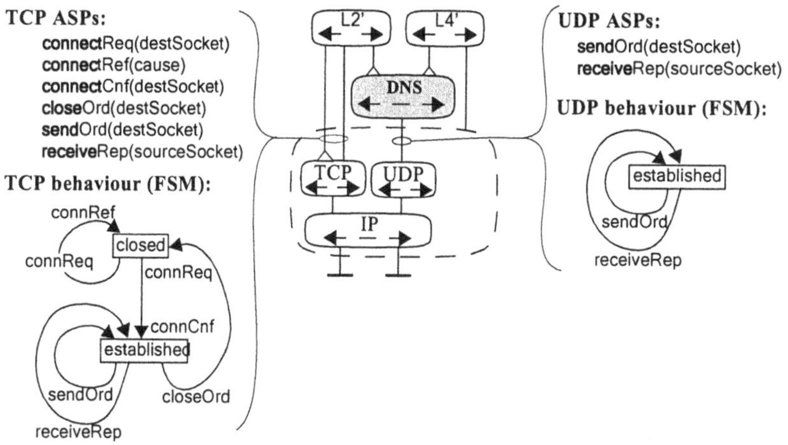

Fig. 4.6 Layer-interface specifications. Cnf=confirm; FSM=finite-state machine; Ord=order; Ref=refuse; Rep=report; Req=request

Note that while UDP defines two operations only (send and receive), TCP defines another two (connect and close). The behavior description for datagram services (such as UDP) is also so elementary that (normally) no behavior is explicitly described for such interfaces. The difference in complexity lies otherwise between control points (CPs) and connectivity service points (CSPs):

- CP interfaces are normally more complex than CSP interfaces, both as to the number of ASPs they define, and the behavior they assume. In general, an interface language (such as AMLs or the OSI ROSE notation[1]) must be used for CP layer interface specifications (which in this case would be used also for specifying the CP that DNS resolvers offer to their users).
- From an ASP definition point of view, all CSP interfaces are very similar, normally defining only a send and receive operations (if duplex transport is supported). More complex CSP interfaces exists, however. For example, some CSP interfaces may also report failures of data transfer (a faultReport ASP) and use flow-control commands (a congestionReport ASP).

4.3.2 Refining Layer Interfaces

Let's now discuss the last step in horizontal partitioning: layer interface refinement. The way a layer interface in a layer structure is specified is deliberately chosen not to assume any particular realization of the interface. That is also why layer interfaces are often called *abstract* interfaces, the primitives *abstract* service primitives,

[1] See the recommendation ITU-T X.219

and the interface language an *abstract* language (at least we do that in AML). To realize a layer interface, therefore, implies translating the abstract description of it to a description in a design language, which may be a language for software interfaces (such as IDL[2]) or hardware interfaces (such as VHDL). When solving this problem you have to consider at least three different situations:

1. The two layer elements that use a mutual layer interface will always exist on the same *processing platform* (or in other words, in the same derived node). In this case, assuming that the platform supports communication between software elements but does not define an interface description language, you may define a number of message handling procedures (one for each operation) and store them in a library as the *application programming interface* (API) for the layer interface.
2. It must be possible to run the layer elements in different derived nodes. At low network levels (i.e., within a local network), it is sometimes possible to use a network platform that supports distribution of software elements to different nodes. An example is the CORBA[3] platform that also defines IDL as its interface language.
3. The two layer elements may not be realized as software, however. Even if they are, network platforms such as CORBA may not be suitable. The limited real-time properties they offer and the processing power they require may make them unsuited for the actual layer interface.

In those cases we must transform the layer interface into a protocol that can carry the ASPs of the layer interface. It is important that the model can tell that such protocols are realizations of layer interfaces, which is why we call them *interlayer protocol* (to distinguish them from the *intralayer protocols* that are a result of vertical partitioning). An interlayer protocol must also be supported by an additional connectivity stratum that is defined internally in the node that defines the layer interface. The principle of this transformation is depicted in Fig. 4.7.

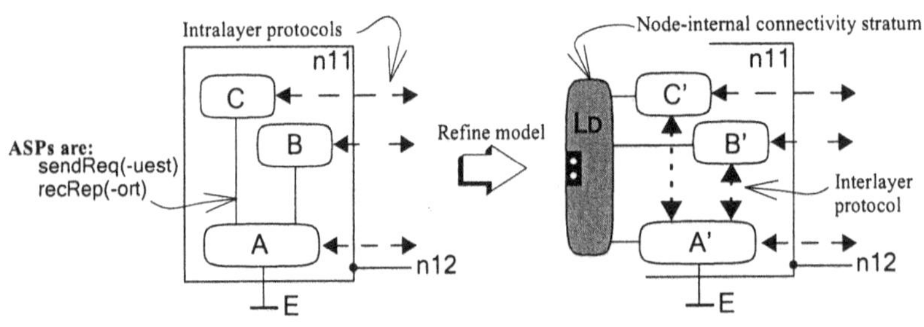

Fig. 4.7 The principle of realizing a layer interface in terms of an interlayer protocol.

[2] See CORBA/IIOP specification, Chap. 3, IDL Syntax and Semantics. www.omg.org/

[3] See CORBA/IIOP specification (www.omg.org/) or Orfali (1996).

In this model, A, B and C are layer elements that handle intralayer protocols and layer interfaces, A', B' and C' are layer elements that handle both intra- and inter-layer protocols, and E is a layer element that offers connectivity within nw1 for node n11.

It is obvious that the decision that is made about how to realize layer interfaces affects the design of layer elements (i.e., A, B and C turn to A', B' and C' in Fig.4.7). Since different decisions in this respect may be taken for different nodes of the same logical network, it is important for a layer element to be designed so that the essential part of it (which might be its actor or its agent) can be reused in different nodes. In case of the model in Fig. 4.7, we should therefore try to isolate parts of A', B' and C' that handle interlayer protocols from parts that handle intralayer protocols. We use the actor–agent separation principle to solve this problem (see Fig. 4.8 which is a refinement of the model in Fig. 4.7).

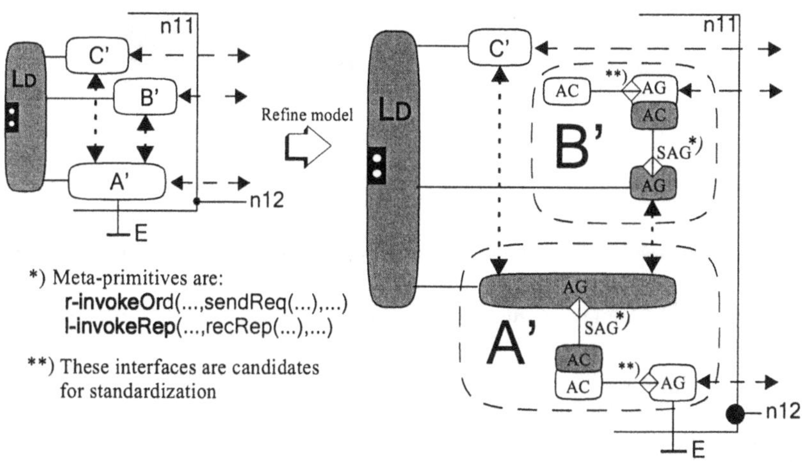

Fig. 4.8 The transformation of a layer interface to an interlayer protocol. Shaded parts are the effects of horizontal partitioning. AG=agent; AC=actor; A',B',C'=elements of layers LA, LB and LC respectively; LD=local connectivity stratum

In this model we introduce different actor–agent interfaces for intra- and inter-layer protocols, which leaves the intralayer actor of element B' and the intra-layer agent of A' independent of how the layer interfaces are realized (which may differ between nw1 nodes). Instead we define a local *common* agent layer inside n11 that supports meta primitives (i.e. r-invokeOrd and l-invokeRep), which in this case are used for carrying the primitives sendReq and recRep that were defined for the layer interface of A' in the original model.

Note that the idea with this refinement is to support maximum reusability of the same functional elements in different realizations of layer interfaces (i.e refinement models). This is achieved by:

- Replacing sendReq and recRep primitives defined by A with meta primitives carrying those primitives.
- Standardizing intralayer actor–agent interfaces (i.e., specify r-invokeOrd and l-invokeRep).
- Specify a common agent layer protocol.

The introduction of this common agent layer changes the original layer structure according to Fig. 4.9.

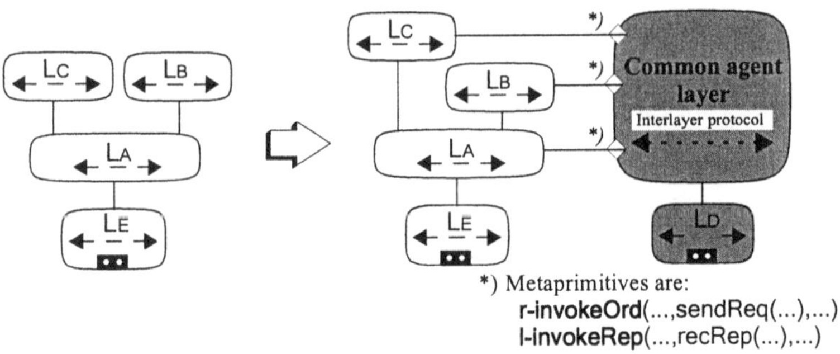

Fig. 4.9 The effect of introducing interlayer protocols on the layer structure

The existence of actor–agent interfaces in this model does *not* tell that it represents a separation of a control structure from a connectivity structure, since LA through LC still are layers interconnected over connectivity layer interfaces. The new layers only define a way for *some* layer interface of realizing these layer interfaces (which we applied on node n11 in Fig. 4.8).

Be aware of that the distinction between inter- and intralayer protocols is never made in standardized layer structures. Furthermore, protocol relations are always drawn horizontally in models, which further hides this distinction (see Fig. 4.10).

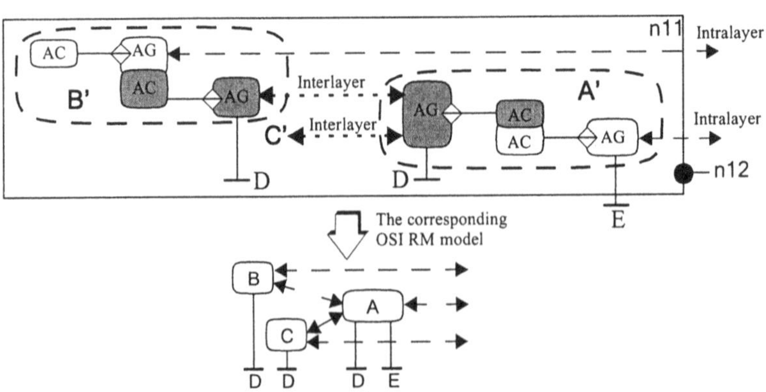

Fig. 4.10 The OSI RM way of modeling hides the intra- and interlayer protocol distinction

The upper model is equivalent to the one to the right in Fig. 4.8. Below the corresponding model drawn the "OSI RM way": logical nodes and routes are not defined and the actor–agent separation is not shown. Consequently, the model does not tell if protocols are for inter- or intralayer distribution.

4.3.3 Horizontally-Partitioned Logical-Network Structures

In this chapter we will discuss how realization of layer interfaces as interlayer protocols affects logical-network structures. Let's start by looking at a scenario in the original model of Fig. 4.7 and how this scenario changes when the model is refined. Figure 4.11 shows a scenario in the original model on network level 1, where layer element B receives a message.

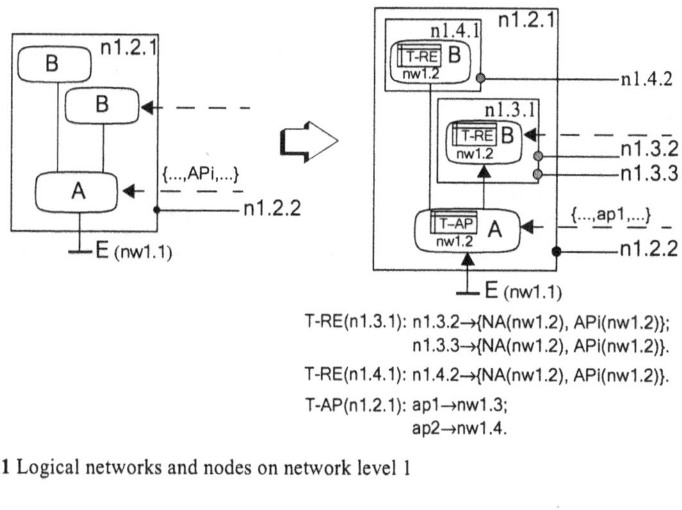

T-RE(n1.3.1): n1.3.2→{NA(nw1.2), APi(nw1.2)};
n1.3.3→{NA(nw1.2), APi(nw1.2)}.

T-RE(n1.4.1): n1.4.2→{NA(nw1.2), APi(nw1.2)}.

T-AP(n1.2.1): ap1→nw1.3;
ap2→nw1.4.

Fig. 4.11 Logical networks and nodes on network level 1

For the sake of this discussion, we name logical networks after the network level on which they are defined (e.g., nw1.1 is defined on network level 1 and has the network number 1). Consequently, the node n1.2.1 is a node on network level 1 of network 2 and has the node name 1 within that network.

We assume here that layer element A is part of a switching layer. The hosting node n1.2.1 is a terminating node of the switching logical network nw1.2 that realizes socket routes for nodes of nw1.3 and nw1.4. As usual, the hosted nodes n1.3.1 and n1.4.1 handle T-RE tables that provide relevant address parameters of nw1.2. Layer element A in n1.2.1 handles a T-AP table that relates APi values to network identifiers of hosted networks (assuming that n1.2.1 hosts only one node per hosted network).

So far there has been nothing new. Let's now realize the layer interfaces of this model as network interfaces, i.e., comprising routes and interlayer protocols. This implies that we introduce a completely new view of logical networks and nodes on network level 2 (see Fig. 4.12).

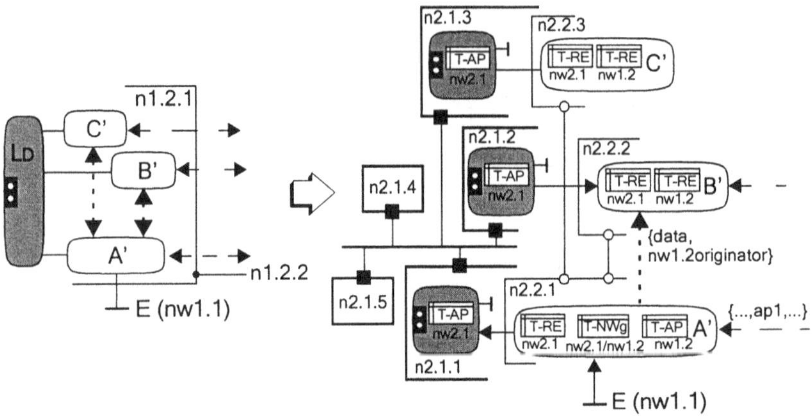

Fig. 4.12 The network level 2 view of a horizontally-partitioned network level 1 node

This view is only defined inside node n1.2.1, i.e., it is not visible from any other network level 1 node. We therefore say that this node structure is *orthogonal* to the node structure on network level 1. We have assumed that the network level 2 view inside n1.2.1 comprises a switching network nw2.1 (e.g., a LAN) that is a hosting network for, e.g., an internal network nw2.2. This network realizes the network level 1 node n1.2.1 by including each of the layer elements A', B' and C' in an nw2.2 node.

The logical nodes of nw2.2 are connected through switched routes that are realized by nw2.1. Consequently, all nodes of nw2.2 now also need a T-RE for accessing nw2.1. Node n2.2.1 need in addition a new type of table, the T-NW table (NWi→REi). The reason is given by the depicted scenario:

- When a message arrives at A' from the outside, the received APi is first translated by the T-AP for nw2.1, which in this scenario returns the network level 1 network identifier nw1.3 (see Fig. 4.11).
- The only way to deliver the associated data to B' is by using the route to n2.2.2, which is defined in the T-RE for nw2.1. A table that can translate between a network identifier (on network level 1) and a route (on network level 2) is therefore needed. We call this table T-NWg (NWi→REi). The suffix "g" indicates that this is a table in a hosting node. Hosted nodes may also need T-NW tables, if they have alternative choices of network level 1 hosting networks (although not in this case). When such a table is needed we call it T-NWd, where "d" stands for hosted.
- The node n2.2.1 now produces a message of the interlayer protocol that contains the message data of the B' protocol and information about the nw1.2 originator

of this message (originatorNA and originatorAPi). This message realizes the primitive receiveReport(...) in the layer interface between A' and its users, B' and C'.

- The data of the interlayer message is then sent over nw2.1, referring addressing data for the route to n2.2.2.

Let's now analyze how the relations between the views on network level 1 and 2 look like. Taking the node model of Fig. 4.11 as a starting point, Fig. 4.13 shows a possible logical-network structure on network level 1.

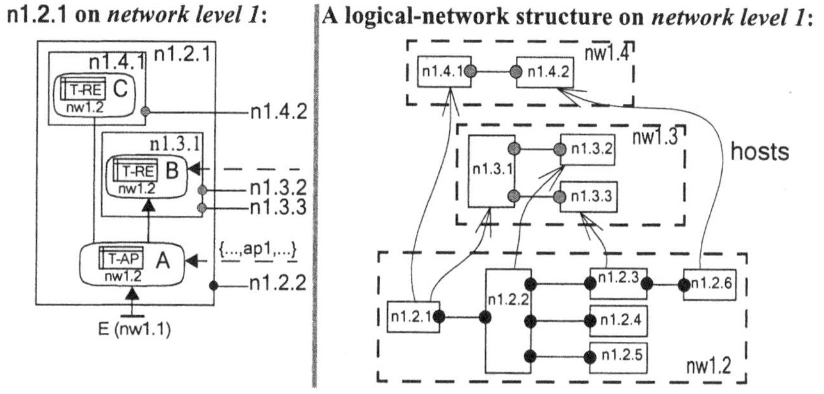

Fig. 4.13 Node and network models on network level 1

The network level 2 view of a node structure in n1.2.1 is completely independent of network level 1. According to Fig. 4.12 it looks like Fig. 4.14.

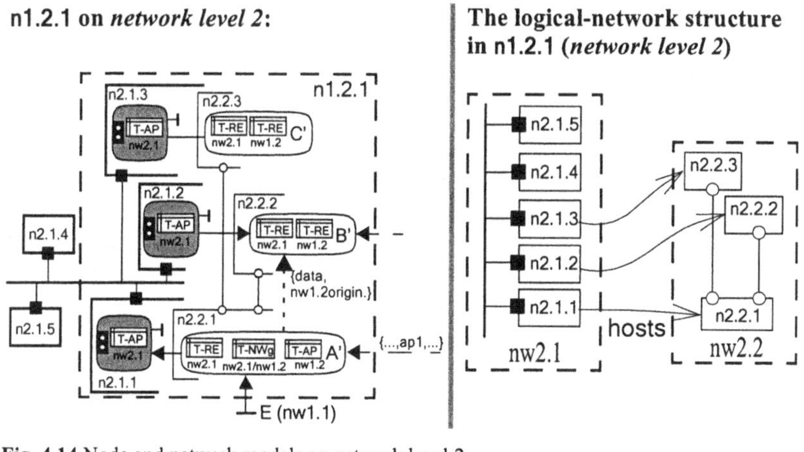

Fig. 4.14 Node and network models on network level 2

A logical-network structure model that covers both network levels can be easily drawn by substituting the node n1.2.1 in the structure of Fig. 4.13 (to the right) with the structure in Fig. 4.14 (to the right), using the sub symbol to distinguish between levels. The result is the model in Fig. 4.15.

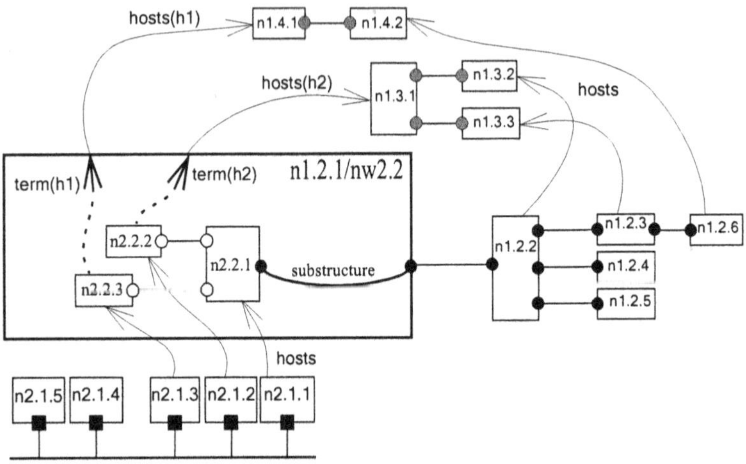

Fig. 4.15 The logical-network structure for the refined model, showing two adjacent network levels (through a sub relation). Vertical partitioning on network level 2 is defined through **terminates** relations

This model reflects the fact that a model that covers both network level 1 and 2 is produced just by adding entities and relations to the original model. Thus, the modeling technique supports seamless refinement of a logical-network structure model over any number of network levels.

As mentioned before, the fact that there is a sub relation between nodes n1.2.1 and n2.2.1 indicates that they belong to different network levels (no matter how nodes are named). A sub relation cannot, however, tell if the refinement implies vertical or horizontal partitioning. If we compare the models in Fig. 4.13 and 4.14, we see that the nodes n1.3.1 and n1.4.1 that are visible on network-level 1 will, after the refinement, exist in nodes n2.2.2 and n2.2.3, i.e. in nodes that are defined first on network level 2. When horizontal partitioning is applied, the relations between nodes on adjacent network levels are obviously not hosts relations, but relations that tell *how hosts relations that are defined on a higher network level are terminated on the nearest lower.* We call such relations **term(-inates)**. A term relation has a hosts relation as its attribute, such as term(h1) in Fig. 4.15.

In conclusion: to make a logical-network structure reveal that horizontal partitioning has been applied between network levels, term relations must be included in the model. If vertical partitioning (i.e., separating layer elements into parts) had been applied in our case study, no term relations would appear in the logical-network structure model. Instead we would insert allo relations between network level 2 nodes and the corresponding layer structure model, as we saw in Fig. 4.2.

We will end this discussion by showing all relations of the network system we have used for exemplification (see Fig. 4.16).

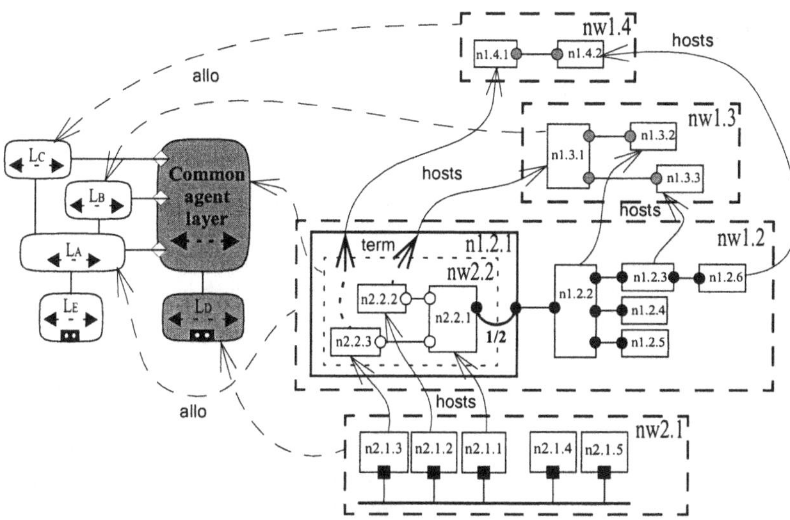

Fig. 4.16 Layer and network structures, and their relations

The complete logical-network structure is built by different types of routes, by sub relations and by the three inter-structure relations hosts, term and allo, where:

- sub defines adjacent network levels within a logical-network structure.
- hosts defines the relations between hosting and hosted logical networks within *the same* network level.
- term defines the relations between nodes on *different* network levels
- allo defines the relations between nodes and layers of a corresponding layer structure.

This model consists of considerable more network and node definitions than if we had not introduced horizontal partitioning and network levels. That does not matter at all, however, since the complexity of a system/model is not measured in terms of the number of elements, but by their relationships. By introducing hosts and term relations we have been able to dissolve an otherwise impenetrable mass of allo relations (compare with Fig. 4.3) into much fewer relations of three different types.

In a sense, this type of model summarizes almost everything there is to say about structures in network systems. Besides being an excellent tool for requirement tracing and network configuration, it gives strong support for the evolution of network systems and networks. For example, if we would like to replace the internal structure of n1.2.1 with a single processor node, the only model editing required is to remove the sub relation in n1.2.1. The modeling tool would be able to draw the following conclusions as a result of such a request.

- The network nw2.2 will disappear from the model.
- As a result, the term relations for n1.3.1 and n1.4.1 will disappear
- All hosts relations from nw2.1 will disappear as well.
- Since the model includes no hosts relations from nw2.1 to nodes of other networks, nw2.1 will also disappear.
- As a consequence the allo relation from nw2.1 to the layer LD will disappear.
- The allo relation from nw2.2 to the common agent will disappear.

This leaves us with a model that comprises network level 1 only in this case. Note, however, that in another case there may exist many sub relations that represent the relation between the two adjacent network levels, in which case all such relations must be removed in order to return to the network level 1 model. This calls for defining a **substructure relation identifier** (SRi) that, for each relation, identifies the adjacent network levels. In the model of Fig. 4.16 we have annotated the sub relation between n1.2.1 an n2.2.1 as "1/2" in order to tell that the relation goes between network level 1 and 2.

Using SRi annotations in complex models implies that a model user can just refer the network level (or levels) to be displayed, and the modeling tool will do the rest. In order to facilitate model maintenance it is also wise to keep the network level structure of models strictly hierarchical, i.e., only allow sub relations between adjacent network levels (e.g., an SRi=1/3 should not be allowed).

5 Management and Traffic Systems[1]

5.1 Introduction

The structuring principles that have been defined in previous chapters are applicable on every network system. When applied on a model, they result in:

- The separation of layer structures from node structures.
- The separation of control structures from connectivity structures.
- For control layers, the separation of layer-state machines (LSM) from layer protocol machines (LPM).

One separation principle is however still missing: *the separation of management functions from traffic functions*. No network can function without the existence of management functions that supervise its performance and manages (under control of operators) the resources (tables, devices, etc.) used by traffic-handling functions. Identifying management functions separately from the traffic functions that deliver services to users is therefore also a general modeling principle (although with many different solutions). In this chapter we deal with the management functions, their relations to traffic resources, how to model this in AMLn, and, to some extent, how existing standards deal with the problem.

5.2 Two Systems[2] and Two planes

We define **traffic system** as a network system that provides network services for end-users. In previous chapters we have seen many examples of traffic systems structures, and also touched on some management functions (e.g., "layer management functions"). So far we have not mentioned the system that defines all management functions for a traffic system, the **management system**. The number of functions of a management system often exceeds the number of functions in the traffic system that is managed.

---

[1] Most structural features for management systems that are described in this chapter are taken from the TMN concept (the ITU–T reference model for management systems), but modified and adapted to AMLn modeling concepts.

[2] In TMN, these systems are named "managed system" and "managing system," respectively.

Since a traffic system is a network system, it follows that a management system is a network system as well. Therefore, the two systems each define their own planes of L dimensions and N dimensions, as depicted in Fig. 5.1. The **traffic plane** is defined as the LT–NT plane, the **management plane** as the LM–NM plane. An L dimension defines all layers and all types of connectivity service points (CSP), user control points (uCP) and resource control points (rCP) in the layer structure of the system (as defined in Sect. 2.1.1). The N dimension defines the nodes and routes of networks that build up the system.

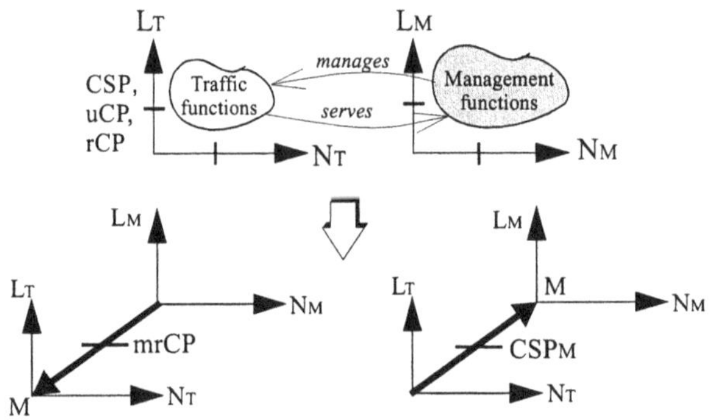

Fig. 5.1 Functions of a management plane (M) manages resources of a traffic plane (T), and functions of a traffic plane may offer connectivity services to functions of the management plane

Thus, each of the planes defines its own layers and nodes. There are two types of relations between the planes:

1. One is a "manages" relation. On the level of functional elements, management functions control managed resources (tables, devices, etc.) of layers in the traffic plane. Conceptually we model this by **managed resource control points** (mrCP) between elements of the planes.
2. The other is a connectivity service relation, denoted CSPM in the model. This relation is used by management functions that run on connectivity layers of the traffic plane.

The structure of mrCP and CSPM between the planes define a completely new, orthogonal dimension, the **management dimension** (M), as depicted by Fig. 5.1.

In order to be able to create models that relate traffic plane elements to management plane elements, we must first define the distinction between the behavior of a layer element in the traffic plane (which is described by an element state machine (ESM)) and the resources on which that behavior relies since only the latter are the things to be managed. We touched on this subject already in Sect. 2.3.2.1. As an example, let's assume a layer element that relies on a local routing table (T-RO, which is a typical managed resource in a control layer). Figure 5.2 shows the refinement of that layer element.

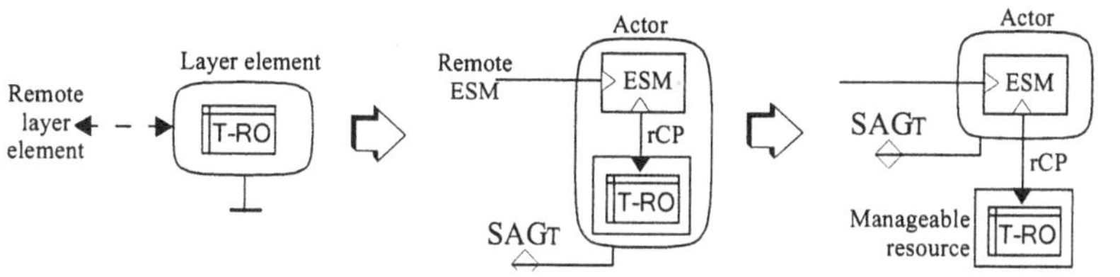

Fig. 5.2 Refinement of a layer element in the traffic plane into actor, agent and manageable resources

1. The model to the left shows how we model this resource in the layer structure of the traffic plane.
2. In the middle we separate the layer element into an actor and an agent. The SAG is called SAGT, where "T" denotes that it connects to an agent layer of the traffic plane. We also introduce the ESM of the actor and define a separate abstract machine that handles the table. This machine allows the ESM to use information stored in the table to control traffic system functions. The operations available to the ESM are defined by the rCP.
3. Since local resources are not visible to remote elements, we can move the resource out of the actor, as shown to the right, thereby making it an abstract machine that can also be accessed from the management plane.

Let's now add a function that can maintain the table resource (see Fig. 5.3).

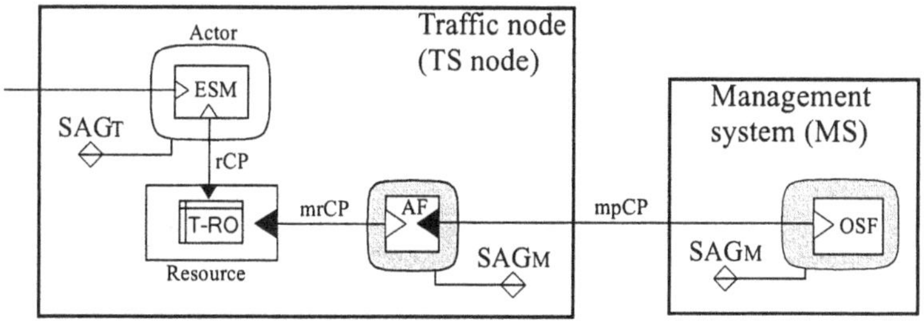

Fig. 5.3 The principle for managing traffic system resources from a management system. AF=adaptation function; CP=control point; ESM=element state machine; mpCP=management plane CP; mrCP=managed-resource CP; OSF=operations system function; rCP=resource control point; SAG=specified agent interface

This function is assumed to be part of the management system, and therefore an **operations system function** (OSF). It controls the contents of the table over a (local) mrCP in the traffic node, via an **adaptation function** (AF) in the same node. The AF is the ESM in an actor that uses a management plane agent layer for communication with the OSF. The SAG is denoted SAGM, where "M" denotes that it connects to an agent layer of the management plane (which may, or may not, be common for the two systems).

The model shows that we must distinguish between mpCPs that describe the interaction between management elements in the two systems, and mrCPs that are local in traffic-system nodes and describe the interaction between an AF and local resources.

- The purpose of an AF is to adapt to interfaces of local physical resources on which traffic system actors rely. These mrCPs consists of control and test points of resources that are first defined in implementation models, and are therefore proprietary (i.e., not standardized).
- mpCPs between the two systems are subjects for standardization[3], since one wants a single management system to be applicable to all types of traffic systems

Similar to all control points, mpCPs are functional interfaces, which means that they can be described as abstract protocols. In all control points, one element (the OSF here) is the controlling party, the other (here the AF) the controlled party. This means that an mpCP abstract protocol is fairly simple, defining (more or less) a number of operations that an AF can perform, mostly upon request from an OSF. This paradigm can be supported by the AMLn–SAG, as well as by TCAP and ROSE agent layers. What complicates things in mpCPs is that one wants to be able to refer not just to operations but to *operations on particular managed resources*. These resources are therefore specified as **managed objects** (MO), as shown in Fig. 5.4. Since the AMLn–SAG supports object references (see Appendix B), it can also support mpCPs. TCAP and ROSE are, however, not sufficient in this respect, which is why ITU–T had to define additional functionality for supporting MOs (i.e., CMIS/CMIP, see Sect. 5.3.3).

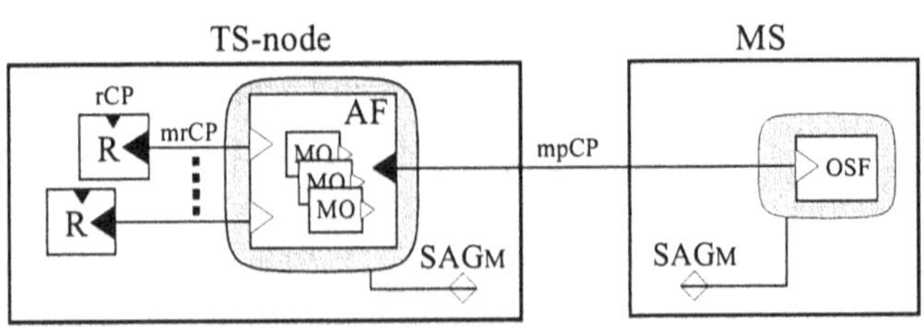

Fig. 5.4 In communication between the management and the traffic system, managed objects represent real traffic system resources

In this model we have used the AMLs notation for the relation between abstract machines and **abstract objects** ("managed objects," MO here). Each MO type defines its own operations, as parts of its interface specification. The individual MOs exist in an AF. In communication over the mpCP, each MO instance has a

[3] In TMN, these mpCPs are defined as the CMIS/CMIP protocol. The corresponding protocol in the Internet is SNMP (more about that later)

unique identity (the "object identifier" in AMLn invokes). The realization of MOs in an AF are object instances that can handle actual mrCPs.

The model also makes the distinction between MOs and real resources (R). This distinction allows for a single resource to be represented by several different MO types (or the other way around), i.e., there is not necessarily a one-to-one correspondence between an R and an MO. For example, a T-RE table (a single physical resource) may define route endpoints that are realized by two different connectivity strata, and therefore have different types of addressing parameters. This would require definition of two different MO types, both associated to the same real resource. It is the role of an MO implementation to relate an operation defined by an MO specification to invocations on some part of a real resource. AFs and MOs are therefore elements that must be realized by implementors of traffic systems.

So far, our discussion tells only part of the management story, however. In standardized management concepts (such as TMN and SNMP), all specific management functions are regarded as *centralized*, i.e., OSFs. In AMLn we also regard management functions that are *not* part of the management system as parts of the management plane. We regard these functions as *decentralized* and call them **autonomous management functions** (AMF). Examples of such functions are the IP routing protocols (that autonomously maintain routing tables), IPCP (an Internet function that assigns IP addresses to hosts), the TEI assignment function in LAPD (that assigns terminal identifiers to ISDN terminals), and all functions which autonomously monitor and maintain the traffic system's behaviour and performance. Figure 5.5 shows the addition of such a management function in a traffic node.

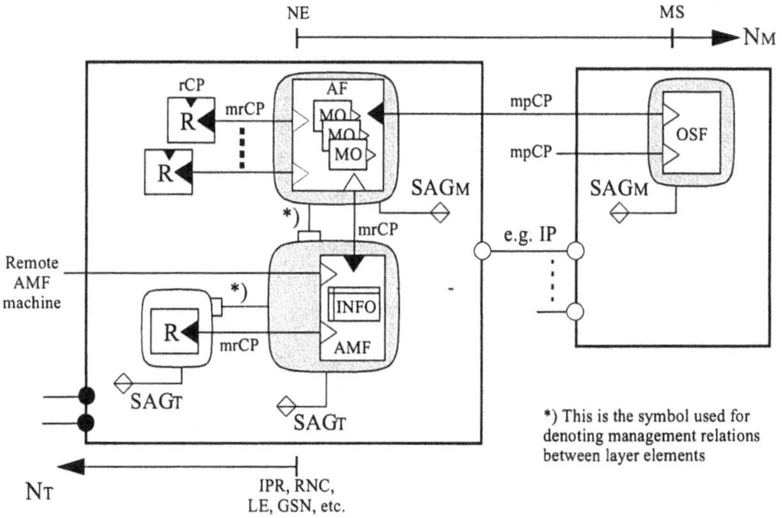

Fig. 5.5 Traffic nodes and the management system in the N dimension. AMF=autonomous management function; GSN=GPRS support node; IPR=IP router; LE=local exchange; MS=management system; NE=network element; RNC=radio network controller

Note that we use different symbols for denoting management relations in layer-structure models and in LSM models. In the former, all relations between layers are some kind of layer interfaces, the latter defines no layer interface at all. Instead all interfaces are interfaces between abstract machines, either of a **user-provider** type (as in Fig. 5.5) or of a **peer** type (you can read more about this in Appendix D).

The AMF element in Fig. 5.5 manages a resource of a traffic system layer element over an mrCP. It is assumed to do that in collaboration with AMF elements in other traffic system nodes, which is why it uses a traffic system SAG. Once and a while, or upon request of the AF, an AMF may deliver information ("INFO" in the model) to the management system (MS) over an mrCP between the AMF and the AF. This may be performance information, data for charging, or for statistics analysis in the MS. AMFs may also spontaneously report alarming conditions to the AF. The AF will define MOs for covering all this kind of information as well.

Since the model in Fig. 5.5. shows nodes, it allows us to analyze the distinction between planes and network systems in this context.

- The M dimension is realized inside traffic system nodes since it is defined by mrCPs. Thus, the boundary between a traffic plane and a management plane does not coincide with the boundary between a traffic system and a management system. The boundary between the planes cannot be standardized (realization of mrCPs is an implementor's business). The boundary between the systems is defined by all network interfaces between the management system and the network elements it controls. This interface (TMN calls it Q3) must be standardized, however, if a common managing system is to be used.
- Since the boundary between planes lies inside managed nodes, these nodes have an existence in both planes. In the management plane, all traffic nodes are viewed as **network elements** (NE). In the traffic plane the same nodes are normally given names after the traffic system type and the role they play in the particular system (such as IPR, RNC, LE and GSN, see the model).
- These names are of interest to the management system only as naming of MOs are concerned since every traffic system defines its own particular resources. A single management system can therefore be used to manage several different traffic systems in parallel, as long as it knows which MOs and which routes belongs to which network. All MO specifications that belong to a particular traffic system are therefore sometimes referred to as a **management information base** (MIB). To support the identification of standards that describe MOs across all standardization bodies, all network systems, strata and layers, the ITU–T developed the concept of a global classification system for MOs, identified by a structured OBJECT IDENTIFIER (a predefined data type in the ASN.1 standard).

5.3 The Management Plane Control Point (mpCP)

5.3.1 Introduction

The abstract protocol of an mpCP is defined primarily by the operations that are defined by the MO classes that are allocated in an AF. This is an open set of MO definitions, some of which may be standardized, some may be not. Basically, which MOs are defined depends primarily not on the actual traffic plane, but on what one wants to manage over an mpCP.

In general, management over mpCPs is also not restricted to using MO-based protocols (which are designed for short requests, responses, and reports). A number of other, non-MO-based methods are used as well, primarily for configuration of traffic nodes (which is an area where MO-based control of resources may mean too much overhead, or may be too unreliable). For example, Internet management also comprises (besides management over SNMP) the following:

- Telnet is used for connecting to remote processors.
- Simple industry standard command line interfaces (CLI) are used for node configuration.
- DHCP (Dynamic Host Configuration Protocol) is used for host configuration on LANs.
- LDAP (Light-weight Directory Access Protocol) is used for configuration via a central server (based on the ITU–T X.500 directory standard) that contains MO information.
- FTP (File Transfer Protocol) is used for bulk data transfer (or FTAM, in OSI environments).

In MO-based mpCPs, there are a number of common LSM events that will exist, depending on the fact that management activities go via MOs. MO instances must be possible to create in an AF, e.g., when a new device is installed on the traffic plane. Consequently, an OSF must also be able to order the removal of MOs from AFs. For all traffic planes of concern, spontaneous events will occur and must be reported from AFs to OFSs for analysis, etc. It may therefore be worthwhile to define a common abstract protocol for mpCPs, based on a number of generic operations. Examples of such protocol specifications are the CMIS/CMIP (defined by ITU–T), SNMP (defined by IETF), and CORBA/GIOP (defined by OMG). To this we could add an AMLn protocol specification based on AML–SAGs, which we will refrain from, however. As Tanenbaum (1996) puts it: "The nice thing with standards is that you have so many to choose from; furthermore, if you do not like any of them, you can just wait for the next year's model."

Such generic operations will depend on how the actual management system defines the concept of "managed object." There is so far no common view on this regarding the standards mentioned. Consequently each management standard defines its own mpCP abstract protocol.

In AMLn we do not provide any generic operation definitions for mpCPs since they would not differ much from existing standards (e.g., CMIS/CMP and SNMP). Let's instead look more generally on which type of operations may be needed in MO-oriented mpCP protocols. An AF will at any time include a large set of object instances (MOs) of different object classes. The things you (as an OSF) may want the AF to do can be listed as follows (this list corresponds more or less to the definition of "services" in CMIS):

- GET information about the state of traffic system resources. Since real resources are represented by MOs, this information is represented by a single MO or by a set of MOs. The request-ASP must therefore include parameters that identify either an individual MO (by an object identifier) or a set of MOs. It also defines the attributes of the MO class it wants to read. The operation may provide information about the time when the information was read, and when a certain attribute was last changed.
  GET operations must be operations of the AMLs operation class REQUEST. Note that GET operations do not change anything in the traffic plane, i.e., the state of the traffic plane is not changed. In AMLs we call such operations **selector** operations.
- SET attribute values of existing MOs. This operation may be used for initiating MOs, for specializing MOs according to attributes in the MO class, or for setting an MO in a particular state. A SET operation may be of class ORDER or REQUEST. A SET operation also does not change anything in the traffic plane. In AMLs we call such operations **initiator** operations.
- CREATE a new MO instance. A CREATE operation must be of class REQUEST. A CREATE operation increases the range of management performed via the management system, but in itself does not change anything in the traffic plane.
- DELETE an MO instance. A DELETE operation must be of class REQUEST. This type of operation decreases the range of management from the MS, but in itself does not change anything in the traffic plane.
- Report EVENTs that happen in the traffic system, particularly alarming events. In AMLs we define such reports as operations of the operation class REPORT. Such reports will normally inform about spontaneous state changes in the traffic system, which have already happen.

None of these operations can be used for creating state changes (i.e., affect any real resource) in the traffic system. If we want to do that, we must use specific operations on an MO instance. In AMLs we call such operations **modifier** operations. These are specified for the corresponding MO class. For example, a switching element may be defined by an MO. The OSF can create a new path through that switch by inserting the information elements of a connectRequest(inlet,outlet) event as parameters in an invokeOrder in its SAGм, also including the actual object identifier and possibly an invocation identifier. Thus, since the AMLn definition of invokes allows object identifiers to be transported in agent layers, there is no need for a generic operation for modifier operations in an mpCP.

However, one may have to define other parameters than just the object identifier for a modifier operation, e.g., information for access control that is confined to the particular MO class, MO instance, or MO operation. That might be a reason for defining a separate ACTION operation in mpCPs. Such an operation will take over the meta-primitive role of the AMLn invokes, as operations on MOs are concerned, since the information elements of MO events will be carried as parameters of ACTION primitives. Instead the AMLn invokes will carry the information elements of an ACTION event. Thus, we get a structure of meta-primitives on two levels, as exemplified in Fig. 5.6 for the mentioned connectRequest(inlet,outlet) primitive.

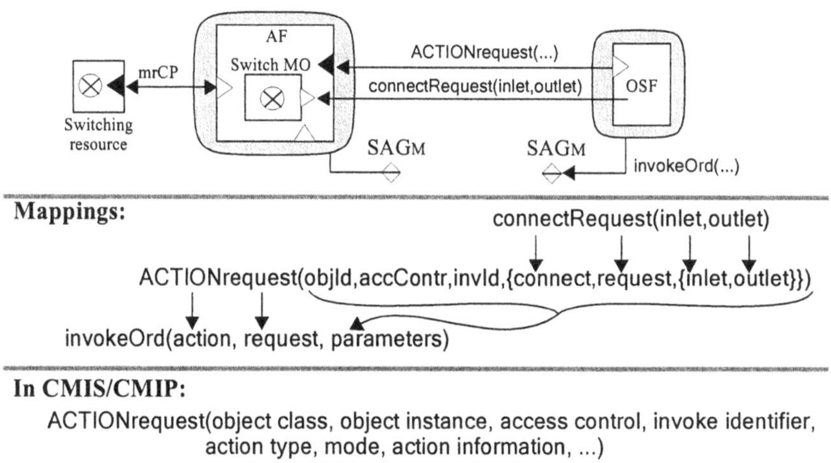

Fig. 5.6 Defining an ACTION operation as part of an mpCP

A possible drawback with this method may be the extra steps of packing and unpacking that has to be done by AF and OSF. Note that the AMLn invoke will carry the information elements of the ACTION event, using neither the object identifier nor the invocation identifier parameter.

CMIS/CMIP is an example of an mpCP that is defined in this way (see Fig. 5.6). The creators of this standard obviously had problems with the terminology (since they did not use the concept of a meta-primitive): the operation ACTION was called "service," the operations of MOs were called "action type," their primitive types "mode" and their parameters "action information." The primitive types of an ACTION are the OSI request, indication, response and confirmation, since the CMIS/CMIP places the AF–OSF actor layer as an OSI layer on top of ROSE and ACSE.

CMIS/CMIP, SNMP, CORBA/GIOP, and OMAP are all methods that can be, and are, used as abstract protocols for mpCP. They are far from equivalent, however. Sometimes several of them are used in managing the same traffic plane, which, besides the costs and inconvenience for an operator, may require complex management gateway functions for translating management information between management systems. Differences between methods can be identified in three areas:

1. The view on MOs and how they are specified (see Sect. 5.3.2).
2. The method used for providing connectivity for the mpCP, modeled as the SAGM in AMLn, and whatever supports this agent layer (see Sect. 5.3.3).
3. The operations defined in the mpCP, i.e., the "management protocol" (see Sect. 5.3.4).

5.3.2 Managed Objects in mpCP

Issues related to MO specifications are:

- How to *classify* and *identify* MOs, considering the problem of managing the development and maintenance of a large set of MOs.
- How to *describe* MOs.
- How to *manage* MOs (create, maintain, store and distribute MOs to, e.g., AFs).

The number of MOs for an operator network becomes very large, considering all the strata, layers and resources that have to be managed. For managing the TCP/IP connectivity plane alone, some 180 MOs are presently designed. *Classification* of MOs is therefore needed. In general, this can be done according to the following criteria:

- The actual network system.
- Logical networks, if defined.
- The functional structure of the network system. The basic classification may reflect the interlayer structure of the network, possibly described over several network levels. The MOs can then be further aggregated according to a possible interstrata structure.

Considering that an MO that is specified for a particular resource may be usable for a whole class of resource implementations, and that MOs are specified by many suppliers of traffic network products, a global identification of MOs is very valuable for building a management system. A suitable identification system is then the global object identifier system that we already mentioned. This system is part of the ASN.1 standard. It defines such identifiers in a classification tree that can comprise all standardization bodies as well as any supplier who produces ASN.1-based specifications.[4] Both TMN–MOs and SNMP–MOs are identified by the ASN.1 data type OBJECT IDENTIFIERs. SNMP also use object identifiers to identify their interlayer structure for the TCP/IP connectivity stratum (presently consisting of the IP, ICMP, TCP, UDP, and EGP layers).

A common approach to network management requires that all network systems specify their MOs in the same way. Since an MO is an "object," and, for the most part, implemented as a software object, it seams appropriate to specify it in an object-oriented (OO) way. The language to be used should be independent of any software design language, which implies the use of UML, AMLs or the CORBA approach.

However, MOs have been described for a long time already with TMN, SNMP, and lately also with CORBA specific methods. TMN and SNMP both use ASN.1 for data specifications (as AMLs does), while CORBA use C++ typing (as UML does). The similarities end with the use of common data types, however:

- *TMN* describes its MO-specification language as "Guidelines for the Definition of Managed Objects" (GDMO). A TMN–MO is a "real object," i.e., specified as to its operations, data structure, behavior and inheritance relations. The syntax for TMN objects is defined as an extension of ASN.1, using the "macro facility" of that language.
- *SNMP* describes its MO-specification language as "Structure of Management Information (SMI)." An SMI is divided into three parts: module definitions, MO definitions, and notification definitions. MOs are stored in a virtual "management information base" (MIB). Collections of related objects are defined in MIB modules that are written using an adapted subset of ASN.1.
  SNMP–MOs are not objects in the same sense as in TMN. SNMP–MOs are described in terms of variables that may have states. The only operations that can be performed on these MOs are therefore some kind of reads and writes.
- If *CORBA* is used as a platform[5] for mpCPs, IDL (the "interface definition language") will be used as the MO-specification language. IDL is a language that relies on how CORBA views objects in general. This view is very similar to how you specify objects in C++, although generalized in order to facilitate adaptation to different software design languages. Mapping specifications between IDL and languages such as C, C++, ADA, JAVA, and SMALLTALK are provided. CORBA objects are stored in an "interface repository" (the correspondence to a MIB).
- *OMAP* ("operations, maintenance, and administration part") is an ITU–T management framework defined for SS7 management. This framework comprises management methods and functions specifically for measurement initiations and data collection, and for initiation of tests. It extends the TMN approach by defining a separate management system inside an SS7 network. This system may be connected to an TMN management system as well, in order to allow operators to

[4] Note that specifications that are identified by ASN.1 object identifiers are not necessarily MOs. Any ASN.1-based specification can be identified in this way. Organizations that want to make their specifications publicly available have to request that from an ISO authority who creates a new organization code in the object identifier tree.

[5] CORBA is developed for being a general platform for distributed OO software applications, i.e., not only for MOs.

initiate testing activities from a central point, for example. The OMAP system will then act as an AF in some mpCPs, defining MOs the TMN way. Internally, however, communication between OMAP elements is by operations only (supported by TACP), which is equivalent to an OMAP actor handling a single MO (since TCAP does not support an object identifier). Thus, OMAP does not rely on MOs inside an SS7 network.

5.3.3 Connectivity Structures for mpCP

Management plane control points exist, e.g., in network interfaces between a traffic system and a management system, which require the support of a connectivity stratum. This stratum may be completely separated from any traffic-system stratum, or it may be a stratum that is also used by traffic system layers.

- Using a separate stratum (in TMN called the "TMN data communication network") is typical for circuit-switching traffic systems. The solution is to install a LAN in a network element and to allocate its AF in a special server on that LAN. This server may then be connected to a (remote) node in the management system over some data communication network (X.25, FR, or IP), running the AF on available transport layers (UDP or TCP in IP networks). This implies that traffic system nodes and management system nodes will be nodes of *different* switching networks.
- If the traffic system includes a suitable packet-switching strata, it will most likely be also used as the stratum in mpCP network interfaces. This is how OMAP operates in an ISDN and how management planes are realized for IP, X.25, and FR traffic systems. Note that this implies that traffic system nodes and management system nodes will all be nodes of *same* switching network (which can have severe consequences if that network goes down).

In AMLn we identify the needs of abstract protocol operations for mpCPs and provide alternative usages of the AMLn–SAGM. How to realize the network interface between traffic and management systems is the choice of the designers of a network system for management, however. Since most of the existing management concepts are not built on the actor–agent separation and LSM–LPM separation principles, we will briefly discuss their connectivity solutions here.

The TMN solution *CMIS/CMIP* is based on the fact that the connectivity stratum is an OSI RM stratum, which unfortunately creates a very complicated structure and set of specifications. This structure is depicted in Fig. 5.7.

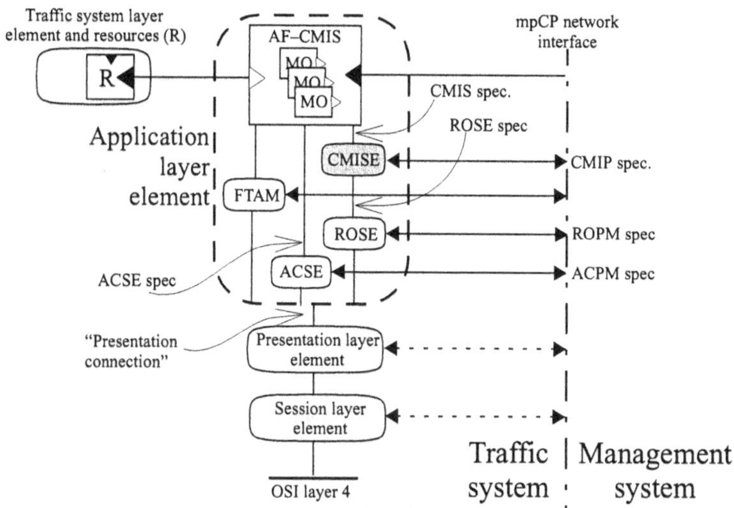

Fig. 5.7 The mpCP connectivity stratum in TMN (note: this is not an AMLn model). ACPM=association control protocol machine; ACSE=association control service element; CMIS=common management information service; CMISE=CMIS element; CMIP=common management information protocol; FTAM=file transfer, access and management; ROPM=remote operations protocol machine; ROSE=remote operations service element

The structure shows that an AF must be made part of an OSI application layer element that includes a CMISE, an ACSE, and a ROSE, and possibly also an FTAM element (for bulk data transfer). The actual structure defined by the layer interfaces in the application layer element is the author's best guess, since the standards are somewhat vague in this respect. The structure indicates that an AF must use the other service elements in a particular order, since they all operate over the same "presentation connection."

Every element in the structure, except the AF, is specified as a layer, i.e., by a "service definition" and a protocol specification (the AF falls outside what the OSI RM defines). Since actually neither of the elements are real layer elements in the sense that they create and interpret messages (messages are first created by the presentation layer), and therefore define no LPM, a large set of complicated specifications and artificial state machines have been produced that exist only because the OSI RM upper layer architecture says so. According to our discussion in Sect. 6.2, CMIS, ROSE, and ACSE are all functions of an agent layer and therefore belong to the same LPM. AFs and OSF are actors of such agents and define an LSM that is supported by that LPM.

This becomes particular obvious by the protocol specification CMIP: as can be expected, the only thing the CMISE has to do is to map primitives of CMIS operations on primitives of ROSE operations (i.e., act as a translator between two levels of meta-primitives). We can therefore expect "CMIP messages" to be identical to CMISE primitives which will be mapped one-to-one on ROSE-primitives, which is exactly what happens in the CMISE (although CMISE renames its "messages," presumably in order to make the effort of specifying the protocol worthwhile). Thus, we can conclude that the OSI upper layer architecture was forced on network designers as a (doubtful) specification method, but is not very helpful when it comes to specifying distributed applications.

TMN has been upgraded today and is no longer restricted to rely on the OSI upper layer architecture alone. TMN has incorporated SNMP and CORBA connectivity structures as realizations of mpCP network interfaces as well (see ITU–T Q.811 and Q.812).

Since *SNMP* runs in the very network it manages (an IP network), it relies simply on an TCP/UDP/IP stratum. The nodes of the SNMP management system run on TCP/UDP port number 161. SNMP does not make any distinction between actors and agents, so there is nothing similar to CMIS/CMIP in SNMP (which partly explains its popularity over TMN). Thus, an SNMP element is a single layer element that includes an element state machine (ESM) of either the AF or OSF type, that handles SNMP–MOs.

The *CORBA* platform, being a general platform for communication between distributed objects, is designed to be possible to run on different types of connectivity strata. The connectivity structure for an mpCP realized in CORBA is therefore somewhat more complex then for SNMP (see Fig. 5.8).

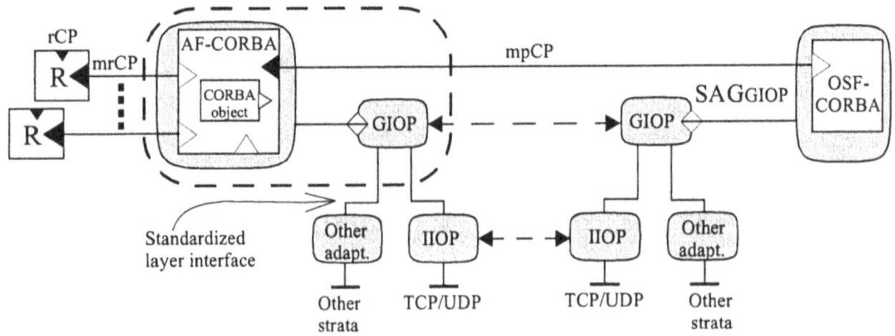

Fig. 5.8 Realizing the mpCP by a CORBA connectivity structure

The structure is well-aligned with the AMLn view: The CORBA layer is separated into an agent layer defined as GIOP ("general inter-ORB protocol"), where CORBA applications[6] are connected over GIOP–SAGs. The GIOP layer includes all the functions that in OSI/TMN are associated to ACSE, ROSE and CMIS/

[6] A CORBA application consists of user defined objects and a number of CORBA defined objects and services that are available to all user defined objects. For example, user defined objects do not know in which CORBA system they exist (in this case either in an AF or in an OSF). The CORBA platform (including GIOP and IIOP) makes object distribution transparent to user defined objects.

CMIP. Although the invokes in this SAG are not those of the AMLn–SAG, they are well defined by a request-reply semantics, defined as part of the CORBA standard. GIOP is designed for working over connection-oriented services (CON). In order to adapt to different strata, adaptation layers that offer a CORBA standardized CON layer interface to GIOP are defined. The layer for running CORBA on an IP network is the IIOP ("Internet inter-ORB protocol"). IIOP runs on TCP/UDP port number 535.

As already mentioned, the *OMAP* framework for management implies that an OMAP management system may exist without being connected to a TMN management system. This is because an OMAP system is designed for running on TACP, i.e., (similar to SNMP) relying on the very connectivity stratum that it manages. In case OMAP activities are to be initiated from a TMN management system, special OMAP–MOs must exist in some network elements where the OMAP and the TMN management planes meet. In such cases, the OMAP management function will rely on an SS7 connectivity stratum and on a TMN connectivity stratum (i.e., OSI, SNMP or CORBA).

5.3.4 The mpCP Protocol and Spontaneous Events

The AMLn and CORBA approach to mpCP suggests that the management layer between a traffic system and its management system can be described in three parts:

1. Specification of all MOs that are used in the interface between AFs and OSFs.
2. A number of general MO-handling operations (such as the CREATE, DELETE, and may be ACTION operation we discussed in Sect. 5.3.1).
3. An agent layer that defines a suitable set of invoke operations that support the abstract protocol.

In any mpCP specification, bullets (2) and (3) should be described as parts of the management layer specification, while operations on MOs (bullet (1)) should be described separately.

Looking at CMIS/CMIP, SNMP, OMAP, and CORBA, we may expect that (except for OMAP) they support similar but not identical general MO-handling operations, since they have been developed by different standardization bodies at different points in time. The differences are mainly due to how they view MOs. While CMIS/CMIP identifies the operations to be performed on MOs, SNMP defines only read and write operations on variables, for example:

- Get-request, an operation that requests the value of one or more variables.
- Set-request, updates one or more variables.

Note that these are SNMP messages (not primitives) since SNMP defines no SNMP–SAG.

One thing they all have in common, however, is that they define and handle spontaneous reports from the traffic system in ways that are different from how they define any other MO and operations. The reason is: the view that standards take on interactions between distributed objects is confused with how objects communicate in OO software systems. In AMLs we have no problems with that, since AMLs objects are abstract objects that are not restricted to being software objects. AMLs operation classes can therefore comprise several forms of operations where an object spontaneously generates events (se Appendix D). Figure 5.9 shows an AMLn scenario of this kind.

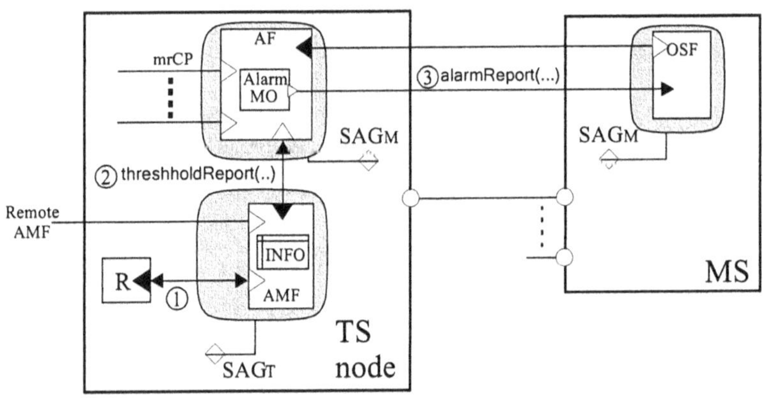

Fig. 5.9 Spontaneous events, the REPORT operation in AMLn. MS=management system; TS=traffic system

In this model we have assumed that the AMF autonomously monitors and maintains some resource R. In an earlier phase (not shown here), the OSF has told the AF that if a certain threshold value on some property of R is reached, the AF must send an alarmReport to the OSF. As a result, the AF informs AMF about the threshold value (also not shown here). The following scenario tells what happens then:

1. Every time the AMF changes R, or detects a state change of R, it compares the new state to the stored threshold value.
2. If the threshold value is reached, the AMF will alert the AF via a local threshold-Report event.
3. The AF may realize a special alarm MO, or define a REPORT operation as part of an MO that represents R. In any case, the MO generates an alarmReport event to the OSF in the mpCP.

Since LSM events are only mapped on the parameters of invokes in an AMLn–SAG, this event can be handled as any other LSM event. CMIS/CMIP as well as CORBA do, however, perform mapping of LSM event on SAGs in a different way:

- *CMIS/CMIP* is locked to the ROSE view on SAGs, which implies that the primitive type of a spontaneous events (Report in AMLs) must be mapped on the primitive types of a ROSE invoke. ROSE does not, however, define any primitive type for Report, since ROSE did assume that an object never generates any spontaneous event and only handles requests from the object user (which may or may not require an answer). Thus, spontaneous events must be mapped on a set of SAG primitives that are not suited for them. This complicates the handling of spontaneous events.
  The CMIS specification defines two "notification services" for CMIS users, a confirmed M-EVENT-REPORT and an unconfirmed one by the same name. In the CMIP specification is then defined how to map these services on "CMIP messages" and then on primitives of the ROSE–SAG.
  Besides the extra complexity caused by the CMIS/CMIP architecture, and the way mapping is done on the ROSE–SAG, the author feels it is rather strange to map an event report from an object in an AF on a primitive (the ROSE RO-INVOKE) that normally is used to carry requests for a remote party to execute an operation. This is especially peculiar since the event-generating object in the AF does not have a clue of what the OSF is going to do with the information.
- *SNMP* calls spontaneous events "traps", and defines a special message called SnmpV2-trap that is sent by an AF to an OSF when something unexpected happens in the traffic system. The trap mechanism is rather simple: the message just says that something happened in the sending node but gives no details. The OSF must find out what it is by issuing queries on SNMP–MOs in AF by generating GET-requests. The trap mechanism comprises no confirming message from the OSF that it has received and logged the message. Since the used connectivity stratum is unreliable (if UDP is used), an SNMP–OSF should also poll the traffic nodes it manages occasionally.
- *CORBA* defines an "event service" paradigm that can be used for reporting spontaneous events in management applications. The CORBA event is defined as "an occurrence within an object," i.e., it could be anything. That occurrence is specified by using IDL for the particular object type. A GIOP message that contains information on events that have happened are called "notification".
  Since CORBA is an open platform for distributed objects, it defines a mechanism that allows objects to register their interest in particular events, and for distribution of notifications to one or many objects. This mechanism is called "event channel," which in reality is a standard object that is available to all CORBA applications, such as the AFs and OSFs of our management model in Fig. 5.8. It could be applied in a management application as in the following scenario:

  - An object in the OSF that is an "event-consumer" object, tells its "event -channel" object that it is interested in a particular event of a particular "event-supplier" object (in this case an object in the OSF and an MO in an AF).

- The event-channel object registers that interest (event, event supplier, event consumer).
- It then tells the actual MO (via GIOP and the event-channel object in AF) to report the particular event.
- When the event happens, the MO reports the occurrence to the event-channel object in AF, which looks up which remote CORBA systems have registered for that particular event. It then uses GIOP to inform those systems, in this case the OSF.
- In an OSF, the event-channel object looks up which local event-consumer objects have registered for the event and informs them.

This is called the "push scenario", and is supported by four event-channel operations that any object can use. The "pull scenario" defines another four operations. In pulling, the consumer must actively ask the event channel object if any event of interest has happened, and a supplier object must announce its presence, as well as the events it can report, to the event channel object.

The picture we have given on how spontaneous events are handled in different management concepts is scattered, and (as can be expected) comprises different solutions, concepts, and terms. To use the AMLn actor–agent separation and AMLs operation classes (in particular such classes where the "event supplier" initiates operations, i.e., REPORT, OFFER and INVITE), could improve this situation considerably.

5.4 The Management System

5.4.1 Introduction

Since a management system (normally) has to manage a large set of distributed network elements, it is a network system in itself. It can therefore be modeled as any other network system, i.e., by a layer structure and a logical network structure.

From a functional point of view, it has a simpler structure than traffic systems because all management functions (AFs, OSFs and the agent layers on which they rely) belong to a single **management layer** in the management plane. This layer extends into the network elements that are managed. Since the whole layer is a control layer, we can simplify its model by showing its layer state machine (LSM) representation, thereby eliminating agent layers and layer protocol machines (LPM). The LSM structure, in its simplest form, would then look like Fig. 5.10, consisting of a single OSF and a strictly hierarchical structure of element state machines (ESM). The ESMs that concern the management system are the AFs, the OSF, and the WSF (which are ESMs that are also defined in TMN). We added the operator function (OPF) and therefore also the work station function (WSF) to avoid forgetting that there are users of the layer who have the overall control of everything that happens (or shall happen) in, or related to, the traffic system.

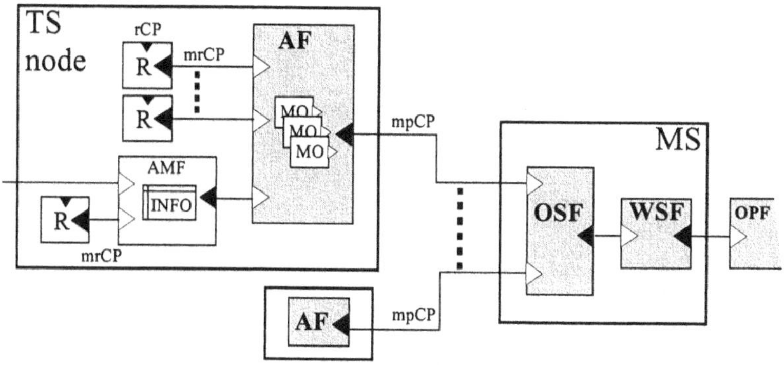

Fig. 5.10 A single OSF managing all traffic system nodes. Shaded ESMs are parts of, or known by the management system. OSF=operations system function; OPF=operator function; WSF=work station function

This model is extremely simplified, considering what we know about the complexity of traffic systems: a system that may consist of many strata, layers, and a more or less complex structure of logical networks. In addition to maintaining this complex system in an optimal condition, the management system may also have to handle many functions that are not directly related to maintenance, such as traffic analysis and network dimensioning, subscribing and connecting new customers, charging, accounting, security and overall business aspects, which altogether can make a management system functionally much more complex than the traffic system itself.

Managed objects (MOs) are developed and standardized for each type of traffic system. The number of MOs for a traffic system becomes very large, as already mentioned. Classification of management functions (i.e. MOs) is therefore needed. If a network system model exists for the traffic system, MOs should be classified and identified (using the OBJECT IDENTIFIER system) according to the structure of that model. If the model is an AMLn model, the basic classification structure will be defined by its layer structure and logical-network structure. The principle is shown in Fig. 5.11 for management functions that concern resources of a particular layer. In a similar way, configuration of layer elements to each other in traffic system nodes (described by the **hosts** relation of a logical network structure model) can be supported by logical node MOs.

Note that the way the OBJECT IDENTIFIER system is used today indicates that traffic system models are not made up or used for classifying MOs. Only SNMP–MOs have identifications that are related to some extent to the Internet layer structure.

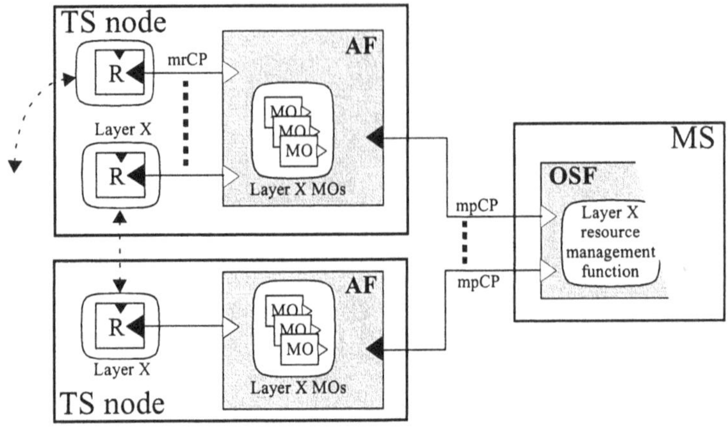

Fig. 5.11 MOs should be classified according to the structure of traffic system models

5.4.2
The TMN Management System

The OSF (the user of MOs) is not standardized. However, because of the functional complexity of the OSF itself, TMN (which was intended for public network operators) deals to a great length with classifying all functional requirements on the OSF. Although the TMN classification may be overkill for other public networks (e.g., the Internet), its function classification approach is interesting as an exercise in managing functional complexity. TMN suggests a separation of the complete OSF in management functions that are relatively independent and have well understood relations to each other. The classification system has two dimensions: "functional levels" and "functional areas."

The functional level classification is based on an assumed or possible network structure of logical nodes in a management system (MS). The ESMs of those nodes are called "managers". The structure suggest that each OSF functional level can be allocated to managers of its own. Four types of managers are suggested that deal with *network element* management (EM), *network* management (NM), *service* management (SM), and *business* management (BM) (see Fig. 5.12). In this model we make a distinction between mpCP and msCP, the latter being control points inside and between management systems (thus invisible to traffic systems). We have also indicated the possibility for different management systems to interwork (more about that later).

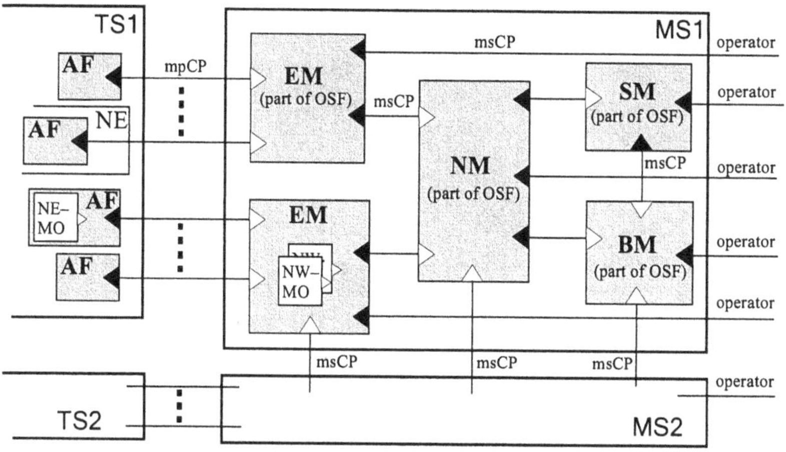

Fig. 5.12 The structure of OSF managers in a management system. BM=business manager; EM=element manager; NM=network manager; SM=service manager

1. *Element manager (EM).* TMN suggests that a manager connecting to network elements (NEs) will deal with logging, backup and maintenance of implementation elements (hardware and software) in the network elements it connects.
   In addition it can deal with threshold detection and alarm generation versus network managers. This implies that an element manager is intended to deal with many types of alarms from AFs itself, without bothering its network manager. It also implies that there can exist network-MOs (NW–MO) in an element manager that rely on network element MOs (NE–MOs) in AFs, where the NW–MOs are used by network managers. Thus, an element manager will behave as an OSF versus AFs, but as an AF versus network managers.
2. *Network manager (NM)*: A network manager that connects element managers will deal primarily with network configuration (setting data in routing tables, etc.). This is the management area in which AMLn logical-network models and configuration tables would be specially helpful. The manager may also deal with testing and traffic analysis.
3. *Service manager (SM)*: Deals with definition, administration, and charging of services.
4. *Business manager (BM)*: This manager deals with business oriented aspects, such as analysing usage trends, service quality and producing accounting reports.

The functional area classification assumes that, for each manager type, several or all of five types of problems may have to be handled (although, for each manager, constrained to resources that are characteristic for the type of manager):

1. *Fault handling*: These are functions that deal with fault detection, fault isolation, and fault reporting and logging.

2. *Accounting*: These are functions that deal with collecting, storing, and delivery of charging information.
3. *Performance handling*: These are functions that deal with collecting, storing, and delivery of operations statistics, and with optimization of the traffic system within the limits of existing resources.
4. *Configuration*: These are functions that deal with installation and initiation of resources and with network configuration, for example.
5. *Security handling*: These are functions that deal with authentication, with multiple access of management functions and with protection against access attempts via the traffic network.

In the application domain of "public networks," the network structure depicted in Fig. 5.12 for the TMN management system is still too general. TMN deals with three additional problems:

1. *Managing simple resources*: There are always simpler network elements, such as line terminals, multiplexers, and regenerators, for which installation of an MO handling AF with a protocol stack for communication with an element manager, is not justifiable. TMN therefore suggests a management function, a "mediation function" (MF), that can act as an AF versus element managers, but connects a number of such resources over a some type of local communication network (LCN). This network could be a separate LAN, but also just a control channel in the traffic system part. For example, both SDH and PDH define built-in control channels to which an MF could connect.
2. *Adapting existing management systems*: The traffic system (or part of it) may already be managed by a non-TMN management system. It is then of interest to be able to connect that system to a common TMN system, for creating a unified management environment for some management functions and/or possibly for enhancing the range of management functions. The TMN solution is to install a special device in larger network elements, a "Q-adapter" (QAF), that can handle MOs and can act as a gateway between the mpCP network interface that the TMN system defines, and whatever network interface the existing management system defines. The QAF will translate commands and reports between the two management layers, possibly using different connectivity structures on the two sides.
3. *Interworking between management systems* (as indicated in Fig. 5.12): When introducing a TMN system, it does not immediately take over the management responsibility of the whole traffic system. For some time one can expect two or more MSs to manage different parts of a traffic system in parallel. One MS may even have to manage resources that are accessible only over another MS. This is a matter of shared responsibility that requires that the MSs can cooperate about resources of common interest. A similar situation exist between management systems of different interworking operator networks.
   Two interworking management systems may both be TMN systems, which facilitates interworking. Often they are not (e.g., one may be a TMN system, the

other an SNMP system), which creates a more difficult interworking situation because they specify MOs differently, for example.

Interworking between management systems can comprise anything from interworking between element managers to interworking between service managers. If two MSs must interwork around a shared resource (e.g., a route, a charging record) that is represented by MOs, they must also have a common understanding of that resource. This common understanding can be specified as something called **virtual MO** (vMO) in AMLn. The set of all vMOs that are agreed upon between two MSs will act as a contract between them that defines which information in one MS is available to the other, and which resources in one MS may be managed by the other.

In their communication, the two MSs will refer operations and reports concerning virtual MOs. Locally, however, they must translate such information to operations and reports of MOs defined for their own traffic system (or part of a common traffic system). To support this paradigm, a TMN manager that interworks with another MS must include a translation function, as indicated in the model in Fig. 5.13 (here exemplified as interworking between network managers). Note that interworking between management systems requires attention to authentication and other security issues, such as eaves-dropping and unauthorized intrusions.

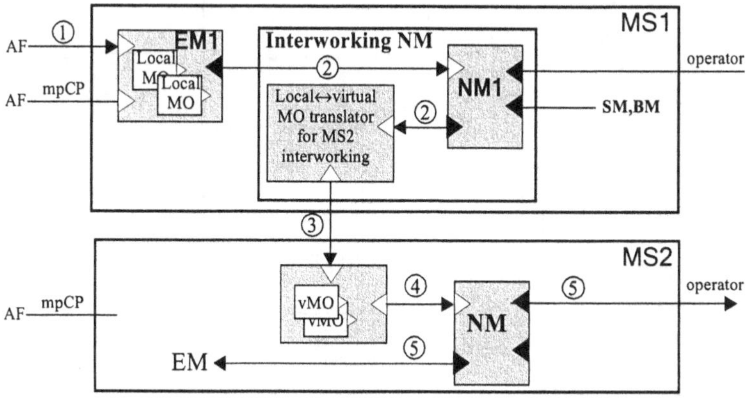

Fig. 5.13 Interworking between management systems can be specified by virtual MOs (vMO)

A scenario is included in this model (only one way is shown here, but it works both ways, of course):

1. A network element reports an event in the traffic system to its EM (EM1 here). EM1 analyzes the event. Depending on the type of event, it is either dealt with without notifying NM1, or stored in a local MO for reporting to NM1 later, or immediately reported to NM1.
2. NM1 may find that this event is related to an MO that is shared with MS2 (i.e., a "virtual MO"). It therefore tells its translation function for MS2 about the event.

3. The translation function translates this information to an event of a particular virtual MO, which can be more or less complicated, if the vMO is not specified as a TMN–MO. The translation function then reports to its peer translation function in MS2 that the particular event has happened, referring the vMO.
4. The receiving translation function in MS2 generates the vMO event towards its NM.
5. This NM may just store the report for later use, order an EM in MS2 to perform some operation on an MS2-MO, or generate an alarm to be dealt with by an MS2 operator.

If we include the TMN functions that adapt a TMN to different types of network elements, as well as interworking between a TMN and other management systems, we get the full picture of the TMN management plane/layer in its LSM representation as shown in Fig. 5.14. Note that TMN calls an AF that can handle TMN–MOs and the TMN defined mpCP (called Q3) a "network element function" (NEF).

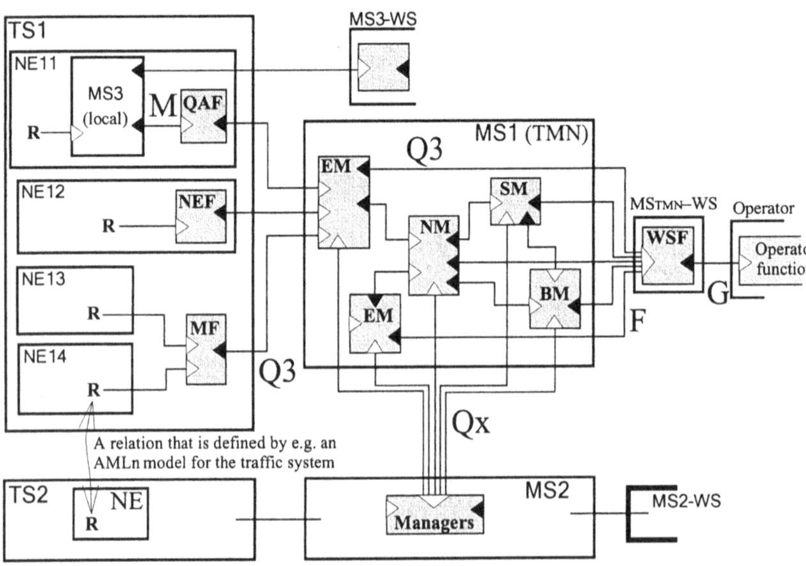

Fig. 5.14 The complete picture TMN gives of the management layer LSM

TMN defines this structure by giving names to the different control points. To be more precise: TMN does not define "control points," but "reference points," which is more a type of a network interface, since these points also include connectivity structures, such as the CMIS/CMIP protocol stack, according to Fig. 5.7.

- The mpCP and msCP of TMN are both referred to as Q3.

- Interworking control points are called Qx, where "x" indicates that the interworking system (MS2 in the model) may be something other than a TMN system.
- The interface between a QAF and an existing management system (MS3 in the model) is called M, but is "outside the scope of TMN."
- Managers interface work station functions (WSF) over an interface called F.
- Work station functions support the TMN operator's access to the TMN over an interface called G, and perform presentation functions for information available in the TMN. It also includes functions for creating, editing, loading, and administration of MOs.

The logical network structure for a TMN have many different configurations since there are many types of connectivity strata that may be used, as described previously. In Fig. 5.15 we show the logical node that performs mediation functions (MF) if SNMP is used in Q3. Note that TMN calls the hosting physical node "mediation device" (MD)

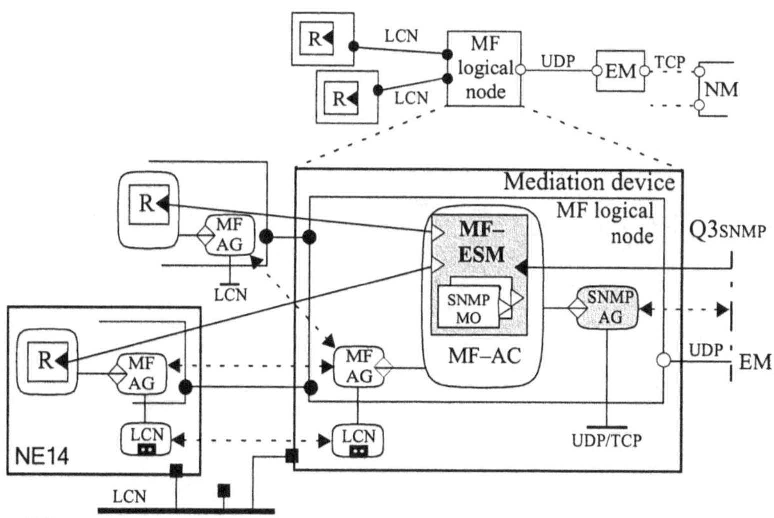

For Q3OSI:
Replace the SNMP agent with CMIS/CMIP, ACSE, ROSE, L6 and L5.
Replace SNMP–MOs with TMN–MOs

For Q3CORBA:
Replace the SNMP agent with GIOP and IIOP.
Replace SNMP–MOs with CORBA objects

Fig. 5.15 Part of the TMN management logical network, showing how an MF logical node may be hosted

On the element manager side (to the right), there are at least three alternative ways in which a Q3 network interface can be realized: using the OSI upper layer connectivity layer structure; using SNMP over TCP/UDP; and using a CORBA interface (GIOP, IIOP or other adaptation layers but IIOP). Note that each Q3 variant defines its own abstract protocol between MFs and element managers, as well as methods for defining MOs. On the traffic system side, an LCN connects the MF logical node to a corresponding node in NE14, which allows the MF-ESM to access resources (R) of simpler network elements.

Figure 5.16 shows another example: a network manager (NM) logical node, that interworks with another management network. This NM uses OSI Q3 interfaces in its own management system, but CORBA (a Qx interface according to TMN) when interworking with the other MS. This implies that the vMOs in the translators will be specified differently.

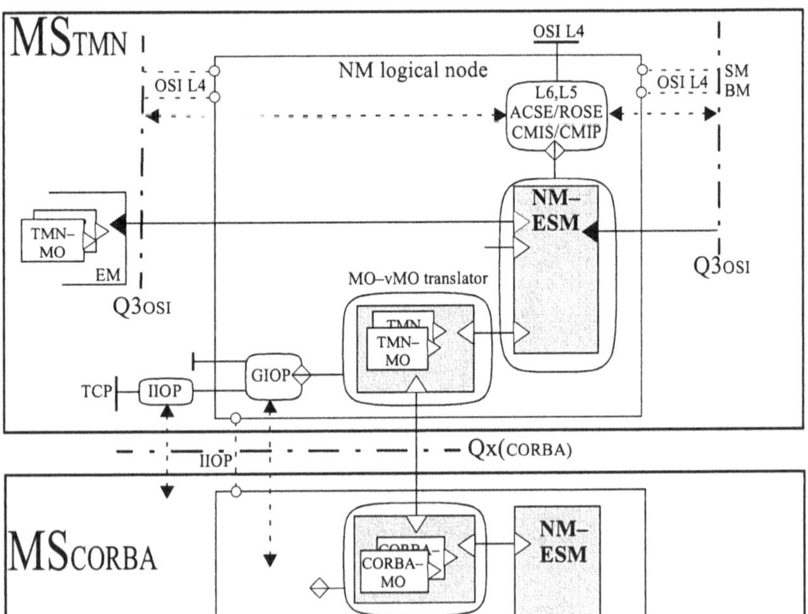

Fig. 5.16 Part of a TMN management logical network, showing how an NM logical node that interworks with another management system may be hosted

5.4.3 The SNMP Management System

The Internet was originally based on *decentralized* management, trying to do automatically as much as possible in traffic network nodes. For example, while in the PSTN/ISDN, routing tables are normally created and managed off-line the traffic nodes, the Internet handles this through a number of different routing protocols within the nodes of the TCP/IP traffic network. However, we now see a trend towards more and more *centralized* management functions in IP-based networks as well, especially when IP is used in public network systems such as UMTS. It is often noticed that today's telecom networks (basically circuit-switched) are migrating towards packet-switching and other Internet characteristics. However, if we look at management principles only, it is the Internet that is migrating towards telecom principles, not the other way around.

SNMP, defined by IETF, was developed to support centralized management functions, i.e., for building management systems for IP-based networks. The basic principles applied to the structure of the SNMP management system is the same as for TMN. The terminology is similar, although not identical, since TMN and SNMP have been developed by different standardization bodies and at different points in time. The SNMP management plane is defined in considerably fewer details than the corresponding TMN model, primarily because SNMP

- Does not consider interworking with any other type of management system and
- Leaves to builders of management systems to distribute the TMN functional levels and functional areas as they prefer.

The management system model that corresponds to the TMN model in Fig. 5.14 is shown in Fig. 5.17.

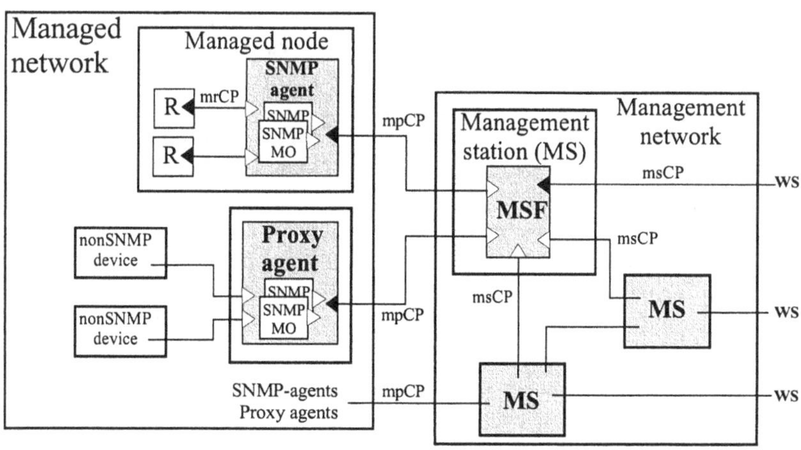

Fig. 5.17 The picture that SNMP gives of the management layer LSM. MSF=MS function; WS=work station

The SNMP-defined node concepts are included here in a "managed network" (corresponds to the TMN traffic system) and a "management network" (corresponds to the TMN management system). Other differences between the TMN and the SNMP management systems are:

- All mpCP and msCP of SNMP use UDP and/or TCP for connectivity.
- The correspondence to a TMN–QAF is not defined in SNMP.
- Terminology differences:
  - The correspondence to a TMN-NEF is called "SNMP agent".

- The correspondence to a TMN-MF is called "SNMP proxy agent." This agent is used to connect older devices or devices not originally intended for being IP nodes ("nonSNMP device" in the model, e.g., printers, bridges).
- The correspondence to a TMN–OSF is called "SNMP management station".

5.5 Using AMLn Models in Management Systems[7]

We have already, when we discussed TMN, indicated the potential of an AMLn model when MOs for the resources of a traffic system are to be specified. The structures of the model (layers and logical network) can be used for classifying MOs that exist in network elements. For example, if a logical-network structure model is defined, MOs for different logical networks should be kept apart. The configuration tables that are defined for layer elements in the layer structure model are obvious candidates for MOs in NEs. Tables that belong to different layers should also be kept apart, etc.

An AMLn model can, however, be used for defining high level MOs as well so that in a network of NEs, element managers (EM) and network managers (NM), operators can operate through MOs. The basis for defining such MOs is the network level structure of an AMLn model (as discussed in Sect 3.1). According to this structural feature, an AMLn model will normally be partitioned into a number of model fragments that (in the complete model) are related over substructure relations. Fragments on higher network levels are candidates for high level MOs, as depicted in Fig. 5.18. Here we have aligned the TMN functional level view to network levels.

The top three network levels (0..2) cover what normally is standardized, and therefore would be candidates for standardized MOs. Implementors may refine these fragments further into any number of lower network levels (taken as a basis for proprietary MOs defined by implementors).

Each fragment is a "small" AMLn model, i.e., described by a layer structure and a logical network structure. Therefore, for any particular network level, the fragment can be taken as the basis for at least two network MOs, one for its layer structure, another for its logical network structure. Following the TMN classification, we may separate NW–MOs into EM–MOs and NM–MOs. For every NW–MO, a number of network element MOs (NE–MO) can be identified, specified and implemented.

[7] In this section we discuss the potential of AMLn models in network management, presenting some ideas about how such models can be used in that context. Actual applications of NW–MOs in management systems cannot be presented in this book (but hopefully in a later book).

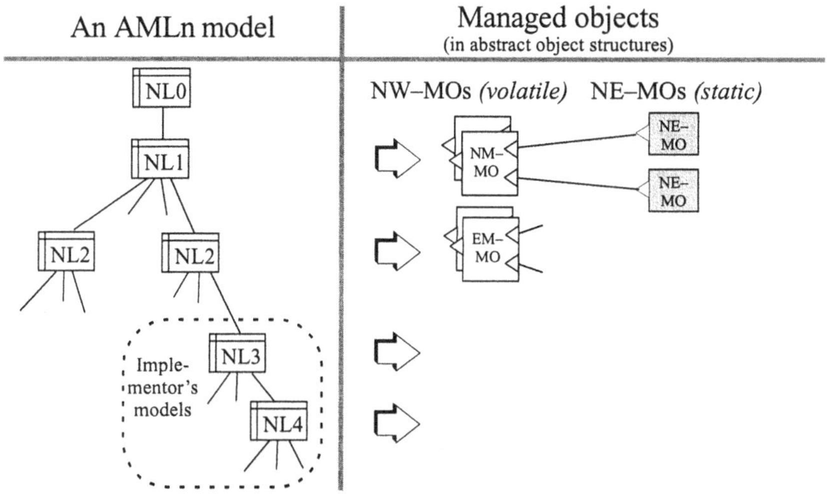

Fig. 5.18 Fragments of an AMLn model are candidates for managed objects. EM=network-element manager; MO=managed object; NE=network element; NL...=network level; NM=network manager

Note that the role of a network MO is to translate operations on the very MO to invocations on lower level MOs, which normally will be MOs in one or several NEs (i.e., NE–MOs), but could be operations as well on one or several MOs in EMs (EM–NOs). A model that shows dependencies between objects (MOs or other types) is called **(abstract) object structure** in AMLs, see Appendix D.

NE-MOs are *static* in the sense that they exist (in NEs) whether or not a management action is performed. This is, however, not so for EM–MOs and NM–MOs, which becomes revealed if we consider how such MOs are used. Figure 5.19 depicts an MO in an NM. This MO supports management of either the layer structure of NL1 or of its logical-network structure.

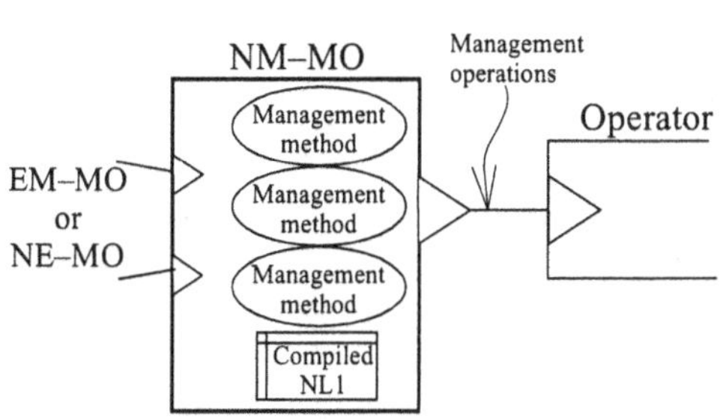

Fig. 5.19 The components of a network MO

The NM–MO is created by compiling the actual NL1 fragment into a data structure for the object (for a software object it is the equivalence to its *attributes*). An interface to the object user (which in this case would be an operator), consisting of a number of management operations, is defined and corresponding *methods* are implemented for the object.

Let's look at a simple scenario of how network MOs can be used (see Fig. 5.20).

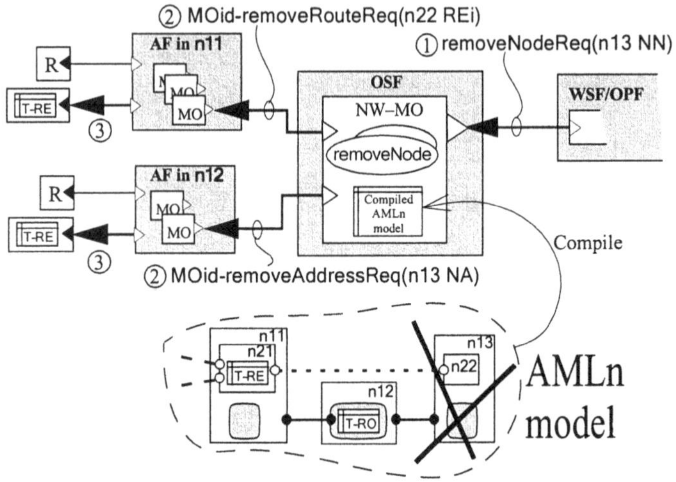

Fig. 5.20 A scenario, showing the removal of a logical node from a traffic network via an NW–MO. NA=network address; NN=node name; REi=route endpoint identifier

The model shows the LSM of a management layer with an OSF that, e.g., maintains tables in network elements of a traffic system. Two network elements are indicated by the AFs. These AFs include MOs that are used to update T-RE and T-RO tables in the two network elements. The network elements are defined as logical nodes in an AMLn logical-network structure of two logical networks: a hosting network nw1 that hosts nw2, as depicted. This model is available as a data structure in the OSF. This structure is used to create network MOs on operators' requests. The model shows a scenario where an operator uses the NW–MO to remove (logically) the logical node n13 (part of a terminal network element) from the traffic network.

1. The WSF/OPF requests removal of n13 ("n13" is the name of a logical node (NN) in the AMLn model here).
2. The NW–MO sees from the AMLn model that the T-RE table of n21 and the T-RO table of n12 must be modified. The NW–MO knows which network elements contain these nodes and which MOs to refer. Consequently it generates suitable requests to the AFs of these elements:
   - removeRouteRequest(n22)
     affects the T-RE in n21, where the actual route endpoint (REi) is identified by the name of the remote node, in this case n22.

- removeAddressRequest(n13)
  affects the routing table in n12, where n13 is no longer a possible network address (NA).
3. The referred MOs in the AFs execute these operations by operating on the tables in the traffic plane by proprietary means.

An NW–MO will most likely be of a *volatile* object type, created (on-demand) by an operator request, since all operations that could be performed on the NW–MO most likely are not allowed for all users/operators. In order to protect the management system from accidental or unauthorized actions, operators will have different access rights to different operations on the same part of the managed system (in this case the NL1 part). Most operators will have view access to most of the information in the NL1 model, while *modifier* operations (i.e., operations that can change the state of the traffic system) will be selective, according to some classification system for operator categories.

As a consequence, an operator requesting access to an NW–MO will have to provide the identity of the NL fragment, an access code that identifies the operator category and (possibly) define what kind of management operations are to be performed. The network manager (in this case) will check which operations on the fragment that operators of the category have access to, and create a corresponding NW–MO. This might not be the NW–MO the operator wanted, but it is the only MO the system will open as a result of the operators original request.

Thus, network MOs can easily be defined and used by operators of a management system, provided that the traffic system is described as an AMLn model (or by any equivalent modelling language). If network MOs are to exist,

- Compilation of the model into data structures must be supported, which implies that a formally defined language (such as AMLn) must be used for modeling.
- Furthermore, a management function must be added to allow managers to create network MOs according to operators' access rights.
- Work stations must know the language's graphical notations so that model fragments can be displayed.

5.6 Summary

We can summarize the discussion about traffic versus management systems by drawing generic layer structure models. In Fig. 5.21 we show the two planes, assuming a simple traffic system that consist of four layers. This model is an applied version of Fig. 5.1, extended with what we know about service and control points and other concepts used to model management functions.

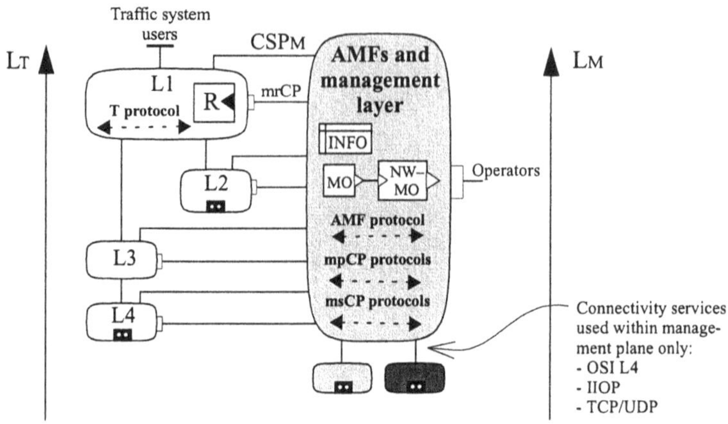

Fig. 5.21 The combined layer structures of the traffic and management planes

- We define as management plane all functions that deal with management in some form, whether they exist in the management system or the traffic system, and whether they are just management agents (AFs, parts of the management layer) or autonomous management functions (AMF).
- Two types of relations exists between the planes: mrCPs and connectivity service points for AMFs (CSPM).
- The layer structures of the traffic plane and the management plane are completely separate and independent of each other, since all relations that the traffic plane has to the management plane terminate either in the single management layer or in AMFs.
- Connectivity services that are used especially for the management layer are those we have discussed: OSI layer 4 if mpCP protocols rely on CMIS/CMIP; IIOP if they relay on CORBA and run on TCP; and TCP/UDP if the mpCP protocol is SNMP.

A corresponding intralayer structure for the management plane is shown in Fig. 5.22. Here we choose to show two nodes: the management system is one node; the other is an imaginary traffic-system node that takes part in all traffic layers and handles service users as well. We also indicated the separation of MOs in AFs according to the traffic system layer structure.

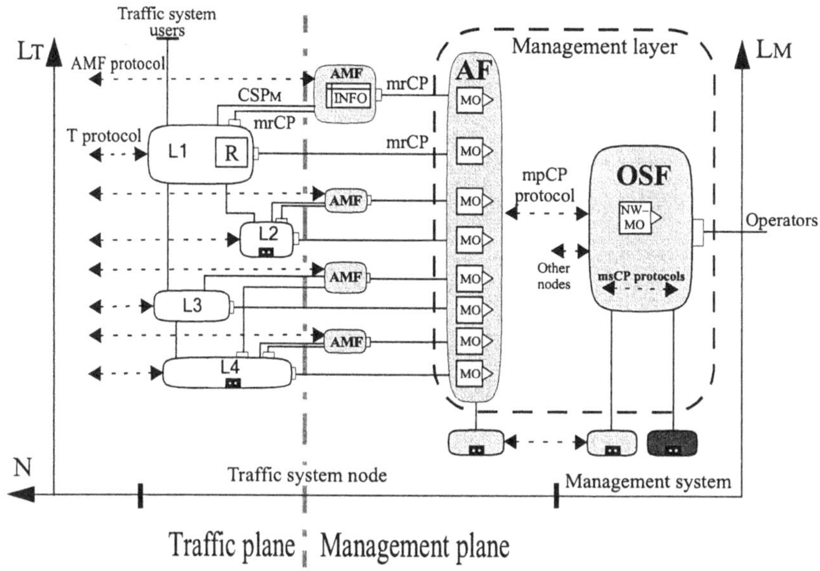

Fig. 5.22 The intralayer structure, seen over two nodes: the management system and a traffic system node that takes part in all traffic layers

Note that this model is a reference model only. When applied to actual traffic systems, all traffic system nodes will not handle all layers (some may only handle one or two low level layers). Furthermore, only some layers may be managed by AMFs, and only some AMFs may deliver information to AFs. Also, as shown in Fig. 5.22, an AMF may use the connectivity services of one layer (here L3) but manage resources in another (here in L4). An example of that is the routing protocol BGP (see Appendix A) in the Internet that runs on the TCP layer and manages routing tables in the IP layer.

6 Applying AMLn

6.1 Introduction

Basically, AMLn can be used in two ways: for specifying network systems or for analysing network systems that are not described in AMLn (which are all systems that exist today).

- Used as a *specification* tool, the resulting AMLn model is an exact model of the system to be implemented. Once implementors get used to AMLn, such a model is easily understood (compared to understanding thousands of standard pages). The model is also easily accessible for every individual in the development organization (compared to the standards libraries of today). Furthermore, whether AMLn is used in producing standards or only in implementors' development processes, AMLn gives strong support for validation through simulation (provided that model creators describe the behavior of interfaces and elements, e.g., using AMLs).
- The purpose with using AMLn as an *analyzing* tool may be to produce a model that can explain a network standard or an implementor's system documentation. AMLn is therefore also an excellent tool for re-engineering. Furthermore, an implementor who supports and documents the evolution of his system in an AMLn model will experience a considerable increase in system quality and length of life. The model can also be a basis for code generation as well as for network configuration, which can result in a considerably shorter time-to-market for new system functions and services.

The author has developed and applied AMLn for almost a decade. A few co-workers have been exposed to AMLn ideas over the years. AMLn is, however, completely unknown to a broader audience. The purpose with the case studies in this chapter is therefore to demonstrate AMLn as a modeling language by analyzing some well known network standards. In doing so, two effects have become very obvious.

1. *The problem of describing a model on a serial media*, such as a book. Using AMLn the way it is intended means that an information model is created in a database. Such a model is a more or less hierarchical structure of a large number of information elements. The intended way of getting to know the system is by browsing the model, i.e., accessing different information elements through the model itself in a user-selected fashion. There is very little need for adding eluci-

dating text to an AMLn model. We cannot however *store a model* on the pages of a book, we can only *describe a model*. This implies that (a) a particular order of description had to be chosen and (b) a lot of text had to be produced that is not part of the model itself. The descriptions of AMLn models in the present chapter cannot therefore give full justice to real AMLn models and to AMLn as a modelling language.[1]
2. *The shortages of today's standards*. The models behind these case studies belong to a set of models that were initially produced as the author's private means for verifying AMLn as a language. An effect was that many common shortages of today's standards became revealed. When describing these case-studies in the present chapter it is unavoidable that these shortages become exposed, since the actual purpose with this chapter is to demonstrate the usefulness of AMLn as a modeling and analyzing tool.

The reader should, however, bear in mind that AMLn was not known when these standards were produced, and that no modeling approach was available to the creators of standards at the time. Standards could only be produced as natural language descriptions, more or less enforced by pictures.[2] Therefore, the fact that the model descriptions in this chapter reveal many shortages in existing standards should not be taken as a criticism that standard workers have been doing an inferior job, and that all standards are poorly written. There are many standards that are well written with the readers' needs in mind (although there are as many that are not).

Having said that, let's summarize all kinds of problems that were encountered when browsing through standards with AMLn as a torch:

1. *No common terminology*: This problem refers primarily ITU–T standards, since ITU–T deals with a large set of different network systems. IETF and OMG are better off since they each deal with a single architecture (the Internet and CORBA, respectively).
   The complete absence of common network terminology (already commented on in Chap. 1) is probably the most annoying thing with ITU–T standards. It makes

[1] Real AMLn models will be published on the Internet as soon as an AMLn modeling tool becomes available. In an upcoming book (expected to be published in 2006) the author will also focus on AMLn modeling in practice by including more full scale and complete example models.

[2] This statement is correct as far as architecture descriptions are concerned. Certain formalisms are sometimes used, primarily for message specification (ASN.1, defined by ITU–T), for interface specification (ROSE, defined by ITU–T; IDL defined by OMG) and for protocol test specifications (TTCN, defined by ITU–T). Functional structures and behaviors are sometimes described as SDL models and MSC diagrams, both defined by ITU–T. SDL is, however, a common-value language, not an added-value language. When used in standards, it is not for substituting the text and pictures by a model, but to create realization models as a means for explaining features of the standard (which in reality increases the complexity of already-complex standards).

it extremely difficult to apply one's knowledge of one network system on to another (or in other words, knowledge management is not a part of ITU–T activities). In reality, almost every single standard reinvents the same concepts, and, probably in order to be semantically accurate, provides definitions and names of its own.

AMLn proves that a rather small set of concepts can be used for describing any network system on the added-value level. When using AMLn as an analysis tool, it will therefore be implacably revealed that, e.g., TCAP describes an AMLn common agent layer but defines no actor layer identifier (which creates transparency problems on the SCCP level), that MTP L3 use the term "signaling point code" for network address, that SCCP use the term "subsystem number" for access-point identifier, while MTP L3 prefers to call it "service indicator". We could go on like this and create a large list of synonyms for the few concepts that AMLn defines.

2. *No common notation*: ITU–T standards in particular contain quite a few pictures. Some basic common graphical notations could have supported the needs of users who must read and work with many standards. With the exception of efforts in the transport network area (see ITU–T G.805) and the frequent use of SDL for implementation models (see bullet 6 below), no common notation is used. At best, an explanation is given for notations used in a particular standard document.
3. *WHAT but not WHY*: Each single standard describes, down to the bit level, how network functions and elements are supposed to function. This is fine (and necessary), but for everybody trying to understand a network system (or part of it), the first question is "Why?", i.e., what problem does this function solve, and why is this particular solution needed (especially if solutions to the particular problem exist in older systems, which is most likely the case)? As ITU–T standards are concerned, they give, almost without exception, no answers to such questions. For others who are not the inventors of a standard, this implies that other sources of information have to be searched for in order to acquire the deep knowledge that makes it possible to understand the standard (a profitable business for many authors of tutorial books, including the author of this book).
4. *Fragmentation but no overview*: In the absence of a modeling technique that can help in managing the complexity of network systems, its complete specification is normally broken down to a large number of standard documents (not always according to the picture of layer and network structures that, e.g., OSI RM and AMLn support).

   Over the years, the author has also noticed a tendency to break down a single document into a large and deep paragraph structure, where some paragraphs may be as small as a single sentence. Behind this approach lies probably a misguided expectation that system's complexity can be managed by fragmenting text. Fragmentation of a whole requires defined relations between the parts, however. Regarding standards, this means that they are crowded with references between paragraphs and between documents, which is rather annoying for the reader (to say the least), and almost certainly guarantees an early inconsistent and erroneous view, as the specification of the network system as a whole is concerned.

5. *Too many functions in a single layer or standard document*: Some standard documents represent a reasonably sized part of a whole network system, no matter which conceptual modeling view is applied. There are however too many that do not, resulting in single documents that may contain several hundreds pages each.
   An examples of this is the SCCP standard. This layer comprises a functionality that, compared to the Internet architecture, integrates DNS, TCP, and UDP functionality in a single layer.
   Another example is the V5.1 standard that manages to describe six layers in a single document.
6. *Creating implementation models as explanation models*: The lack of intelligibility in many standards is sometimes dealt with by describing a possible implementation model (sometimes, but not always, using SDL as the modeling language). However, such models are not substitutes for anything in the standard. In the author's opinion they merely increase the complexity of already complex standards, and should be taken as a sign of that an added-value modeling language is badly needed.
   In this chapter, the TCAP sublayer model is such an example (see Sect. 6.3). Block diagrams are used for the same purpose in, e.g., SS7 standards (e.g., MTP L2 and MTP L3) and ATM standards (e.g., SSCOP). For example, an AMLn model of MTP reveals that this architecture comprises two traffic layers (MTP L2 and MTP L3) and two (related) management layers. The architecture is described by two documents. This looks acceptable until one realizes that the description of the layers is almost randomly spread over the two documents (which is the result of the L2/L3 layer interface not being specified). The authors of these document obviously realized the problem, but, instead of breaking up the document into a number of separate documents for the layers and layer interface specifications, they produced extensive diagrams of an assumed implementation of layer elements (in terms of block diagrams and SDL flowcharts). These diagrams were *added* to the two existing documents, thereby considerably increasing the size and complexity of the standard.
7. *Erroneous use of, or interpretation of the layer concept*: This comprises two issues: defining "layers" that are not layers, neither in the OSI RM nor in the AMLn sense; violating the requirement that anything inside a layer should be transparent to other layers.
   The most flagrant violation to the transparency requirement is the OSI presentation layer and the ACSE (see the discussion in Sect. 6.2).
   TCAP manages to define two layers, one of which handles no messages and is therefore no layer (see Sect. 6.3).
   An unnecessary coupling between SCCP and TCAP is also created by letting SCCP access-point identifiers (SSNs) identify actors that are supported by

TCAP. Thus, which actor layers exist on the TCAP level is not transparent to SCCP.

ATM manages to define five layers in its signaling AAL architecture (SAAL), of which only two handle protocols (see Sect. 6.4.4).

MTP exhibits a rather blurred definition of the layer interface between MTP L2 and MTP L3, as already mentioned.

There are many other examples of this shortage. It seems that the OSI RM-layer concept, as well as the distinction between "protocol" and "layer interface," clearly defined in the OSI RM, has not always been understood by authors of standards.

8. *Integrating a traffic function and its management in a single layer*: Almost without exception, protocol standards integrate the description of a traffic function with some management functions (often called "layer management") in the same document. From an AMLn viewpoint this is not acceptable for two reasons:

   - The time perspective for the traffic function and the management function are completely different (on-demand and as needed, respectively).

   - The resources on which the traffic function operates are completely different from the resources on which the corresponding management function operates. The former operates on other traffic functions, the latter on tables and other resources of the traffic function itself.

   In AMLn, the traffic function and its management functions are specified as different layers/protocols, interconnected by management control points (mrCP, as defined by AMLn). The way standards integrate these functions often creates a very complex picture of something that is rather simple to describe,

   An example of such an unnecessarily complicated standard is the V5.1 standard. An AMLn model of V5.1 reveals that it comprises four traffic layers and two management layers, the specification of which are all squeezed into a single standard document. Another example is the MTP architecture, as already mentioned.

9. *Implied software implementation of protocols and interfaces*: Probably the most outstanding example of negative effects in this respect is the OSI upper-layer architecture. The standard confuses OO software implementation of OSI application layers with a network's functional structure and node structure. Structural concepts in this architecture also comprise pure software concepts, such as the "application process". The architecture therefore has a restricted applicability, and is also unnecessarily complicated. To a large extent this explains why OSI failed as a reference model (for data networks). You can read more about this in Sect. 6.2.

10. *Lack of important separation principles*: The actor–agent separation principle and its consequences (primarily the meta-primitive concept, the separation of control layers from connectivity layers and the division of control layers into an LPM and one or several LSMs) are very strong principles for keeping complex functional structures simple, intelligible, and maintainable. With the exception

of ACSE, ROSE, and TCAP, standards use nothing of this kind. As a result, too many standards and protocols are very complex, difficult to understand, difficult to validate, and difficult to maintain.

As ACSE, ROSE, and TCAP are concerned, these standards are unnecessarily complex and too restricted as to their applicability, due to that the LPM–LSM distinction is not applied. In addition, ROSE exhibits a method for describing behavior ("linked operations") that is more damaging than useful. ROSE and ACSE also suffer from misplaced OSI layers (layer 5 and 6), while TCAP has managed to avoid this problem. You can read about this in Sects. 6.2 and 6.3.

6.2 OSI Upper-Layer Architecture

6.2.1 Introduction

During the 1980s, OSI specialists (including the author) believed that the OSI RM and its standards would prevail over other "open systems" concepts, in particular in its competition with the Internet architecture. For example, in Rose (1990) we could read:

> The... OSI model will join and eventually displace the Internet protocol suite as the off-the-shelf commodity of choice.... The OSI suite eventually will dominate, then eclipse and finally make the Internet suite "immaterial."

The fact that the evolution of open systems has not only gone completely in another direction (the Internet way), but has also made the OSI RM an almost forgotten architecture, is not only due to the fact that there are millions of TCP/IP-based systems deployed today. It is as much the architectural solution that is represented by the OSI RM, which is too complex, too inefficient, and above all, too implementation-oriented. Nobody should apply it today.

The discussion here is of interest not only because it demonstrates the power of AMLn, but also because it demonstrates the negative effects of applying a software engineering approach to network architecture, and because parts of this architecture (ACSE and ROSE) represent an original initiative of defining common agent layer functions[3] (and has inspired AMLn in this respect).

The OSI architecture for layers above its transport layer (layer 4) is (in principle) not possible to model in AMLn. Besides that AMLn defines a generic solution for modelling any layer (the actor–agent separation principle) while OSI RM

[3] The OSI upper-layer architecture is described by the OSI RM. ACSE ("association control service element") and ROSE ("remote operation service element") are generic parts of this architecture. A third generic part is called RTSE ("reliable transfer service element"). RTSE seems to be a function that compensates for the lack of transparency in the OSI application layer, as regards the OSI presentation and session layers. Since we believe that this is a reason as good as any for abandoning these layers completely, we can disregard RTSE in this discussion.

defines a generic layer model only for its application layer, a number of architectural principles differ. In summary:

- The actor–agent separation principle does not exist in OSI.
- The identification of something similar to the abstract protocol in AMLn is defined only for the OSI application layer (layer 7), which (in OSI) is a *software* layer by definition.
- Encoding is dealt with in different ways. In AMLn it is a task that is performed by agent layers within *every* (OSI) layer. In OSI it is a task performed by a *separate* OSI layer (the presentation layer, layer 6) on behalf of only one specific layer: the application layer.
- Dialogue handling is performed in the OSI session layer (layer 5) and in ROSE (layer 7). In AMLn, the corresponding function is regarded as a part of the layer state machine (LSM) of a control layer since dialogue handling is assumed to be more or less unique for a specific application. Allocating LSM dialogue handling in the layer protocol machine (LPM), as is done in ROSE, is avoided in AMLn, to make LPMs more generally applicable.

In the AMLn sense, the OSI application layer is a control layer. There is no simple way to map between an AMLn control layer and the OSI RM upper layer architecture, however. An intuitive comparison is made in Fig. 6.1.

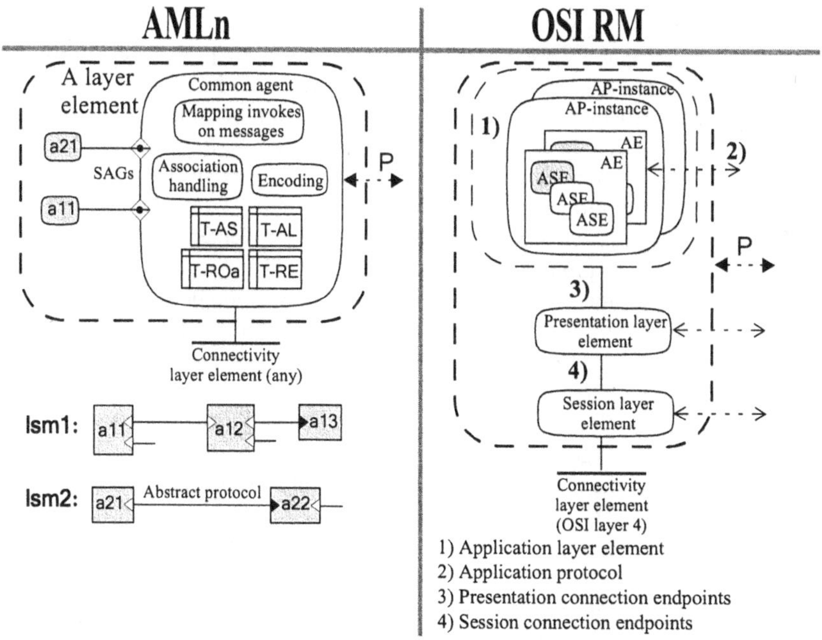

Fig. 6.1 Comparing AMLn layering with the OSI RM upper-layer architecture. AE=application entity; AP=application process; ASE=application service element

The AMLn model to the left depicts a layer element that includes a common agent supporting actors of two actor layers: {a11,a12,...} and {a21,a22,...}. The LSMs (lsm1 and lsm2) of these layers are indicated below.

An important difference is that an AMLn agent includes everything defined by the OSI RM upper-layer architecture, *except actors*, while the OSI application layer includes everything of the AMLn layer element, *except encoding* and *dialogue handling* (which are functions placed in layers of their own). The correspondence in AMLn to a presentation layer function is the mapping performed inside an agent between invokes in actor–agent interfaces and messages between agents. The correspondence in AMLn to a session function is performed by actors and specified as a part of the abstract protocol (i.e., as a part of the element state machine, ESM, of an actor). Therefore, applying an AMLn layer architecture means that the complicated OSI presentation and session layer interfaces need not be implemented (and of course neither do the presentation and session protocols). In the OSI RM, these interfaces are defined (in analogy with real connectivity interfaces of lower OSI layers) by artificial "connect," "disconnect," and "data" primitives, used for establishing "presentation connections" and "session connections," and for "sending" and "receiving" data over such connections. The way properties of such connections are defined also becomes complicated and rather strange, comprising the definition of things like "presentation context," "synchronization points," "tokens," etc., which the author is sure of that very few designers have ever understood and applied.

Another important difference is the way that the unique part of an application protocol is defined. In AMLn we isolate this part in terms of an LSM and its abstract protocols, which we specify (using AMLs techniques) independent of the LPM that will be used. In the OSI RM, the correspondence to an actor is one of a number of "application service elements" (ASE) that exist in an "application entity" (AE), that in turn exist in an "application process" (AP, a typical software concept). Sometimes such unique ASEs are called "user elements," but when specifying an "application protocol" all ASEs are treated alike.

The OSI application protocol defines the communication between AEs. It is a specification that comprises not just a "user element" ASE, but all other ASEs as well (primary ROSE and ACSE ASEs). The specification technique is based on an extension of basic ASN.1, where an ASE is described by an *ASN.1 macro* that is an aggregate of other ASN.1 macro that specify operations. An application protocol is then specified as an aggregate of ASE macron, as an *application context* macro. To further complicate the matter, an AE may handle alternative application contexts (which necessarily requires additional negotiation procedures in association handling).

This complex architecture can only be understood if we know the purpose with it, which is to facilitate code generation and management of interface specifications, based on the particular specification technique that is represented by ASN.1 macro for operations, ASEs, AEs, and application contexts. Another stated purpose is to support OO software implementation of applications (the ASE being close to the specification of a software object type). For example, in the ROSE standard (see ITU–T X.219) you find the statement: "The notation is based on established object-oriented programming principles." The standard does not define which these principles are. What were "established" OO principles back in 1984 are unfortunately no longer so. The statement also indicates that the OSI application-layer architecture is not suitable as a general architecture for control layers, unless you choose to implement them in software. These conclusions indicates the danger with letting implementation concerns direct the way system architectures are defined.

Furthermore, in order to support this specification technique, and as an effort to create a common residence for all types of encoding algorithms (an objective that concerns only software realization and management of network functions), OSI RM defines encoding of ASN.1 data as a layer of its own (the presentation layer), instead of what it actually is: encoding algorithms for ASN.1-typed data (the existence of other data types and alternative algorithms were initially assumed, but never realized in any network system). If the creators of the OSI RM had realized that, network designers would never have been exposed to the strange layer services, such as those offered by the OSI presentation layer.

Thus, the OSI upper-layer architecture is the result of an engineering approach that aims at code generation of software implementation of control functions in networks, based on a complicated specification technique. It does not reveal necessary insights in network systems and how to model them. For network system modeling, we prefer a technique that is applicable in all cases, i.e., whether network control functions will be implemented in software or hardware, and whether the ROSE specification technique (ASN.1 macron) or some other technique is used. Furthermore, the technique must be useful for all types of control layers, not just for what OSI regards as "application layers."

6.2.2 ACSE, the Association Control Service Element

The OSI RM approach to association handling is defined by the ACSE standard. ACSE is one of the common ASEs that are defined in the OSI application layer. If we use ACSE in AMLn, it would become a functional part of an agent in the AMLn common agent layer model. In Fig. 6.2 we have replaced the AMLn association-handling function (as described in Sect. 2.3.2.3) with an ACSE to show the similarities between association primitives defined in a specified agent interface (SAG), with association primitives that ACSE defines.

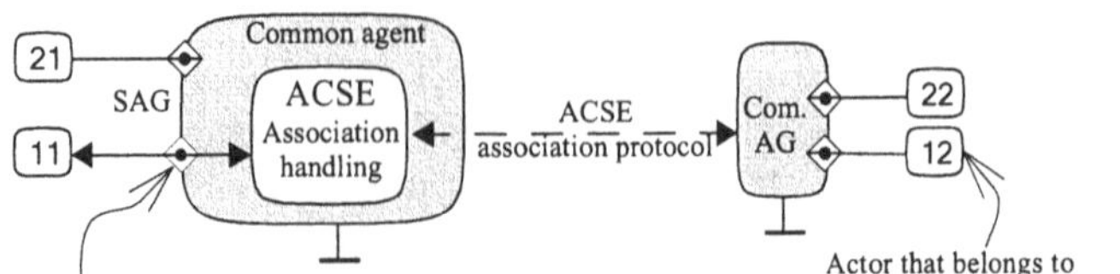

ACSE primitive names:	AMLn primitive names:
A-ASSOCIATErequest	acAssRequest
A-ASSOCIATEconfirmation	acAssConfirm
A-ASSOCIATEconfirmation	acAssRefuse
A-ASSOCIATEindication	agAssOffer
A-ASSOCIATEresponse	agAssAccept
A-ASSOCIATEresponse	agAssReject
A-RELEASErequest	acRelOrder
A-RELEASEindication	agRelReport
A-RELEASEresponse	-
A-RELEASEconfirmation	-
A-P-ABORTindication	agRelReport

Fig. 6.2 ACSE primitives, compared to AMLn association primitives

ACSE and AMLn association operations have similar purposes. The difference in number of primitives and primitive names lies primarily in the way AMLs and OSI looks at the operation concept:

1. In OSI, operations are always some kind of global functions. For example, the operation A-ASSOCIATE extends between two remote actor–agent interfaces and is defined through four primitive types (request, indication, response, and confirmation). AMLs operations, on the other hand, are always bound to a single interface type. Therefore the ACSE operation A-ASSOCIATE is broken up in two operations in AMLn: acAss (request, confirm and refuse) and agAss (offer, accept and reject).
2. The difference in names of primitive types indicate the difference in operation classes in AMLs and OSI RM. An important difference is that AMLs always makes a distinction between positive and negative outcomes of a request, in order to provide better semantics for behavior descriptions. OSI does not, which is a typical software language approach to remote operations. An example is the ACSE primitive A-ASSOCIATEconfirmation in Fig. 6.2 that is represented by two primitives in AMLs: acAssConfirm (positive outcome) and acAssRefuse (negative outcome). The OSI operation classes are defined in ROSE (see Sect. 6.2.3).

The similarities between ACSE and AMLn end here however, due to the difference in architectural models. A minor difference to AMLn is that ACSE does not define an association identifier (ASi). A major difference is how element identification is done when associations are established and used. In AMLn we need two parameters: an actor layer identifier (ALi, for common agent layers only) and an actor suffix (As), and possibly an association identifier (ASi) when an association exists. This is reflected by the tables in the model to the left in Fig. 6.1.

The model to the right in Fig. 6.1 indicates which types of identifiers are needed in creating ACSE associations: since an association is established between two AEs, and an AE is contained in an application process, AP (which is the element that must translate application layer identifiers to presentation layer addresses), both an "application process title" (APtitle) and an "application entity qualifier" (AEqualifier) must be given in ACSE messages. Since no ASi is used, all application protocol messages must include these parameters both for the "calling" and "responding" parties. The APtitle is the closest we get to the AMLn actor layer identifier. Note however that ACSE does not know which APs belong to the same application. This implies that every AP that is called by another AP must check if it is an illegal call or not, by checking the calling APtitle against stored information about which APs belong to its own application.

We still have not covered all identification needs in the OSI architecture, however. As Fig. 6.1, shows, multiple instances of application processes is also supported, which requires an additional set of identifiers: "AP invocation identifier" and "AE invocation identifier". The author was lost at this point when reading the ACSE standard: the rationale for handling process instances is explained as a matter of being able to recover previous processes after an association breakdown. This is a matter of how to realize network reliability requirements in the software implementation domain, but it does not explain why AE instances must be referred as well.

Thus, addressing and naming in ACSE is a complex matter and affected by the software engineering approach to the OSI upper layers. ACSE also reveals that the session and presentation layers represent an artificial division of a layer since the general transparency requirement for layers is completely abandoned in the OSI upper-layer architecture. A user element that uses ACSE must also deliver all presentation and session layer parameters as parameters of ACSE primitives. This approach not only excludes the possibility of replacing the presentation and session layers with other layers, it also increases the complexity of the ACSE interface considerably.

6.2.3 ROSE, the Remote Operation Service Element

Besides ACSE, ROSE is the only general application service element (ASE) of interest in an application entity (AE). ROSE was later reused, extended, and modified in the TCAP standard (see Sect. 6.3). ROSE is the OSI RM approach to remote operation handling, which in AMLn is defined through the meta-primitives l-invokeReport and r-invokeOrder in SAGs, as described in Appendix B. If we would use ROSE in AMLn, a ROSE–ASE becomes a functional part of an agent in the AMLn common agent layer model. In Fig. 6.3 we have replaced the AMLn invokes with primitives defined by ROSE, in order to show the difference between ROSE primitives and AMLn meta-primitives.

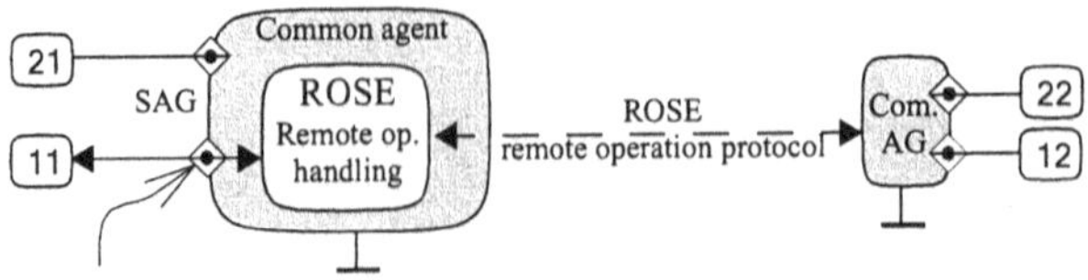

ROSE primitive names:	AMLn meta-primitive names:
RO-INVOKErequest	r-invokeOrder(ex.: LSM request-event)
RO-RESULTrequest	r-invokeOrder(ex.: LSM confirm-event)
RO-ERRORrequest	r-invokeOrder(ex.: LSM refuse-event)
RO-REJECT-Urequest	r-invokeOrder(LSM generalFaultReport event)
RO-INVOKEindication	l-invokeReport(ex.: LSM request-event)
RO-RESULTindication	l-invokeReport(ex.: LSM confirm-event)
RO-ERRORindication	l-invokeReport(ex.: LSM refuse-event)
RO-REJECT-Uindication	l-invokeReport(LSM generalFaultReport event)
RO-REJECT-Pindication	-

Fig. 6.3 ROSE primitives compared to AMLn meta-primitives for remote operation handling

The model shows that the primary difference between ROSE and AMLn is that ROSE does not define meta-primitives. Instead, primitive types of LSM events that are carried by AMLn invokes are defined as ROSE primitive types. For example, an AMLn r-invokeOrder that carries a request-type of LSM event, must be carried by an RO-INVOKErequest if ROSE were used. An AMLn r-invokeOrder that carries a confirm type of LSM event must be carried by an RO-RESULTrequest, etc.

Besides that ROSE needs five operations (RO-INVOKE, RO-RESULT, RO-ERROR, RO-REJECT-P, RO-REJECT-U) where AMLn copes with two (r-invoke, l-invoke), the major disadvantage with ROSE is that the primitive types of LSM events are made visible in SAGs. A ROSE-based common agent layer can therefore only support abstract protocols that apply the ROSE operation classification system, while the AMLn invokes support any type of operation classes, including ROSE and AMLs.

Note also that the ROSE remote operation protocol must define many message types due to this solution, while the corresponding AMLn protocol defines a single message type for invokes (see Fig. 2.26). Thus, by not realizing the need for meta-primitives, a ROSE-based common-agent layer is not only restricted in its applicability, but also considerably more complex than the AMLn common agent layer.

The ROSE solution to primitives in SAGs would make some sense if the ROSE ASE did verify the validity of LSM events. For example, if an agent received a message that identified a duplicate of a previously received LSM event, it could just discard it; if it received a message that contained an LSM event of type Request when an event of type Confirm or Refuse was expected, it could send the other agent an error message without bothering its own actor/user. This is however not so. A ROSE agent delivers any LSM event it gets in messages which is syntactically correct to its actor/user, which must perform such validity checks itself. That is also why ROSE defines the additional operation RO-REJECT-U to be used by an actor/user to inform its partner about formal LSM event errors (in an AMLn actor layer represented by the generic abstract primitive generalFaultReport (defined by AMLs).

The last ROSE primitive, RO-REJECT-Pindication, has no correspondence in AMLn. It is used by a ROSE agent to inform its actor/user if it cannot recognize which LSM event is referred to in a received ROSE message. In AMLn we assume a) that the sending agent has already checked the validity before sending, and b) that used connectivity service is reliable. If there still are problems of this kind, the agents should first try to solve it mutually, and, if they cannot, inform the association element that the association no longer works. The association elements will then give notice to their actor/users by the agRelRep (see Fig. 2.33). Thus, there is no need for an RO-REJECT-P operation in AMLn.

The operation class system defined in ROSE is different to the one defined by AMLs. The rationale for the AMLs operation class system is described in Muth (2001), which also comments on some strange features in the ROSE operation class system. We will therefore not repeat these arguments here. Note, however, that the AMLn invokes (see Appendix B) are not bound to use the AMLs. operation class system. They can serve any operation class system, as long as it defines the syntax of abstract primitives as consisting of an operation identifier, a primitive type, and a sequence of parameters, which includes the way operations and primitives are specified in the ROSE specification technique.

Another difference between ROSE and AMLn is that neither ROSE primitives nor ROSE messages include any element identification parameters (such as the {ALi,As} or ASi parameters in AMLn invokes). This is an effect of that "user elements" are aggregated together with ACSE and ROSE ASEs in the same AE, that relies on a single presentation connection, requested by the ACSE function. Thus, in the OSI solution, a presentation connection carries a single association. Since "user elements" are isolated as actors in AMLn, and since the AMLn remote operation solution will work on any type of connectivity layer (including connectionless layers, such as IP), AMLn invokes must include actor-identifying parameters. This also makes it possible for a common agent layer to support multiple associations over a single connection (a common requirement in application networks).

An important difference between how behavior is specified in AMLn and the OSI RM view on behavior, is revealed by the "parent" and "child" operation construct (also called "linked operation") in the ROSE standard.[4]

- The separation of LSM from LPM in AMLn control layers implies that all behavioral aspects of interest are defined by abstract protocols of LSMs. The behavior that an AMLn abstract protocol exhibits can be simple or show any degree of complexity. The general method for behavior specification that is

[4] This concept reveals a strong (but misguided) ambition to create interface definitions that can support software code generation, *once the behaviour is known*! The author had preferred an approach that could support network designers in actually *specifying the behaviour* of control layers (which inevitably requires a state machine approach).

applied in AMLn is therefore state-machine-oriented, e.g., the one that is defined by AMLs. Through the definition of meta-primitives, the AMLn LPM need, and has no knowledge of this behavior. The LPM has of course a behavior of its own, but that does not interfere with the behaviour of the LSMs it supports.

- This is not so in the OSI application layer, however. When an operation is defined in ROSE (by an ASN.1 operation macro, included in an ASN.1 application context macro), it can define other operations as child operations, regarding itself as a parent operation. The idea was to support behavior descriptions of "peer relations" between application layer elements. By recursively applying this principle, one assumed that all possible sequences of operations in an application layer protocol could be defined in an application context specification (an approach that is as far from OO as you can get).

There are two major flaws with this approach:

1. There is no way to define behavior in terms of the *states* of the LSM.
2. It can give a correct description of behavior only for *positive* responses, i.e., only if the response of an RO-INVOKErequest is an RO-RESULTconfirmation that carries a positive response (which, due to the way OSI defines its primitive types, is only apparent if you scrutinize the LSM event that is carried).

Since in reality layer behavior is completely centred around the state of a layer, and since negative responses are the source for the majority of behavioral sequences in most protocols, there is no way that the ROSE parent–child concept can be used for behavior specification. If used, and if code generation is applied on ASN.1 macro that define "linked operations," only a minor part of the behavior controlling code could be generated, which most likely is worse than not generating any at all.

6.3 TCAP, the Transaction Capability Application Part in SS7

TCAP is a common agent layer solution that is used within ISDN and some other ITU–T public networks (e.g., in GSM). In the following eight pages you get a condensed version of the 227 standard pages that describe TCAP, including a comparison between how the standard describes TCAP and how the OSI RM upper layer structure and AMLn deal with the same issue.

TCAP is an evolution from the ACSE, ROSE, and ASN.1 standards of the OSI application layer. TCAP was developed to support remote operations communication between layer elements of call-handling layers, which is why it is also regarded as a part of the SS7 standard (this is irrelevant to the present discussion, however). TCAP relies on the sequencing connection-less service of the SCCP layer of SS7 for its connectivity, i.e., not on the OSI presentation layer (as is the case for ACSE and ROSE).

A relevant question is: why did ITU–T not use its own solution for remote operations and association in the OSI application layer? Part of the answer to this question was already given in previous chapters: TCAP network architects realized the misguided ambitions with the presentation and session layers when they had to fulfill real network requirements. There were other reasons as well:

1. The intended use of TCAP was for short dialogues, often a single request to and a response from a remote actor. Considering that such dialogues were often parts of a call establishment procedure that should take a minimum of time, it was not always feasible to first establish an association for an invocation (and have to release it afterwards). This had to be considered when defining the specified agent interface (SAG) of TCAP.
2. Since TCAP was developed for supporting call handling in public networks, i.e., not any applications on an "open system," it was felt unnecessary to use the identifying parameters of ACSE (AP title, AE qualifier, etc. which we discussed in Sect. 6.2.2). Actor identification in TCAP therefore relies on SCCP addressing methods.
3. TCAP also assumed that two communicating actors would benefit from running concurrent threads of invocations, called "dialogues". The standard does not explain why. A qualified guess is that two call-handling actors can use this facility to relate each ongoing call to a single dialogue. For example, an ISDN call-handling layer element in a local exchange uses the parameter CallRef in the DSS1 L3 protocol to identify a call. If this element starts using TCAP, it can define a separate dialogue identifier value in TCAP to relate to the CallRef value. Remember that, if AMLn invokes had been used, the TCAP dialogue identifier would be regarded as an association identifier (ASi).[5]
4. TCAP had an ambition to support the sending of several LSM events (called "components") in the same invocation. As is the case in most standards, no explanation is given to why and in which situations this would be beneficial. Furthermore, the solution to this facility was to define two "sublayers" within the TCAP agent layer: the "component sublayer" and the "transaction sublayer", of which only the latter is a layer whether you use the OSI RM or the AMLn definition of what a layer is.

The TCAP standard is a good example of how one can complicate a rather simple network function by aligning it to a less suitable architecture (i.e. the OSI RM) and specification method (i.e. ASN.1 macron, defined the ROSE way). In addition,

---

[5] Considering the fact that neither ACSE association procedures nor ACSE identifiers are used, it is rather puzzling to read in the TCAP standard Q.771 that: "... TCAP is in partial alignment with ... ACSE." To the author it seems that TCAP has substituted ACSE association handling with something rather different, called "dialogue handling." The only thing TCAP and ACSE have in common is that they both rely on application context identifiers (referring ASN.1 macro, according to the ROSE specification technique). This identifier is used between TCAP users to check that they can both operate according to the same "application protocol."

the inventors of TCAP also managed to violate the OSI RM transparency requirements by dividing TCAP in two "sublayers," which did not simplify things. Figure 6.4 shows how TCAP's layer structure is defined in the standard.

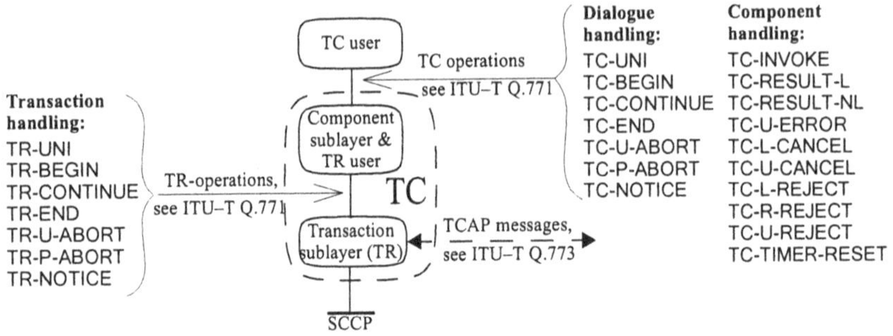

Fig. 6.4 TCAP's layer structure as described in ITU–T Q.771

"Transaction capability" (TC) is the correspondence to an AMLn agent layer (i.e., TC is equivalent to the TCAP layer). An immediate observation, looking at the operations in the interface between the sublayers, is that this interface is not a connectivity service point (...SP); in that case the only operations that would be needed would have been TR-send and TR-receive). It looks more like a control point (...CP). However, since most of the TR operations include LSM event data to be sent between TC user, it is also not a service control point. The standard also shows that none of the component handling operations (TC-INVOKE, etc.) result in any TCAP messages being sent or received (contrary to how the OSI application layer and the AMLn common agent layer work). Thus, the component sublayer is neither a layer in the OSI, nor in the AMLn sense. Note also that the names of TR operations are identical to the names of TC dialogue handling operations (except for the prefix). This also indicates that the sublayer division of TC is aimed at something other than defining OSI layers (or AMLn layers for that matter).

In reality, the TCAP model is an example of a frequently used method in standards, that is to describe the behavior of a layer element (here the TCAP/TC element) in terms of an implementation structure, in this case consisting of an element called "component sublayer" that interworks with another element called "transaction sublayer." The component sublayer handles the user interface and is invisible on the network level, since the component handling operations (TC-INVOKE, etc.) do not result in any messages being sent or received. The "transaction sublayer" does send and receive TCAP messages, however, and is therefore visible on the network level. The fact that this happens through events in the user interface by TC-BEGINrequest, etc., does not alter this conception, since the component sublayer more or less just renames primitives (e.g., TC-BEGINrequest to TR-BEGINrequest).

Thus, the two sublayers together constitute a layer element, in this case an agent of a common agent layer. By applying an artificial sublayer interface between these parts (defined by TR-UNI, etc.), the TCAP inventors managed to create an unnecessary complex model, also reflected by a complex, ASN.1-based TC protocol description, see ITU–T Q.773. If one takes the time to interpret this document it also becomes clear that TCAP defines a single protocol since the document defines a single set of messages called "TC messages". Fortunately, supplier's actual realizations of TCAP normally look different from the model depicted in Fig. 6.4.

TCAP can be modeled as a normal AMLn common agent layer. However, if we want to accurately reproduce the TCAP standard, the best we can do is to map it according to the model to the right in Fig. 6.5.

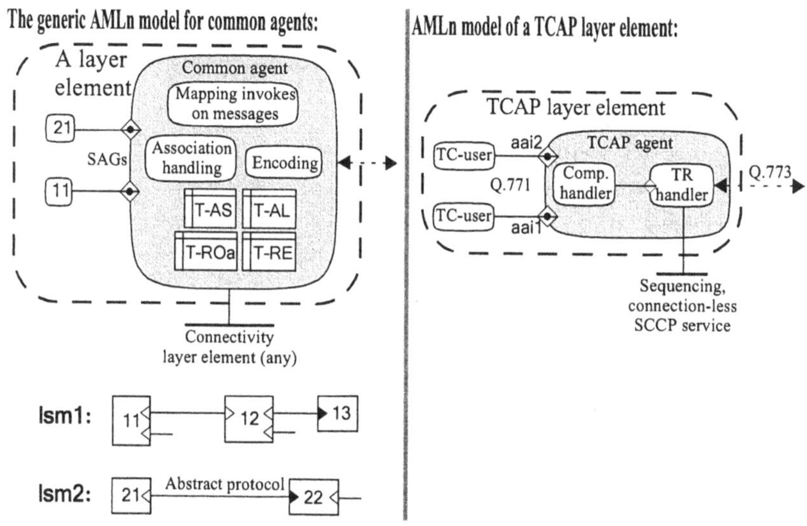

Fig. 6.5 Mapping the TCAP standard on AMLn

The way TCAP is specified restricts how we can model it in AMLn without redoing the standard. For example, the TCAP standard does not distinguish between the TCAP LPM and the LSMs it supports. The reason is that it uses the ROSE specification technique, which, as we saw in the previous chapter, implies that actors are regarded as integrated parts of the specification of OSI application entities, AEs (which is the way an application code generator would like to see it). The sublayer division of the TCAP agent layer is, as we already discussed, outside the scope of AMLn modeling (i.e., modeling on the network level) since communication between TCAP layer elements will work whether or not they implement sublayers and TR operations.

However, in order to satisfy the way the standard describes TCAP, we have included a submodel in the TCAP agent that defines two functional elements: a component handler and a transaction handler (TR handler), interconnected over an internal actor–agent interface. The only way we can explain the TCAP standard is by looking at this submodel and describe the three main phases in sending a TCAP message that the standard implies. The first phase stores "components" in the TCAP agent (see Fig. 6.6). "Component" is a word that does not exist in the OSI RM. It is a TCAP-unique concept that serves the purpose of supporting the way the TCAP protocol is described.

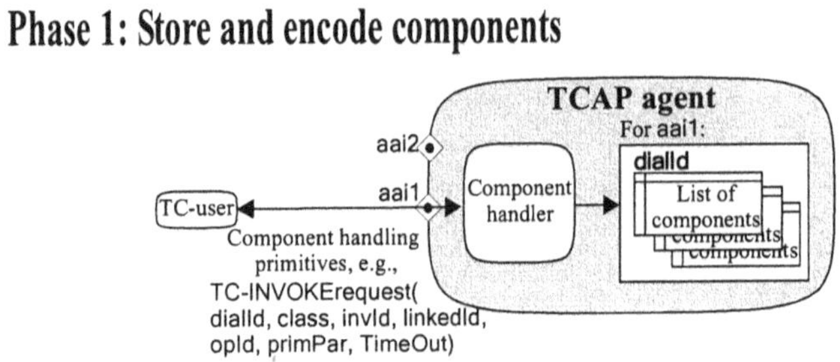

Fig. 6.6 Component handling in TCAP, phase 1

A "component" contains information about an LSM event (an operation identifier and a set of primitive parameters), the type of invoke (e.g. TC-INVOKErequest), and an invocation identifier. The idea is that the TC user can store a whole set of "components," all belonging to the same dialogue, before sending anything. The TC-INVOKErequest therefore also delivers a dialogue identifier (dialId) for each component. Since TCAP applies the strange ROSE "linked operation" concept that we discussed in Sect. 6.2.3, a linked operation identifier (linkedId) is also included in the definition of a "component." The purpose with the class parameter in the TC-INVOKErequest is not explained by the standard. To the author's understanding, communicating TC users should know to which operation class an operation belongs, and the TCAP layer has no reason to know.

This phase is handled by the component handler alone (i.e., it is invisible on the network level), which encodes these data for each components and stores all components that belong to the same dialogue in a common table for the actual TC user (identified by an actor–agent interface identifier (AAIi)).

Whenever the TC user is ready for it, he may order the TCAP agent to send all components of a particular dialogue to a named remote TC user. There are several operations defined for that, but let's just look at the TC-BEGINrequest: the primary parameters of this primitive are defined as: dialId, destination address (destAddr) and application-context identifier (applContextId) that refers the actual "application protocol." Another parameter is the origination address (origAddr), to which we will come back later. A TC-BEGINrequest starts phase 2 (see Fig. 6.7). In this phase, the TCAP message is composed for sending (as user-data for the transaction handler).

As this model shows, besides encoded "component data," the TR user-data also tells the receiving component handler that this was initiated by a TC-BEGINrequest (for some reason coded as dialogueRequest in TR-user data), which TCAP protocol version is referred (i.e., the version of Q.773), and which application protocol the originating TC-user wants to use (coded as applContextIdentifier in TR user-data). In addition, the TC user can add an octetstring of userInfo that is not encoded by TCAP, only transported between TC users.

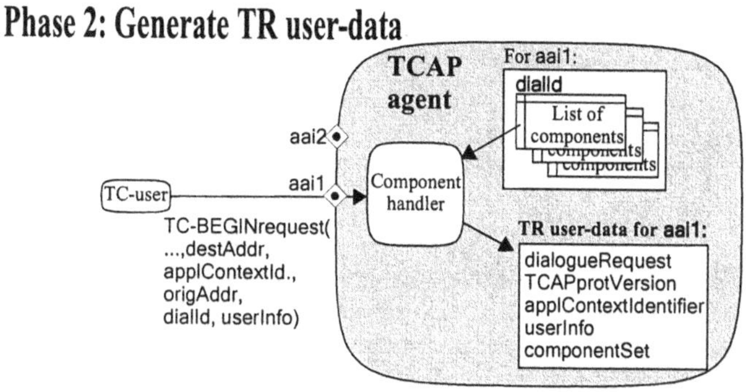

Fig. 6.7 Component handling in TCAP, phase 2

When the component handler has created the TR user-data octetstring, it starts phase 3, which is when something actual happens on the network level (see Fig. 6.8).

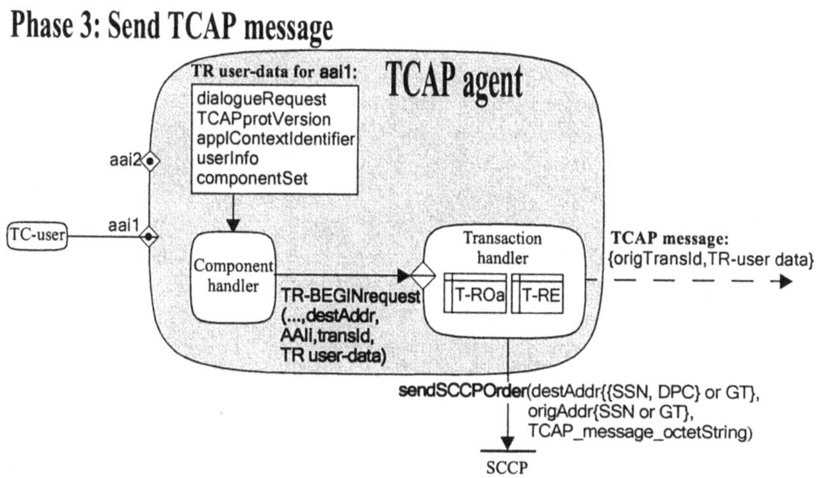

Fig. 6.8 Message handling in TCAP, phase 3. DPC=destination point code; GT=global title; SSN=subsystem number

The transaction handler now starts acting. This handler, including the TR operations it supports and the fact that it interfaces SCCP, shows all features of an agent, which is why we have connected it to the component handler over an internal actor–agent interface. Thus, the component handler acts as an actor in this interface and as an agent in interfaces to TC users (from an AMLn point of view, such elements do not exist on the network level).

Probably in order to uphold the distinction between the component and the transaction handlers, the dialogue identifier (dialId) that defines a dialogue in TC user interfaces is called transaction identifier (transId) in the TR-BEGINrequest primitive (it does however not make life easier for the interpreter of the standard). In the TCAP protocol it is called originating transaction identifier (origTransactionId) which implies that peer TCAP agents use separate identifiers to refer to the same dialogue. This implies that, in further communication within the same dialogue, both an origTransId and a destTransId must be included in TCAP messages.

Let's now look at how TC users identify each other: Both the TC-BEGINrequest and the TR-BEGINrequest include the parameters destination address (destAddr) and origination address (origAddr). From an AMLn modeling point of view, there is no need to constrain the generation of an origAddr to a parameter of these primitives. An alternative method (with some benefits over storing origAddr in the actual TC user) is to store the translation between the actor–agent interface identifier (AAIi) and the address of the actual TC user in T–ROa and T–RE tables in the transaction handler when the TC user is configured to TCAP. Thus the TCAP standard is overspecifying this issue.

The origAddr and the destAddr refer to the addressing system of the underlying connectivity service, which in reality is SCCP, although the standard leaves open the possibility of using other connectivity layers and their addressing systems as well. Thus, TCAP does not use any actor layer identifiers (ALi) and actor suffix (As) that are independent of the addressing system of the used connectivity layer. Since the TCAP agent does support multiple actors, this problem had to be solved by introducing an addressing facility in SCCP that includes a parameter for TC user identification. This parameter is the subsystem number (SSN), defined by SCCP (another possibility for identifying TC users is the global title (GT) addressing mechanism of SCCP. You can read more about that in Sect 3.4.5.2). The way SSN is used assumes that all actors that belong to the same actor layer are identified by the same SSN value. Since SCCP uses the SSN to identify using layers other than TCAP as well, the way the SSN is used when configuring layers on SCCP becomes rather complicated. Furthermore, since TCAP has no addressing system for TC user identification, actor layers that rely on TCAP are actually configured on the SCCP level by binding the identification of TCAP actors to SCCP addressing (a violation of general transparency requirements on layer structures). Note that since the SSN only identifies a TC user within a TCAP layer element, an MTP L3 (the datagram switching layer in SS7) address has to be given as part of destAddr an origAddr as well. This address, called destination point code (DPC), identifies the SS7 node in which the TCAP layer element exists.

We will end this discussion by commenting on the fact that while AMLn needs only two meta-primitives and ROSE nine, TCAP needs 15 (see "component handling operations" in Fig. 6.4) for the same network function. There are three main reasons for that:

1. The ROSE operation RO-RESULT is substituted by two operations in TCAP: TC-RESULT-L (only response or last part of a segmented response) and TC-RESULT-NL (non-final part of a segmented response). This is an effect of that TCAP does not necessarily run on a layer that accepts unlimited size of a user datum (as the OSI presentation layer does). For example, when TCAP runs on SCCP it uses an SCCP service that does not segment user data (allows only about 300 octets per invocation).
   This solution implies that the TC user must know which connectivity layer TCAP uses, which again is a violation of the transparency requirement of layer structures. A more user friendly TCAP had defined an optional segmentation function in the agent layer.
2. There exist three operations that have no correspondence in neither ROSE nor AMLn: TC-TIMER-RESET, TC-L-CANCEL, and TC-U-CANCEL. In relation to ROSE, these are necessary operations. The reason is that TCAP relies on ROSE operation classes, some of which cannot be used without time supervision. For example, in ROSE operation class 3, one TC user generates a TC-INVOKErequest which has no defined result. The TC user expects however to be informed if the operation did *not* succeed. The response time depends on real time factors, including congestion, execution time, and transport delays. If a response does not turn up in a reasonable time, the generating TC user cannot know if that implies that the execution succeeded, or if there are processing and transmission problems. To control this situation, TCAP offers TC users timer and cancel operations, the latter to be used when the TC user is tired of waiting. Timer values are associated to operations and given to TCAP in TC-INVOKEs (the Time-Out parameter in Fig. 6.6).
   AML deals with this issue in a completely different way: the operation classes defined in AMLs define operations from a pure functional point of view, i.e., without any considerations of real time issues. For example, the ROSE operation class 3 is a confirming operation class (class REQUEST) in AMLs where the confirm primitive just does not have any result parameters. If an implementor wants to implement this as a ROSE operation class 3, either a maximum response time value in the corresponding AMLs operation specification may be added, or the TCAP timer operations (or something similar) may be implemented.
3. TCAP defines an extra reject operation, the TC-L-REJECT, in addition to the two that ROSE defines. This is used by a TCAP element to reject its own TC user's invocation request before sending any messages. Thus, this operation is not related to anything that is visible on the network level, which is why it is defined neither in ROSE nor in AMLn.

Let's now summarize the characteristics of TCAP, in relation to the ROSE/ ACSE and AMLn agent layers:

- TCAP has no actor identification system of its own.

- TCAP does not support associations the way AMLn and the ACSE do. TCAP does, however, support a partial association function: a negotiation procedure that checks if peer TCAP layer elements can handle the same application protocol.
- TCAP has an elaborated function for dialogue handling, which combines an application protocol negotiation function with a dialogue between TC users. This feature has no correspondence in OSI.
- TCAP satisfies the first requirement of the bullet list in the beginning of this chapter by a dialogue-handling operation called TC-UNIT. The operation TC-BEGIN has to be used when a TC user wants to establish a dialogue before the TC users start communicating.
- The handling of dialogues in TCAP does not comprise other functions, such as authentication and charging. Thus, TCAP is not designed for "public use."
- Basically, TCAP uses the primitives for invoking remote operations that are defined by ROSE, which implies that the LPM–LSM separation defined by AMLn is not supported by the TCAP standard.
- TCAP compensates for real-time problems that are inherent in ROSE operation classes.
- TCAP also compensates (to some degree) for the fact that it runs on a connectivity layer other than the OSI presentation layer.
- The sublayers of TCAP represent an implementation model, used as a method for explaining how TCAP works. There is no need to realize these sublayers in real TCAP implementations, i.e., any other implementation model will do as well. Unfortunately, the standard does not inform the reader about that. Luckily, however, most implementors do realize that, and develop their own TCAP implementation models.

6.4 ATM Cell Switching

6.4.1 Introduction

Asynchronous transfer mode (ATM) is a stratum that switches *virtual circuits* (VC, in the ATM standard called *virtual channel*). This switching technique can support both data applications and media service (such as voice connections) with the same quality as in PSTN/ISDN (often called "carrier grade" quality). To fulfill such wide requirements, ATM was designed for transporting small fixed length messages (or "packages") of 48 octets of user data and a 5-octet header (compared to the variable and large length of most other types of switched packages). This message type was called *cell*, and the ATM switching technique is therefore called *cell switching*.

To be able to support media streams, an ATM network performs almost no normal link layer functions, such as flow control, error detection, and resending of packages. Such functions (when needed), were to be handled end-to-end by *ATM adaptation layer* (AAL) functions in ATM terminals.

ATM also takes advantage of the development in physical transmission technologies (optical fibers and SDH cross connection), and can therefore offer considerably higher transfer rates than PSTN/ISDN and X.25. The maximum transfer rates that are available for end user services (such as data, voice, video, and multimedia services) depend on the type of core network and access connections, and the type of media encoding. An ATM switching stratum that relies on SDH for physical transport in accesses and in core can switch user data up to over 500 Mbps. Contrary to ISDN, the actual transfer rate for an ATM service need not be selected from a predefined set of levels (64 kbps, 128 kbps, etc.), but could be negotiated for each ATM connection on a continuous scale. However, in order to simplify the realization of ATM switching strata, the standard still defines discrete levels of transfer rates.

When the ITU–T standardized the ATM switching stratum about 10 years ago, it was obvious that the raw cell services provided by this stratum had to be adapted to different types of media service needs, including connectivity services for control signaling. A number of AALs were therefore defined that could be placed in between the ATM switching stratum and using strata (including service controlling strata) and offer a number of different connection types. Thus, an ATM stratum consists of an ATM cell-switching stratum and an AAL stratum, the latter comprising many different types of AALs.

6.4.2 ATM in B–ISDN

An ATM stratum can be used as the switching stratum in any other network system that offers end-user services. One of these could be the B–ISDN[6] ("broadband ISDN") network system. Figure 6.9 depicts the B–ISDN network system, here as a generic derived node. The B–ISDN was intended to substitute the existing ISDN in the long run. Consequently the latter was renamed "narrowband ISDN," N–ISDN. B–ISDN must therefore be able to offer exactly the same services as N–ISDN does. Thus, a B–ISDN operator network with ATM adapted terminals must be able to

---

[6] B–ISDN is an extremely complex system to understand, at least if you try to understand it via its standards. The author made a rough estimate of the number of standard pages that describe just one of its layers (the control layer DSS2 L3 for access signaling). There are in all 54 documents that describe different aspects of this layer. The basic protocol standard, ITU–T Q.2931, comprises 255 pages, so the total number for DSS2 L3 should lie in the range of 2000-3000 pages. The number for B–ISDN as a whole, including ATM standards, is much larger.
Due to the amount of information given by the standards, and to the rather unfriendly way in which it is presented, the author cannot guarantee that the model discussed in this chapter is 100% accurate. This should not disturb the reader too much. After all, the intention here is not to provide a tutorial on B–ISDN/ATM, but to show that the AMLn principles for modeling switching layers are applicable to ATM as well.

interwork with a number of surrounding PSTN and N–ISDN networks when it is put into operation.

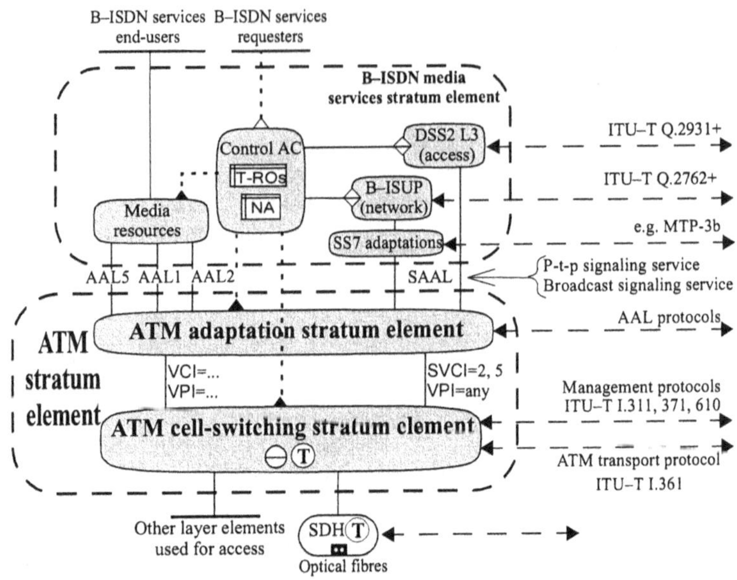

Fig. 6.9 The B–ISDN stratum element. AAL...=ATM adaptation layers; P-t-p=point-to-point; SAAL=signalling AAL; SVC=signalling VC; VC=virtual channel; SVCI=SVC identifier; VP=virtual path; VPI=VP identifier

The fact that B–ISDN is an evolution from N–ISDN can also be seen by the similarities and differences of this model and the N–ISDN model we discussed previously (see, e.g., Fig. 2.54).

- In principle, the B–ISDN network system is created out of the N–ISDN system by replacing its circuit-switching stratum with an ATM cell-switching stratum.
- No special packet-handling stratum as in N–ISDN is needed, since the cell switching layer is already a suitable base for packet services.
- The B–ISDN media services stratum is fairly similar to the one for N–ISDN: it has an access call-handling agent layer, called DSS2 L3, which is a modified version of DSS1 L3; it has a corresponding network agent layer, called B–ISUP, a modified version of ISUP. The modifications take into consideration that a cell-switching stratum has to be used instead of a circuit-switching stratum, and that B–ISDN offers a wider set of end-user services than N–ISDN.
- The structure of connectivity strata used by these agents looks different to N–ISDN. A special stratum, called "SS7 adaptations" in the model, is used by B–ISUP. This stratum relies on the SAAL services ("signaling AAL") of the AAL stratum. DSS2 L3 relies directly on SAAL services.

- The "SS7 adaptations" stratum replaces part of the SS7 stratum. It allows the B–ISUP agent to run on typical SS7 connectivity interfaces (such as those provided by MTP L3, SCCP, and TCAP).
  A relevant question is why the necessary interface adaptations are not provided by the AALs in the ATM adaptation stratum. Obviously the inventors wanted the ATM adaptation stratum to be general enough to be able to support different kinds of signaling systems. They therefore restricted the SAAL to handling signaling links only, i.e., the correspondence to the MTP L2 level in SS7 and the LAPD level in DSS1 (these adaptations are called SSCF–UNI for use by DSS2 L3 and SSCF–NNI for use by the SS7 adaptation stratum).
- A principle difference is how media support is realized. In the N–ISDN model, media resources that are needed for end-user services rely directly on connections offered by its circuit-switching stratum, since such connections are well suited for isochronous media services, such as voice and video.
  The story for media services in B–ISDN is very different: cell-switched virtual circuits are fast and reliable, but are not suited for any particular media service. It was therefore obvious from the start that, in order to make an ATM stratum a platform for all kinds of services, one had to add some adaptation layers (the AALs) on top of the ATM switching stratum, that could offer connectivity services suited for different types of media layers. For example, one such connectivity service was obviously a service that was similar to a circuit-switched connection, often called "circuit emulation." The AAL developed for that purpose was called AAL1. Other standardized AALs are the AAL2, AAL5, and the SAAL (read more about the AAL stratum in Sect. 6.4.4).

6.4.3 The ATM Stratum

We will now turn more specifically to the ATM stratum in Fig. 6.9. In N–ISDN we had to define the N–ISDN switching stratum as interfacing physical media strata since N–ISDN does not define a single or any particular agent layer for data transport (except in accesses). ATM is different: it defines a single, network-wide agent layer that can run on any type of stratum for physical transport. Primarily an SDH cross-connecting stratum was considered, but the development of other technologies (e.g., ATM directly over an optical fiber network) are already in the pipeline. Thus, the ATM switching stratum has no direct interface to physical media (similar to IP). It interfaces other strata, e.g., SDH, over layer interfaces that are connectivity service points. Note that no specific access strata, such as the BRA and PRA in N–ISDN are defined for B–ISDN.

Let's now take a deeper look inside the ATM cell-switching stratum (see Fig. 6.10).

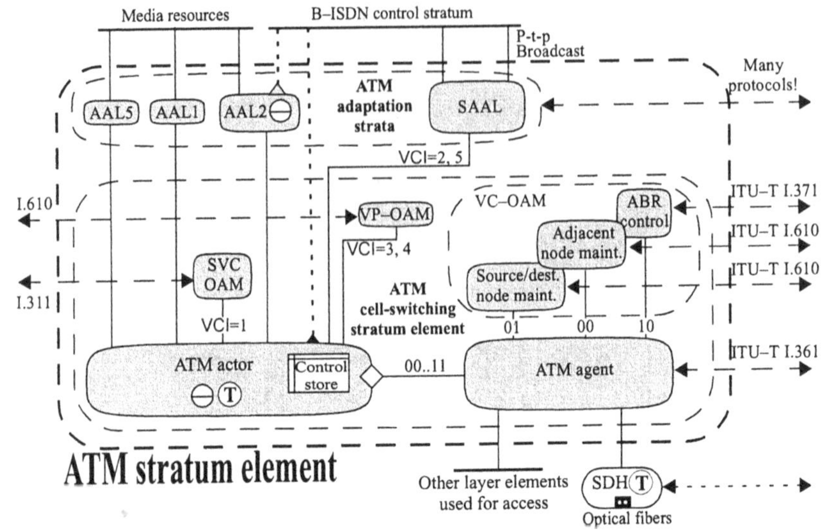

Fig. 6.10 The ATM stratum element. AAL...=ATM adaptation layers; ABR=available bitrate; OAM=operation and maintenance; PTI=payload type identification; SAAL=signalling AAL; SVC=signalling VC; VC=virtual channel; VCI=VC identifier; VP=virtual path; VPI=VP identifier; VC-OAM=operation and management of VCs; VP-OAM=operation and management of VPs

In one sense, the internal structure of the switching stratum is more like the X.25 VC switching model than the N–ISDN circuit-switching model. There is a single agent layer, defined by the ITU–T I.361 protocol, that handles cell transport between ATM terminals and ATM access nodes, and between network nodes. In another sense it is more like N–ISDN in that the switching actor is used for terminating streams for several types of users (the X.25 switching actor supports a single user only). Similar to N–ISDN, layer management (SVC–OAM, VP–OAM, and VC–OAM, see Fig. 6.10 and following page) is also carefully specified.

Virtual channels are offered by the switching stratum. A VC is identified by a virtual channel identifier (VCI). Every VCI must belong to a virtual path, identified by a virtual path identifier (VPI). A VPI can contain up to 65,536 VCs, and the stratum can switch both whole paths and single VCs.

Service properties of the switching layers in N–ISDN and B–ISDN are very different. The N–ISDN switching layer offers services as single duplex synchronous connections with a few discrete levels of transfer rates from 64kbps up to 2 Mbps. The ATM switching layer offers simplex asynchronous virtual channels (i.e., VCs), in order to support different transfer rates in the two directions (many media services require low transfer rate in one direction but high transfer rate in the other).

The circuit-switching layer requires no particular properties to be defined when ordering a connection. On-demand VCs (and VPs) are ordered by the controlling actors of media services strata (such as the B–ISDN control actor, see Fig. 6.9).

Ordering ATM virtual channels is considerably more complex (for a service-controlling actor). Two virtual channels (one in each direction) have to be established for most end-user services. For most channels, a large number of parameters must be given. These are parameters that define the load profile (maximum, minimum, and average transfer rate needed), the burst character of the traffic and a large number of quality of service parameters. The switching layer adapts the resources it allocates (buffers, timing supervision, etc.) to a channel to meet the requested properties. To simplify channel establishment, the standard defines a number of basic types of channels that the switching layer must be able to offer, such as:

- CBR ("constant bit rate" or "circuit-emulating channel"). This is a channel type with good real-time properties (both low delay and jitter), provided that the transfer rate is the same and rather constant in both direction. The behavior of the channel is therefore rather similar to a circuit-switched channel. It is the type of channel that is used for isochronous media services, such as interactive, uncompressed voice and video.
- VBR ("variable bit rate"). This is a channel that can handle large variations of transfer rate with low jitter. It is the type of channel that is intended for compressed interactive services, such as video conferencing.
- ABR ("available bit rate"). This is a channel that guarantees good properties for a certain minimum level of transfer rate, and a best effort for higher rates. If the ATM switching stratum cannot deliver, the source is informed about the congestion situation.
- UBR ("unspecified bit rate"). This is a channel that guarantees only best effort transfer delivery, and gives no information about congestion state. It is the channel type that would be used by an IP network, since its properties are similar to the IP service.

When a virtual channel is ordered and established, it represents a contract between the using-operator network (e.g., an B–ISDN) and the ATM network. In this contract, the ATM network guarantees negotiated properties and the using network accepts the costs (if any).

The configuration of layer elements that is defined by the model in Fig. 6.10 indicates that the stratum provides connectivity for three purposes. Two mechanism are used for that: one is to *dedicate* VCIs in a specific way; the other is to use a parameter in the ITU–T I.361 protocol, called "payload-type identification" (PTI).

1. *Support for AAL1, AAL2, and AAL5.* These strata are supported by regular VCs and VPs that are established by the B–ISDN service control actor over its resource control point to the ATM actor (similar as in N–ISDN and X.25 VC). The VCI for these VCs must lie in the range VCI= 32 through 65,536 and the VPI must not be 0.
2. *Support for the SAAL.* Two VCIs in every VPI are dedicated for that purpose: VCI=2 defines a point-to-point VC; VCI=5 defines a broadcast VC. These VCs are also called "signaling VC" (SVC).

3. *Support for management.* There are three separate management areas defined: operation and management of SVC (SVC–OAM); operation and management of VPs (VP–OAM), and operation and management of VCs (VC-OAM). Both dedicated VCs and the PTI are used as discriminating mechanisms: The three VC–OAM layers are separated by the binary PTI values 100, 101, and 110. Management cells that are addressed to these functions are already extracted by the ATM agent. Cells with other PTI values (000 through 011) are switched by the ATM actor. Cells with VCI values of 1, 3, and 4 are also management cells. If the VCI value is 3 or 4, the cell is terminated in the VP–OAM layer interface. If the VCI = 1, the cell is terminated in the SVC-OAM. A channel with VCI = 1 is also called "meta-channel" because it is used to manage other channels, i.e., SVCs.

VCI values 7 through 31 are reserved for future use. The use of the PTI parameter is quite interesting. Since the inventors wanted to keep the overhead in cells as low as possible, they managed to include four different functions in the three bits (B1, B2, and B3) of the PTI (see Fig. 6.11):

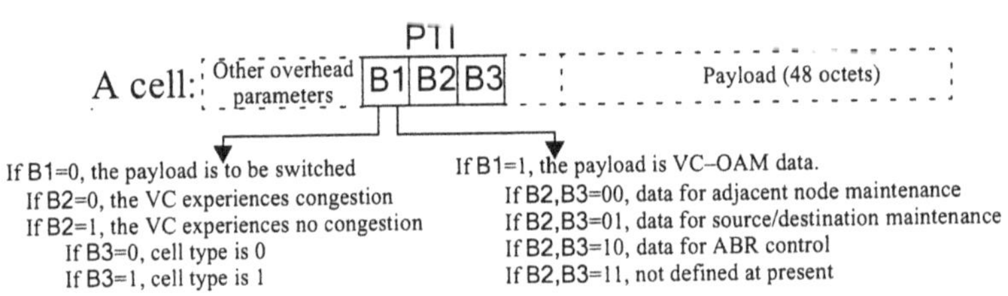

Fig. 6.11 Usages made of the payload-type identification parameter, PTI

This kind of smart coding (although sometimes necessary in order to save capacity) can make it very difficult to understand a standard, unless its authors have made an extra effort to describe the actual mechanism above the bit level (which rarely is the case, also not in the case of ITU–T I.361). Let's therefore try to understand the four functions that are supported by the PTI in AMLn terms.

1. The B1 bit acts as an interface type identifier (ITi) for cells to be terminated by the ATM agent, and cells to be switched by the switching actor.
2. For cells that are terminated by the agent (B1 = 1), the second and third bit together act as an access-point identifier (APi). For example, B2B3 =10 identifies the ABR control layer, B2B3 = 00 identifies the adjacent node maintenance layer.
3. The B2 bit ("congestion state") and the B3 bit ("cell type") in cells to be switched (B1 = 0) has no significance for the switching actor, which only switches cells according to their inVCI, inVPI and the corresponding outVCI and outVPI, that are set in its control store (similar to how the X.25 VC actor works). The meaning of the B2 and B3 bits is first revealed when a cell is terminated in an AAL.

   The B2 bit tells the receiving AAL if the channel is congested (B2 = 0) or not

(B2 = 1). An originating ATM agent sets B2 = 1 when it sends a cell. During transport of the cell through the switching network, any node may change this bit to 0 (to warn the destination ATM agent) if the node experiences a congestion situation for the actual VC.

4. The B3 bit is the encoding of a cell type identifier, CTi[7] (B3 = 0 means cell type 0; B3 = 1 means cell type 1). This bit is end-to-end transparent, meaning that when an AAL element delivers the payload for a cell, it also delivers a CTi value. The CTi will join the cell through the network and finally be delivered to the terminating ATM element at the other end of the VC as a kind of connection identifier. Thus the switching layer encodes the CTi in the B3 bit, but has no knowledge of what the AAL uses it for.
   The CTi implies that in every virtual circuit there can exist two permanent substreams of cells, one cell-type 0 stream and one cell-type 1 stream. The AAL agents may not use this separation, but it can also be used by a discrimination function in a terminating ATM actor. Figure 6.12 shows how this function works in ATM.

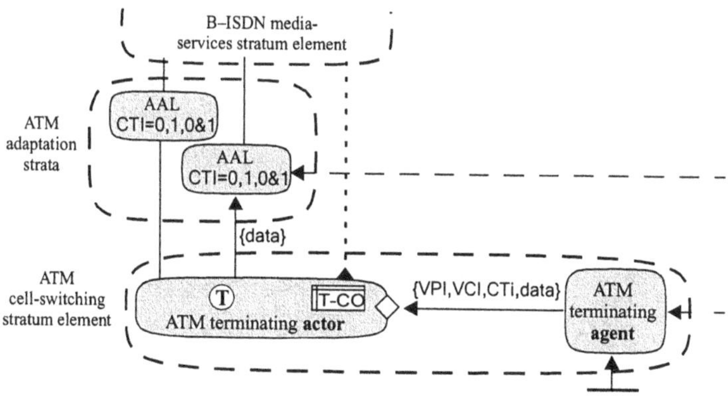

Fig. 6.12 CTi discrimination in an AAL. CTi=cell type identifier

The discrimination function in the terminating actor is dynamically controlled from the B–ISDN call-handling function over a resource control point. For each established virtual channel, the call-handling function tells the ATM actor to which layer interface to relay the substreams defined by the CTi values (information stored in the T-CO table). The substreams can be directed to different AAL elements or to the same.[8]

---

[7] This concept is not defined in ITU–T I.361. We introduce it here to be able to make the distinction between the PTI and the B3 bit.

[8] This is just another example of that the B–ISDN concept has evolved from N–ISDN mainly just by replacing the N–ISDN circuit-switching layer with the ATM cell switching stratum. Most other N–ISDN functions and architectural solutions have been retained, however.

After having read this presentation of how the four PTI functions are represented in a cell, the reader can conclude that a more functional (and thereby more intelligible) encoding is preferable. The author could only agree, especially since it would cost only three extra bits in a cell (two if the CTi was excluded).

6.4.4 The Adaptation Stratum

Let's end this chapter by commenting on the AAL standards: the properties of the connection type that is offered by a particular AAL is a combination of a VC-type (CBR, ABR, etc.) and the functions that are defined by the AAL. For example, the AAL1 stratum was developed to emulate circuit-switched connections, which obviously requires the CBR type of VCs. A basic function that is required in the AAL1 is segmentation and reassembling since the user of AAL1 delivers a continuous stream of bits without any particular message boundary, which the AAL1 must deliver to the ATM switching actor as payload in cells over a particular VC. The AAL1 is not allowed to perform any type of payload error handling since that would corrupt real-time properties of the AAL1 service. However, it is of some importance to detect cells that are lost in the ATM switching stratum. The AAL1 therefore adds two parameters (in the overhead, OH, defined by the AAL1 protocol) as parts of the payload of cells, to be able to detect missing cells. Figure 6.13 shows how an AAL1 element works.

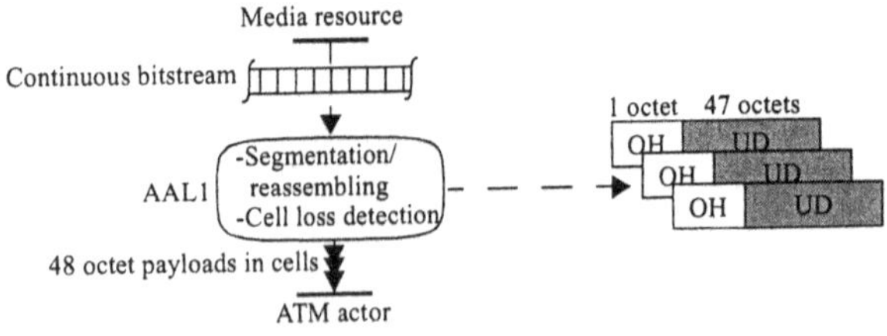

Fig. 6.13 The AAL1 element. OH=overhead parameters in AAL1 messages; UD=media bits

This AAL consists of single layer/protocol that takes one of the octets of the 48 payload octets in an ATM cell and uses it for parameters for cell loss detection. Note that no parameters are needed for sequencing received payload since the ATM switching stratum guarantees sequenced delivery of cells.

Requirements on AAL2 functions are very different to AAL1: users of AAL2 deliver compressed media stream with a variable bit rate. The VC type used is therefore VBR. Contrary to AAL1, it is extremely important that the bits delivered to the receiving media resource are not corrupted since that would create strange results after encoding. The input to an AAL2 element is also not a continuous bit stream, but frames, i.e., the boundaries of frames must be preserved. To cope with these problems, the AAL2 layer adds a number of overhead parameters that, together, take three octets of the 48 payload octets in an ATM cell (see Fig. 6.14).

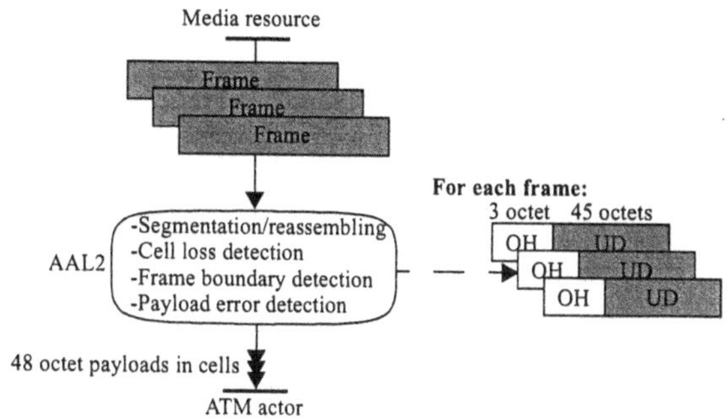

Fig. 6.14 The AAL2 element. OH=overhead parameters in AAL2 messages; UD=media bits

The fact that segmentation and reassembling has to be performed also for ALL2 (and for most conceivable AALs) made the ATM standard committee conclude that this function (generically called SAR) should be defined by an internal "sublayer" in every AAL. The slip in this assumption was only that sometimes segmentation requires no extra overhead, and therefore defines no protocol (as, e.g., for AAL1). Sometimes it does, and therefore defines a protocol (as for AAL2, since frame boundaries must be preserved). Thus, the SAR can be modeled as a layer in the case of AAL2 but not in the case of AAL1. The use of the term "sublayer" in AAL standards has forced some writers (see Tanenbaum (1996)) to statements such as "in the AAL1, the SAR layer has no protocol but in the AAL2 it has," which sounds rather strange, considering that a "layer" that defines no protocol is not a layer (by definition).

Nevertheless, assuming that a generic structure for AALs was possible and meaningful to define, the AAL committee defined an internal *generic* structure based on four generic "sublayers" (also called "protocol" or "functions") and seven structural concepts (see Fig. 6.15). The intention was to use this structure for defining specializations for different user-layer requirements, resulting in a large set of standard documents.[9] Note, however, that the present status of the generic AAL structure has become unclear, since the overview documents have been withdrawn. The depicted model is a best effort model, based on information available to the author.

---

[9] The present set of AAL standards comprises 11 documents in the ITU–T I-series (I.363 through I.366), and an additional number in the Q.2000-series (the latter primarily covering SSCOP variants and SSCF functions).

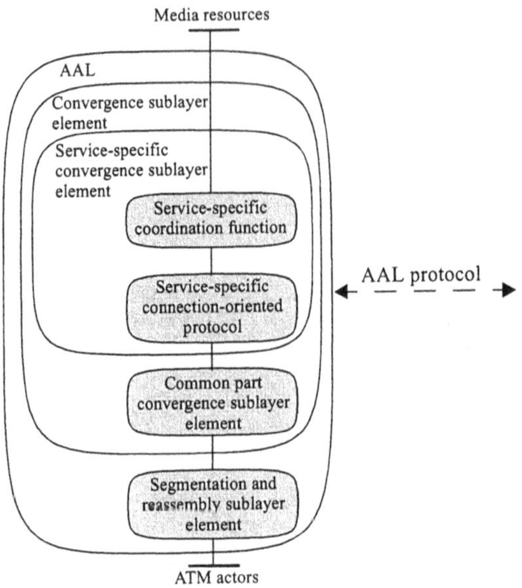

Fig. 6.15 The generic architecture of an AAL element, as described by the AAL standard

Much has been said about this structure and its specializations since it was defined. One of the more moderate comments are given by Tanenbaum (1996): "If the full AAL committee had turned its work in as a class project, the professor would probably have given it back with instructions to fix it and turn it in again when it was finished." The author looks at this structure as the result of the AAL standard committee giving priority to the need of dividing the work of specifying AALs, ahead of defining a useful architecture. In the light of AMLn, we can analyse AAL structures and conclude that:

- Although ITU–T denotes AALs "layer," they are only so in some cases (e.g., AAL1). In other cases (e.g., the SAAL) they are strata, i.e., consist of several layers.
- In AMLn we do not recognize "sublayers." Whenever this concept appears in standards, one can suspect that either they are agent layers, they define actor layer functions, or they are layers of an implementation model, i.e., they are not network layers at all.

The number of structural elements in the internal AAL structures varies. The fact that elements in these structures are characterized differently ("protocol," "function," "sublayer," "sublayer_in_sublayer_in_sublayer"), indicates some hidden semantic differences (most likely the fact that when specializing from this architecture, an element may or may not define a protocol, depending on the actual specialization).

Due to the confused definition of the AAL generic architecture, it is impossible to define a single mapping on an AMLn model. Instead, a specific AMLn model must be defined for each specialized AAL. For AAL1 and AAL2, the correspondent AMLn model defines a single layer, as depicted in Figs. 6.13 and 6.14.

Neither the space available in this book nor the AAL standard inspires the author for an exhaustive analysis of all AALs. We will, however, take the "signaling AAL" (SAAL) as a mapping example of an AAL that defines more than one layer. There are variants of SAALs as well. The variant depicted in Fig. 6.16 is used for supporting the B–ISDN service control stratum. The figure compares the standard's description of the B–ISDN SAAL architecture with a corresponding AMLn model. Note that the standard describes five sublayers although there are only two layers that define any protocol.

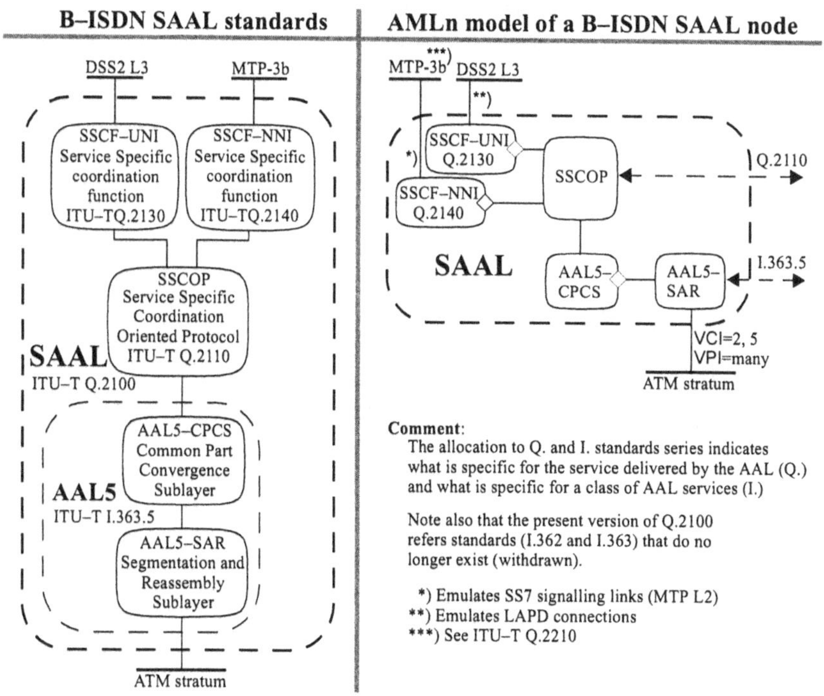

Fig. 6.16 The signaling AAL (SAAL) that is used for supporting B–ISDN control protocols

SSCF is a generic term for actor functions that adapt the connectivity service of an AAL to layer interfaces that are used by existing layers. In this model, the SAAL supports layers of the B–ISDN service control stratum, i.e., B–ISUP and DSS2 L3.

- For the use of DSS2 L3, the SSCF–UNI actor provides a LAPD-like layer interface. Since SSCOP performs basic link layer functions,[10] and the ATM stratum provides the correspondence to LAPD connections, the SSCF–UNI does not do much more than defining a standardized interface.

- B–ISUP is assumed to use ordinary SS7 interfaces, i.e., MTP L3 and SCCP, in order to be able to interwork with N–ISDN networks. The SSCF–NNI actor is therefore designed to act as the correspondence to an MTP L2-type of layer. Note that a reduced variant of the MTP L3 is defined (called MTP-3b, see ITU–T Q.2210), to be used by the B–ISUP agent layer. The MTP-3b defines a reduced variant of the MTP L3 protocol and maps MTP L3 addressing parameters (e.g., DPC, OPC) on the VPI/VCI parameters. Other SS7 adaptation layers exist as well, e.g., for providing SCCP interfaces.

Note, however, that service control layers may use other switched services than AAL for their connectivity (e.g., IP or an ordinary SS7 connectivity stratum). A large number of adaptation standards, both within the context of AAL and of the SS7 adaptation stratum are therefore emerging.

[10]SSCOP is used for purposes other than supporting B–ISDN as well. For example, it is used in an OSI–AAL for supporting an OSI network (according to ITU–T X.213), which seems somewhat superfluous. The SSCF in this AAL is specified in ITU–T I.365.2.

Appendix A: List of Acronyms and Standards

This list contains all standard acronyms and relevant standard references to network systems that have been discussed or mentioned in the book (which still is a very small set of all standards that define those systems).

AAL1	ATM Adaptation Layer 1
	ITU–T I.363.1, B-ISDN ATM Adaptation Layer Specification: Type 1 AAL
AAL2	ATM Adaptation Layer 2
	ITU–T I.363.2, B–ISDN ATM Adaptation Layer Specification: Type 2 AAL
AAL5	ATM Adaptation Layer 5
	ITU–T I.363.5, B–ISDN ATM Adaptation Layer Specification: Type 5 AAL
ACSE	Association Control Service Element
	ITU–T X.217 describes the service. The protocol is describe in ITU–T X.227
AMLn	Abstract Modeling Language, network view
AMLp	Abstract Modeling Language, protocol view
AMLs	Abstract Modeling Language, service view
ASN.1	Abstract Syntax Notation No. 1
	ITU–T X.208, Specification of Abstract Syntax Notation One (ASN.1)
ATM	Asynchronous Transfer Mode
	ITU–T I.361, B–ISDN ATM Layer Specification (the ATM switching protocol specification)
	ITU–T I.311, I.371, I.610, ATM switching-layer management
BER	Basic Encoding Rules
	ITU–T X.209
BGP	Border Gateway Protocol
	IETF RFC 1654, A Border Gateway protocol 4
	IETF RFC 1268, Application of the Border Gateway Protocol in the Internet
B-ISDN	Broadband ISDN
	ITU–T I.121, Broadband Aspects of ISDN, and many other recommendations in the I.- and Q.2000 series
BRA	Basic Rate Access
	ITU–T I.430, ISDN User-Network Interfaces: Layer 1 Recommendation
CMIP	Common Management Information Protocol
	ITU–T X.711, CMIP specification
CMIS	Common Management Information Service.
	ITU–T X.710.
CNL	Connection-less
CON	Connection-oriented

CORBA/IIOP	Common Object Request Broker Architecture and Internet Inter-ORB Protocol OMG CORBA/IIOP specification, see www.omg.org/ The CORBA object model is described in Chap. 1, IDL in Chap. 3
CS	Circuit Switching
DHCP	Dynamic Host Configuration Protocol v4: RFC 2131, RFC 3396 v6: RFC 3315
DIR	The (OSI) Directory ITU–T X.500, The Directory: Overview of Concepts, Models and Services, and other recommendations in the series X.500–586
DN	Domain Name
DNS	Domain Name System (Internet) IETF RFC 1034, Domain Names, Concepts and Facilities
DSL	Digital Subscriber Line. Series ITU–T G.991.1–997.1.
DSS1 L2	Digital Subscriber Signalling System No. 1, layer 2 See LAPD
DSS1 L3	Digital Subscriber Signalling System No. 1, layer 3 ITU–T Q.931, ISDN User-Network Interface Layer 3 Specification for Basic Call Control
DSS2 L3	Digital Subscriber Signalling System No.2, layer 3 ITU–T Q.2931+, ... The B–ISDN access control protocol
DTMF	Dual-Tone Multi-Frequency ITU–T Q.23.
EGP	Exterior Gateway Protocol IETF RFC 094, Exterior Gateway Protocol Formal Specification
FR	Frame Relay Diverse standards in the ITU–T X. and I. series
FTAM	File Transfer, Access and Management ISO 8571
FTP	File Transfer Protocol IETF RFC 0959. Updates, see RFC 2228, 2640, 2773
GDMO	Guidelines for the Definition of Managed Objects ITU–T X.722
GIOP	General Inter-ORB Protocol See www.omg.org/
GPRS	General Packet Radio Service See www.3gpp.org/specs
GSM	Global System for Mobile communication See www.3gpp.org/specs
GT	Global Title
GTA	Generic Functional Architecture (of transport networks) ITU–T G.805, Generic Functional Architecture of Transport Networks. See also ITU–T G.803, which applies G.805 on SDH

ICMP	Internet Control Message Protocol v4: IETF RFC 0792, updated by RFC 0950 v6: IETF RFC 2463
IANA	Internet Assigned Numbers Authority See www.iana.org/
IEEE	Institute of Electrical and Electronics Engineers See www.ieee.org/
IETF	Internet Engineering Task Force See www.ietf.org/
IDL	Interface Definition Language See www.omg.org/, the CORBA/IIOP specification, Chap. 3, IDL Syntax and Semantics
IMF	Internet Management Framework. IETF RFC 1902–1908.
IN	Intelligent Network ITU–T Q.1201, Principles of Intelligent Network Architecture, and many other recommendations in the series Q.1200–1290
INAP	IN Application Part ITU–T Q.1218, Interface Recommendation for Intelligent Network CS-1 INAP is an abstract protocol that relies on the message protocol defined by TCAP
IPA	IP Network Address IETF RFC 1166, Internet Numbers
IPv4	Internet Protocol, version v4 IETF RFC 791
IPv6	Internet Protocol, version v6 IETF RFC 1883
ISDN	Integrated Services Digital Network. This concept refers circuit-switching networks with digital accesses, i.e. both N–ISDN and B–ISDN
ISUP	ISDN User Part ITU-T Q.761–764
B-ISUP	Broadband ISUP ITU–T Q.2762+. Broadband ISUP. The B–ISDN core control protocol
ITU-T	International Telecommunication Union, Telecommunication Standardization Sector See www.itu.in/ITU-T/
LAN	Local Area Network See www.ieee.org (look for 802 standards)
LAPD	Link Access Procedure on the D-channel ITU-T Q.920–921
LDAP	Lightweight Directory Access Protocol IETF RFC 3377, Lightweight Directory Access Protocol (v3): Technical Specification
MAN	Metropolitan Area Network

MAP	Mobile Application Part ETSI GSM 09.02. MAP is an abstract protocol that relies on the message protocol defined by TCAP
MEGACO	Media Gateway Control Protocol IETF RFC 3225, Gateway Control Protocol Version 1. The document is common text with ITU–T H.248.
MGCP	Media Gateway Control Protocol IETF RFC 3435, Media Gateway Control Protocol Version 1.0. This RFC will be replaced by MEGACO
MIB	Management Information Base
MO	Managed Object
MSC	Message Sequence Charts ITU–T Z.120
MSISDN	Mobile Station ISDN Number. ITU–T E.213, Telephone and ISDN numbering plan for land mobile stations in public land mobile networks (PLMN)
MTP	Message Transfer Part ITU–T Q.701, Functional Description of the Message Transfer Part (MTP) of Signaling System No. 7
MTP-3b	Message Transfer Part 3b ITU–T Q.2210, Message Transfer Part Level 3 Functions and Messages Using the Services of ITU–T Q.2140
MTP L1	Message Transfer Part, Layer 1 ITU–TQ.702. Signalling Data Link (part of SS7)
MTP L2	Message Transfer Part, Layer 2 ITU–T Q.703. Signalling Link (part of SS7)
MTP L3	Message Transfer Part, Layer 3 ITU–T Q.704. Signalling Network Functions and Messages (part of SS7)
NGN	Next Generation Network Denotes the evolution direction of today's public networks
NIC	Network Information Center See www.iana.org
N-ISDN	Narrowband ISDN (normally integrated with PSTN) ITU–T I.120, ISDN General Structure, and many other recommendations in the I.- and Q.-series
OMAP	Operation and Maintenance Administration Part ITU–T M.3010, Principles for Telecommunications Management Network ITU–T Q.756, Guidebook to OMAP OMAP is an abstract protocol that relies on the message protocol defined by TCAP
OMG	Object Management Group See www.omg.org/
OSI RM	Open Systems Interconnection Reference Model ITU–T X.200, Reference Model of Open Systems Interconnection
OTN	Optical Transport Network ITU–T G.872. Architecture of Optical Transport Networks

PPP	Point-to-Point Protocol IETF RFC 1661 updated by RFC 2153
PRA	Primary Rate Access ITU–T I.431, Primary Rate User-Network Interface-Layer 1 Specification
PS	Packet Switching
PSTN	Public Switched Telephony Network. This concept refers circuit-switching networks with analog accesses (normally integrated with N–ISDN)
RFC	Request For Comment
ROOM	Real-Time Object-Oriented Modeling See Selic (1994)
ROSE	Remote Operation Service Element ITU–T X.219 describes the service. The protocol is describe in ITU–T X.229
RTSE	Reliable Transfer Service Element ITU–T X.X.218, describes the service. The protocol is describe in ITU–T X.228
SAAL	Signaling ATM Adaptation Layer ITU–T Q.2100. B–ISDN Signaling ATM Adaptation Layer Overview Description.
SCCP	Signaling Connection Control Part (part of SS7) ITU–T Q.700–716
SDH	Synchronous Digital Hierarchy ITU–T G.803, Architecture of Transport Networks Based on the Synchronous Digital Hierarchy. Describes SDH functional structure according to ITU–T G.805. See also other standards in the G.800–899 series
SDL	Specification and Description Language ITU–T Z.100, CCITT Specification and Description Language (SDL)
SMI	Structure of Management Information RFC 1902, Structure of Management Information for version 2 of SNMP
SNMP	Simple Network Management Protocol IETF RFC 1441–1452 defines the SNMP framework IETF RFC 1442, Structure of management information for version 2 of SNMP IEF RFC 1905, Protocol Operations for version 2 of the Simple Network Management Protocol (SNMPv2) IETF RFC 1213, Management information base for MIB-II (additional documents are RFC 2011–2013, and 2963)
SS7	Signaling System No. 7 ITU–T Q.700, Introduction to CCITT Signalling System No. 7 Other SS7 standards, see the series ITU–T Q.701–787
SSCOP	Service-Specific Connection-Oriented Protocol (part of SAAL) ITU–T Q.2110
SSFC–NNI	Service Specific Coordination Function for Support of Signalling at the Network-Node Interface ITU–T Q.2140.
SSFC–UNI	Service Specific Coordination Function for Support of Signalling at the User-Network Interface ITU–T Q.2130

TCAP	Transaction Capability Application Part (part of SS7) ITU-T Q.771–775
TCP	Transmission Control Protocol IETF RFC 793
TMN	Telecommunications Management Network ITU–T M.3000, Overview of TMN Recommendations. See also other recommendations in the M.3000 series ITU–T Q.811. Lower Layers Protocol Profiles for the Q and X Interfaces ITU–T Q.812. Upper Layer Protocol Profiles for the Q and X Interfaces
TTCN	Tree and Tabular Combined Notation ITU–T X.290, OSI Conformance Testing Methodology and Framework for Protocol Recommendations for ITU–T Applications - General Concepts. See also other recommendations in the series Q.290–296
TUP	Telephony User Part ITU–T Q.721–725
UDP	User Datagram Protocol IETF RFC 786
UML	Unified Modeling Language See www.omg.org/
UMTS	Universal Mobile Telecommunication System See http://www.3gpp.org
UPT	Universal Personal Communications ITU–T F.851, Universal Personal Telecommunication (UPT) - Service Description (Service Set 1) ITU–T E.168, Application of E.164 numbering plan for UPT
V5.1	V5.1 interface ITU–T G.964, V-Interfaces at the Digital Local Exchange (LE) - V5.1 Interface (based on 2048 kbit/s) for the Support of Access Network (AN)
V5.2	V5.2 interface ITU–T G.965, V-Interfaces at the Digital Local Exchange (LE) - V5.2 Interface (based on 2048 kbit/s) for the Support of Access Network (AN)
VC	Virtual Circuit or Virtual Channel A channel in a connection-oriented packet switching network, such as X.25, TR and ATM
VP	Virtual Path An aggregation of VCs
WAN	Wide Area Network
X.25 NA	X.25 Network Addresses ITU–T X.121, International Numbering Plan for Public Data Networks

Appendix B: SAG and SAC Operations

B1 Invoke Operations (in SAG)

A SAG ("specified agent interface") defines the operations and primitives that are used over actor–agent interfaces (AAI) in control layers. The abstract primitives of these operations are meta-primitives since they carry information about other primitives (which are the primitives between the element state machines, ESM, of actors). SAG primitives are defined as part of an agent-layer specification.

AAIs are local interfaces in layer elements. Neither these interfaces nor their meta-primitives are therefore visible on the network level. AAIs and meta-primitives are, however, often visible in implementors' solution models. The benefit of defining AAIs and SAGs is that a model of a layer can be separated into two models: a layer protocol machine (LPM) and a layer state machine (LSM) with an unambiguous mapping between these machines. SAG primitives should therefore always be defined, whether or not the intention is to implement them.

AMLn defines two meta-primitives: **r-invokeOrder(...)** and **l-invokeReport(...)**. These primitives have identical sets of parameters. The sets of parameters differ somewhat between two types of agent layers: agent layers that offer no association handling, and agent layers that do. Different AAI/SAG symbols are used in models for these two types (see the figure below).

SAG without associations:

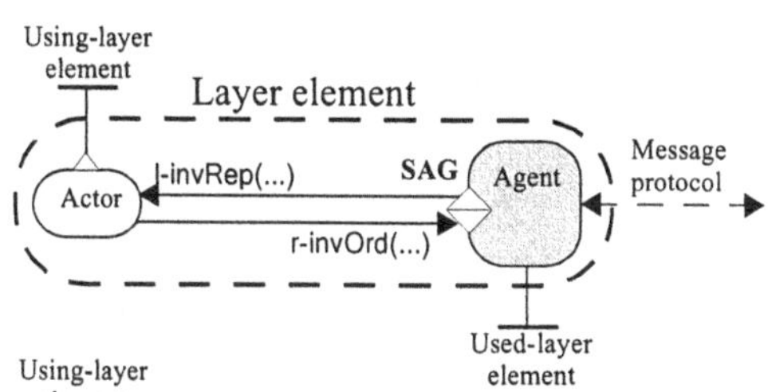

SAG with associations:

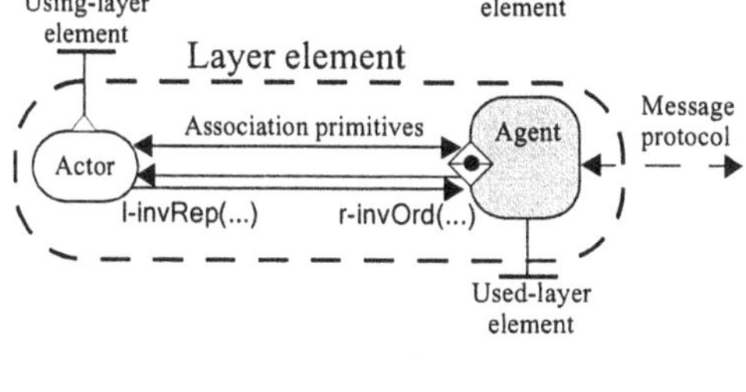

In the following section we apply the ASN.1 style for type assignment[1] to show how the parameters of the meta-primitives l-invokeReport and r-invokeOrder are defined ("OPT" means that the parameter may or may not appear in an instance of the primitive; "SEQ" defines an ordered list of parameters):

r-invokeOrder(InvPar):

InvPar::=SEQ{DetectorReference, invId OPT, LSMeventInfo)

DetectorReference::=CHOICE{AA{ALi OPT, As}, ASi, Other}

--**DetectorReference**: The identity of the remote actor that will handle the LSM event (i.e., the "detector")

--**AA,** actor address: The actor identifier.

--**ALi,** actor layer identifier: Needed only for common agent layers, i.e., agent layers that support more than one actor layer.

--**As**, actor suffix: Defines the actor within an actor layer. The only AA identifier needed if the agent layer supports a single actor layer.

--**ASi,** association identifier: The only identifier needed for agent layers that offer associations, once the association is established.

--**Other**: Any other type of identifier, accepted by the agent, for identifying remote actors, e.g., a network address.

invId, invocation identifier: If an invoke can be generated, referring the same operation as a previous one that is still not finished, this parameter must be used in order to distinguish between the invokes.

LSMeventInfo::=SEQ{objId OPT, event SEQ{opId, primType, primPar{...}}}

--**LSMevent**: Information elements of an instance of an abstract primitive defined in an abstract protocol between ESMs.

--**objId,** object identifier: The two ESMs may optionally associate their operations to (abstract) objects, in which case an object identifier is included.

--**opId,** operation identifier: Identifies the operation that the LSM event refers.

--**primType,** primitive type: Defines the type of primitive of the LSM event, see Appendix D for details.

--**primPar{...}**, primitive parameters: The parameters of the LSM event (in ASN.1).

l-invokeReport(InvPar):

InvPar::=SEQ{GeneratorReference, invId OPT, LSMeventInfo)

--The parameters are the same as for the r-invokeOrder. The only difference is that the actor identifier now identifies the remote actor that generated the LSM event.

---

[1] We trust that the reader has no problem with ASN.1. Otherwise, see the ITU–T recommendation X.208 or some short version of it, e.g., the one given in Appendix 2 of Muth (2001).

The DSS1 L3 example:

How these type assignments are applied depends on the actual layer. Let's apply it to the DSS1 L3 layer. The DSS1 L3 standard does not define the actor–agent separation, and therefore also no abstract protocol and no invokes (with few exceptions, standards do not). We discussed the mapping between abstract primitives (e.g., connectRequest(...)) and the corresponding DSS1 L3 messages (e.g., SETUP) in Sect. 2.3.2.1. What we did not define, however, is how DSS1 L3 actors identify each other. In order to understand how, we must look at the connectivity structure in the DSS1 L3 layer, see the model below (details of the model in Fig. 2.6).

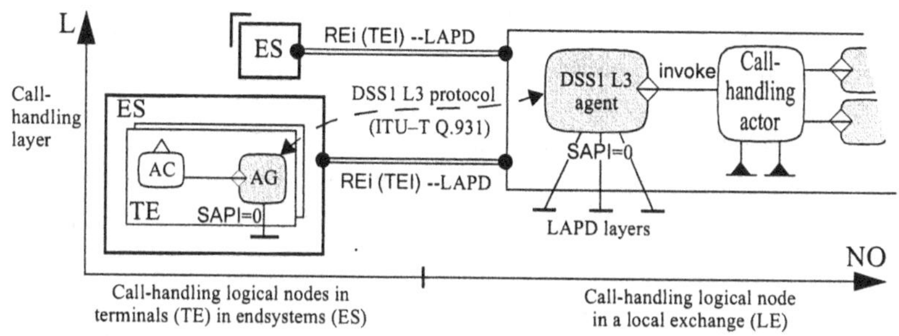

Terminals (TEs) exist in endsystems (ESs), and an ES can comprise many TEs. The DSS1 L3 agent in an LE connects permanently to many ESs over LAPD link routes, using LAPD layer interfaces identified by SAPI = 0 (SAPI is the APi in LAPD). This implies that there is a single link route between the LE and each ES. Since there might exist several TEs in an ES, this route has multichannels, with each channel associated semi-permanently to a particular TE. The channel identifier (CHi in AMLn) is called TEI in LAPD. How these channels are distributed to individual TEs inside an ES is another story, which does not concern us in this discussion.

Let's now discuss which AMLn invoke parameters are needed, considering this connectivity structure:

- DSS1 L3 actors in TEs can only communicate with the LE actor. Furthermore, the DSS1 L3 agent layer is obviously not a common agent layer, i.e., it does not support multiple actor layers. Consequently, when a DSS1 L3 actor in a TE invokes its agent, there is no need for referencing any actor address (AA in AMLn).
- The call-handling actor in the LE can, however, communicate with all TE actors (it may have to deal with hundreds of sessions simultaneously). It must therefore, in its invokes in the actor–agent interface, be able to identify individual ESs by some kind of actor address (only an actor suffix, As, since there is only one actor layer). Furthermore, since there might exist several TEs in an ES, and since any of these TEs can take a call offer, or initiate a call request, a particular

call must be associated to a particular TE by some kind of association identifier (ASi in AMLn) as well.

- The DSS1 L3 protocol specification does not foresee any need to repeat an invoke before a previous one has been dealt with. Consequently, no invocation identifier is needed.

Thus, in summary, all we need is an ASi parameter in the invokes. The mapping of the ASi on a DSS1 L3 message parameter is easily identified as the callRef ("call reference") parameter. An ASi value is, according to AMLn, established as a result of an association procedure. We deal with this issue for DSS1 L3 first in Sect. B2. Suffice here to accept that a callRef exists and is used in LE invokes of the call for addressing actors in TEs. According to the standard, a callRef is unique only within a particular ES. It is, however, reasonable to assume that the LE actor should not be concerned with parameters for identifying ESs, and therefore use an ASi that is unique within the whole DSS1 L3 layer. This requires the LE agent to translate a "global ASi value" that the LE actor knows to a "local ASi value" (i.e. a callRef) that an ES knows. Using some of the tables we have discussed for agent layers (see also Appendix C) we can define the DSS1 L3 agent in an LE as in the model below.

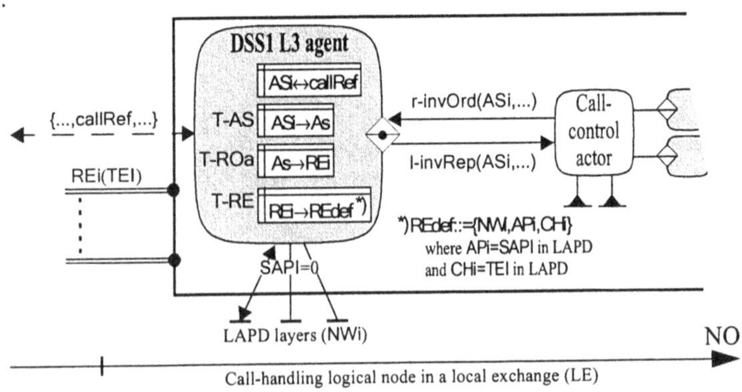

In the send scenario (triggered by an r-invOrd):

- The agent uses the ASi↔callRef table for mapping between the association identifier (ASi) used in the SAG and in the message protocol (the callRef).
- It uses the T-AS table to identify the TE actor (by an As) that handles the actual call (identified by an ASi).
- It uses the As as an entry to the T-ROa table to get the route endpoint identifier (REi) for the route to the actual TE.
- It uses the T-RE table to find which LAPD network to use (NWi), and which address parameters (APi, CHi; in LAPD called SAPI and TEI).

The agents on the TE side of the DSS1 L3 layer need no tables at all.

B2 Association Operations (in SAG)

AMLn defines four operations for handling associations, shown in their context in the model below.

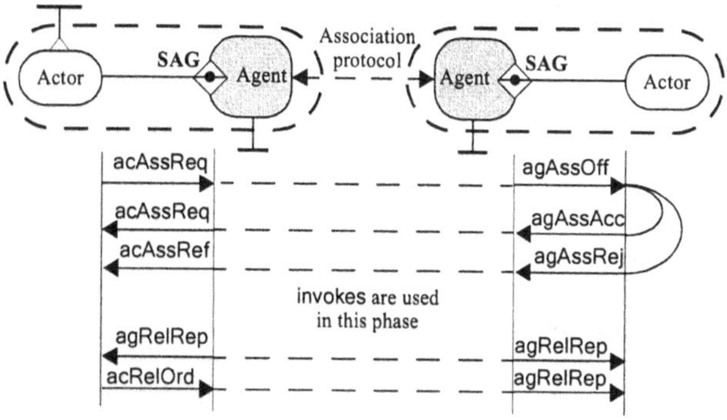

The essential parameters of these primitives are the association identifier (ASi) and the actor address (AA). Since creating an association is a negotiation procedure between two actors, other parameters may exist as well, e.g., parameters for authentication and parameters for abstract protocol compatibility checking. The association identifier is (strictly speaking) a parameter of the association protocol, which implies that local association identifiers may be used in SAGs (disregarded here, however). The four operations are:

Operation is: actorAssociation (acAss)
Primitives are: Request(destAA,ASi,...)
Confirm(ASi,...)
Refuse(ASi,...)

Operation is: agentAssociation (agAss)
Primitives are: Offer(origAA,ASi,...)
Accept(ASi,...)
Reject(ASi,...)

Operation is: agentRelease (agRel)
Primitives are: Report(ASi, causedBy{agentLayer,remoteActor},...)

Operation is: actorRelease (acRel)
Primitives are: Order(ASi,...)

The DSS1 L3 example:

The AMLn definition of association handling assumes it to be a procedure that is executed before any invokes can be generated, and after the last invoke is detected. This corresponds to how a common agent layer would deal with the problem (e.g., TCAP and ACSE). The DSS1 L3 agent layer is, however, not a common agent layer. In such cases it is normal for the association identifier (ASi) to be established at the same time as the first invoke is generated by simply including an ASi value in the first message of a new dialogue. In the case of DSS1 L3 this message is the SETUP and the ASi is the parameter callRef in the SETUP. Logically we can view this as the originating actor generating two SAG primitives before the agent sends the SETUP: one assReq(As, callRef) and one r-invokeOrder(callRef,<the first LSM-event, carried by the SETUP>).

The only questions are: is there any actor identifier parameter (As) in the DSS1 L3 protocol, and what parameters are used for identifying actors in the LE SAG?

- Every event generated by a DSS1 L3 actor in a TE is detected by the single LE actor. Thus, there is no need for a TE actor to identify any remote actor in communication with its DSS1 L3 agent. Consequently, no parameter is needed in DSS L3 messages for actor identification in this case
- In order to make the DSS1 L3 actor in an LE independent of how endsystems and terminals are connected to the operator network, and how ASi values are used over the DSS1 L3 layer, we do not want this actor to know anything about LAPD routes and callRefs to ESs. Therefore, when the LE actor wants to establish a new association to a particular ES, it only refers an idle ASi and an ISDN/PSTN network address, in communication with its DSS1 L3 agent. This agent has a table that associates such a number to a particular LAPD route (i.e., ES). Thus, no actor identification parameter in DSS1 L3 messages is needed in this case either. The DSS1 L3 agent in an LE translates the ASi value to an idle callRef of the actual ES, and performs the association procedure (over the broadcast channel TEI = 127 on the actual LAPD route). Once the association is established, the LE agent uses the tables discussed previously for relaying information between the LE actor and a particular TE actor.

If we take a look at the DSS1 L3 protocol specification, we do find several parameters in the SETUP that are ISDN/PSTN network addresses, however. One is called "calling party subscriber number" and the other "called party subscriber number". It is important to understand that these have nothing to do with establishing associations between actors in TEs and LEs, but are parameters of the abstract protocol of the DSS1 L3 LSM:

- The *calling party subscriber number* is used by a TE user to see who is calling (sometimes referred to as the service "A-number display").
- The *called party subscriber number* is used to support interworking between the operator network and an ES that acts as a gateway (a PABX) of a local subscriber network. Hundreds of TEs may exist in this network, each identified by a separate ISDN/PSTN network address. It is the responsibility of the ES to route calls inside this network.

B3 Operations in SAC

A SAC ("specified actor interface") defines the operations and primitives that are used over actor–agent interfaces (AAI) in connectivity layers, i.e., a layers that transport data between layer interfaces. The abstract primitives of these operations are *not* meta-primitives since a connectivity layer has no LSM to support. The interface symbol used for SACs therefore looks different to SAG symbols. SAC operations are defined as part of an actor specification.

Connectivity layers either realize link routes (such as LAPD), or perform multiplexing, cross-connecting, or switching (different types of switching included). The actors are either relaying only, terminating only, or both. An actor that is relaying is called **common actor** (i.e., connects several agents).

A SAC defines two conceptual operations: **relay** and **terminate**, with the corresponding ASPs **relayOrder(...)** and **terminateOrder(...)**. How these primitives are related to **send** a and **receive** operations in layer interfaces depends on whether the data transport is user-to-user, user-to-system, or system-to-system, as shown in the figure below. Note that an **interface type identifier** (ITi, explicit or by position) is needed in protocol messages if data can be terminated in different ways.

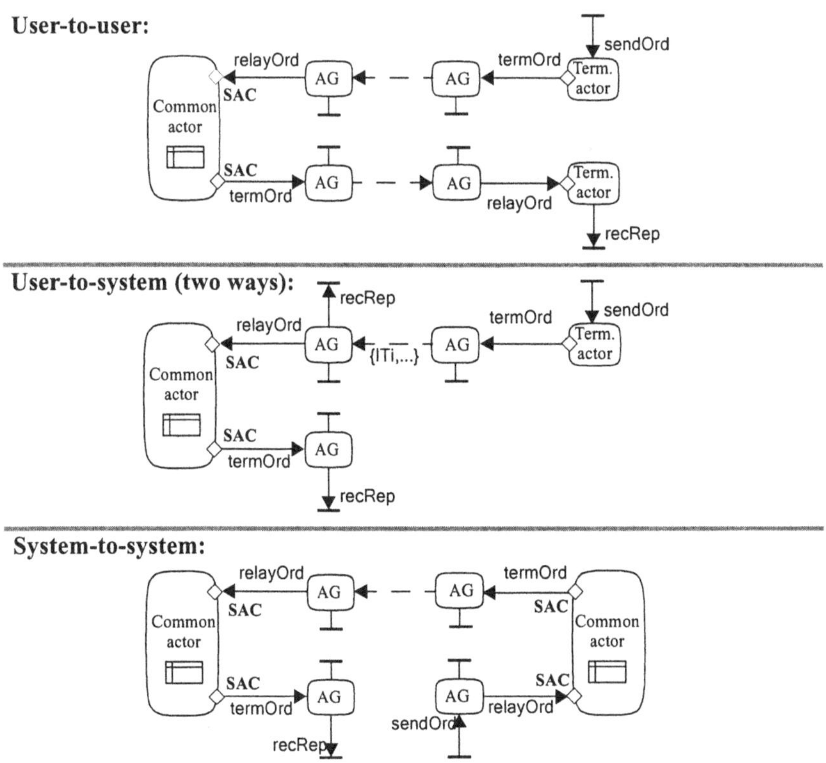

- The SAC primitive sequence in the user-to-user model is based on the principle that agent layers should not know where a data stream is terminated, in order to make agent layers more generally applicable and to confine path control to actors only.
- The sequence in the user-to-system model demonstrates two alternative methods for terminating data in systems: either in a message receiving agent (the method used in, e.g., PSTN/ISDN) or in an agent of a common actor (the method used, e.g., in ATM).
- The sequence in the system-to-system model is typical of how connectivity for communication between control layer elements is provided, e.g., in PSTN/ISDN and in ATM.

As far as SAC primitives are concerned, there is a distinct difference between layers that transport synchronous and asynchronous user-data.

- If synchronous, there will be no implementation of SAC primitives since relaying is defined by the position of bits in frames over a SAC.
- If asynchronous, there will exist explicit primitives since some type of terminator address has to be delivered to the actor, together with the user data. Other parameters may occur as well, e.g. quality-of-service requirements (as in IP) and some type of originator address.

Path control for data streams is performed by common actors through some type of tables, depending on the type of layer.

- Circuit-switching actors (in PSTN, ISDN and cross-connecting systems) and connection-oriented packet-switching actors (in X.25, FR, and ATM) use a control store table (T-CS) that defines how to connect a particular data stream between AAIs, also considered multichannel AAIs. Information in the T-CS is set and removed by a control layer, either in a management system (for cross-connecting systems) or in the traffic system (all other systems).
- Connection-less packet-switching actors use routing tables (T-ROs). Information in the table is either set by a management system (as for X.25 and MTP L3) or managed by special routing layers (as in IP).
- Asynchronous multiplexing actors use a table (T-L) that translates a link identifier (Li) on the multiplexed side to an AAI identifier (AAIi) on the non-multiplexed side, and vice versa. The table is managed by a management system.
- Synchronous multiplexing actors need no tables at all.

How the information in T-CS, T-RO and T-L tables look depends on the type of relaying layer. The model below shows the circuit-switching case. The SAC primitives are just conceptual since channels are defined by their positions in frames. Each AAI comprises many channels, and the actor may connect one or several AAIs.

Circuit switching:

Single-AAI T-CS: CHi↔CHi
Multiple-AAI T-CS: {AAIi, CHi}↔{AAIi, CHi}

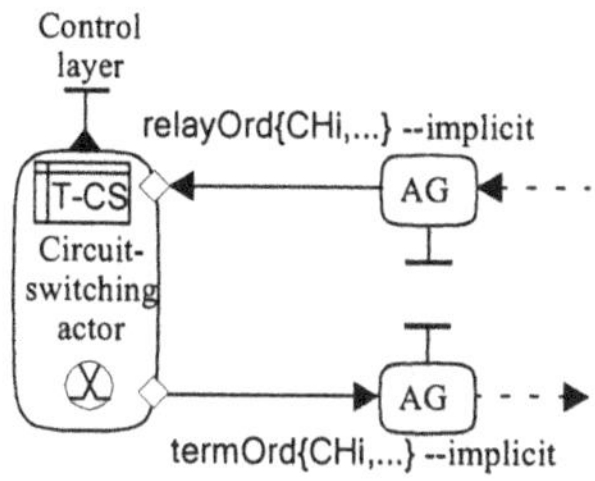

The next model shows the connection-less (or "datagram") switching actor. Note that the layer element is normally both relaying and terminating, and that the agent will know its own network address (NA). The agent is also the terminating element in this case.

Datagram switching:

T-ROs: NA→REi
T-RE: REi→REdef

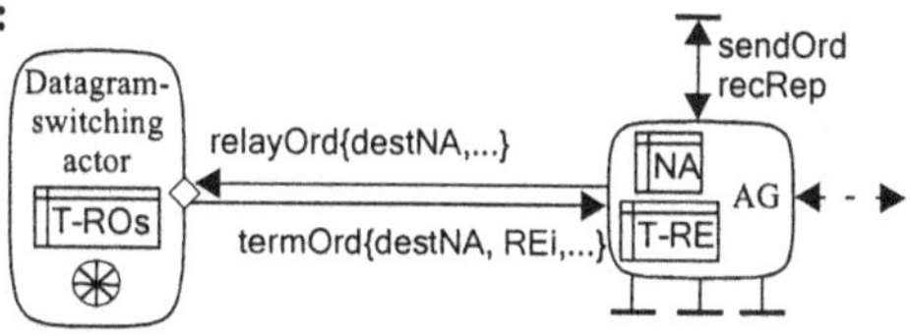

The following model shows the connection-oriented packet-switching actor. The agent may use several link layers (as in X.25) or a single cross-connecting layer (e.g., SDH as in ATM). This model is strictly valid for X.25 and FR only. In ATM, switching can be based both on VCi and VPi.

Connection-oriented switching:

T-CS: {inREi,inVCi}→{outREi,outVCi}
T-RE: REi↔REdef

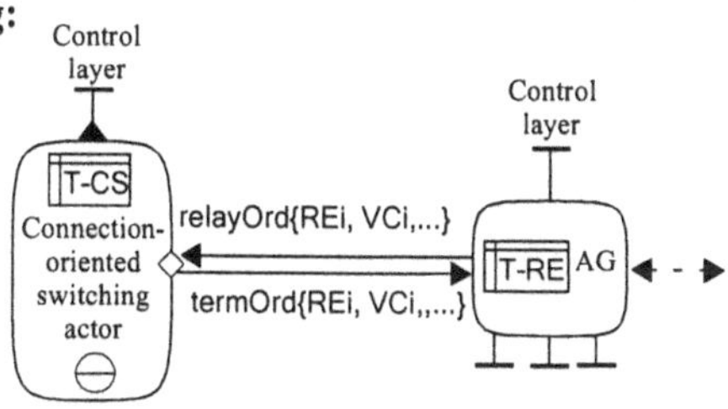

The model below shows the asynchronous multiplexing actor.

Asynchronous multiplexing:

T-L: Li↔AAIi

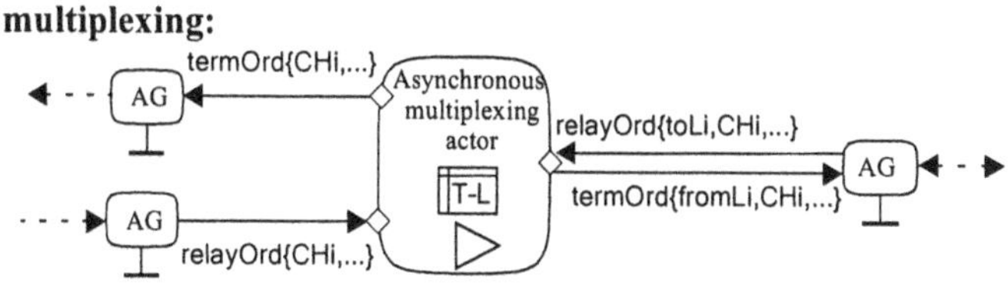

Appendix C: AMLn Configuration Parameters and Tables

C1 PARAMETERS

AA (Actor address):

The identifier used within a layer for an actor. The AA in a common agent layer may be built by the two parameters ALi and As, i.e., AA::={ALi,As}.

AAli (Actor-agent interface identifier):

Identifies an actor–agent interface. Only known by agents and actors inside a layer element. Applicable for both SAGs and SACs

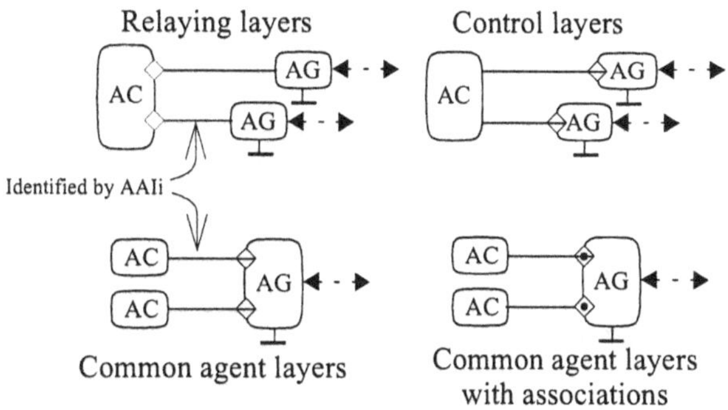

ALi (Actor-layer identifier):

A protocol parameter in an association protocol (of a common agent layer) that identifies an actor layer.

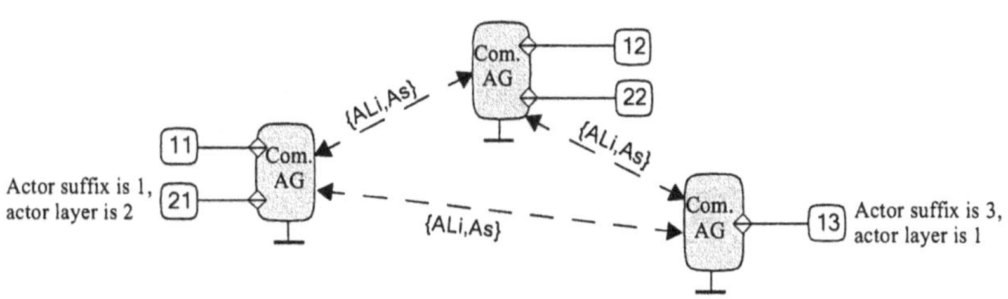

As (Actor suffix):

A protocol parameter in an association protocol (of a common agent layer) that identifies an actor within an actor layer.

ASi (Association identifier):
A protocol parameter in an association protocol that identifies a temporary session (or "dialogue") between two actors.
APi (Access-point identifier):
A protocol parameter that identifies a layer element within a node.

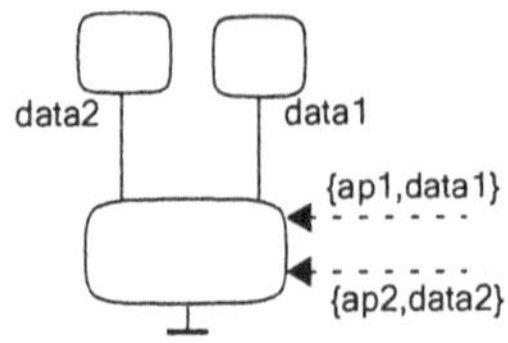

CEi (Connection-endpoint identifier):
An identifier in a layer interface of a connection-oriented packet-switching layer. Identifies a temporary connection.
CHi (Channel identifier):
Identifies one of several channels in a permanent or semi-permanent route.
Ci (Connection identifier):
A protocol parameter in a connection-oriented packet-switching layer that identifies a temporary connection between two (or more) using layer elements. May be translated to a CEi in each layer interface.

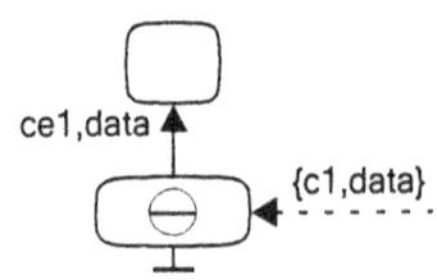

GT (Global title):
An address that identifies a user of a switching layer in a numbering system that is independent of the numbering system of the switching layer. Requires translation to address parameters of the switching layer.
ITi (Interface-type identifier):
A protocol parameter that supports discrimination within an agent between layer interfaces and actor–agent interfaces. May be combined with an APi for layer interfaces.

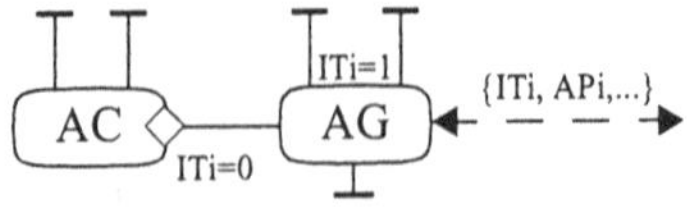

Li (Link identifier):

A protocol parameter that indirectly identifies one of a number of actor–agent interfaces in a multiplexing layer element

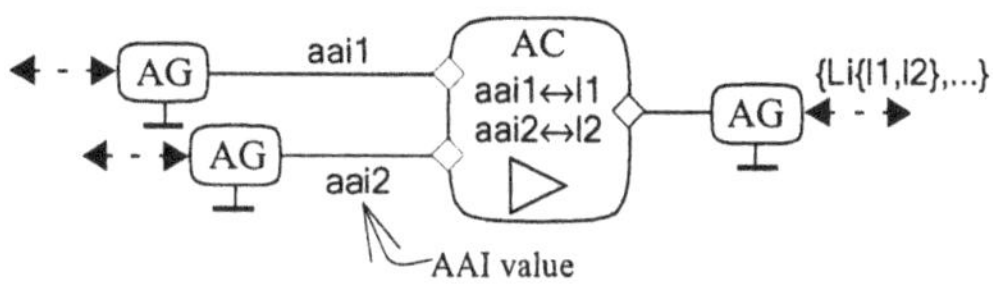

Lli (Layer-interface identifier):

Identifies a layer interface in a layer structure model. This identifier can be used for simulation or translated to addresses of the platform on which the layer elements run.

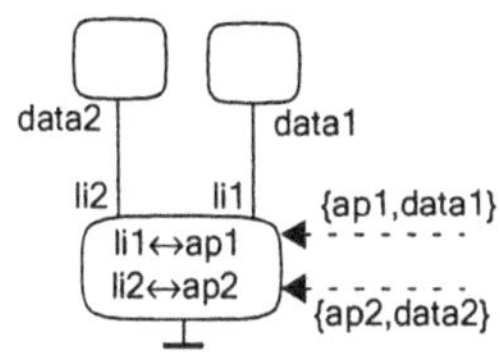

NA (Network address):

Most NAs are defined as concatenations of diverse operator network identifiers and serial numbers (sometimes named "subscriber number"). The international telecommunication numbering system is described in ITU–T E.164, and a number of other ITU–T recommendations for its application on different network systems. IP addresses are assigned by the NIC (Network Information Center). The IP numbering system is described in RFC 1166.

NWi (Logical-network identifier):

This is an AMLn-defined parameter that refers a logical network (hosted or hosting), not to be confused with operator network identifiers in NAs.

PA (Platform application address):

This parameter is used in simulation and execution of a model on a particular processing platform. Two PAs are defined for each layer interface.

REi (Route-endpoint identifier):

Identifies a route endpoint. Only known by the logical node that terminates the route. Must be translated to an REdef before data can be sent over the route.

REdef (Route-endpoint definition):

Defines how a route (and its channels if multichannel route) must be referred in communication with a hosting network over a layer interface. In general, a hosted network can be hosted in more than one hosting network, which is why an NWi might also be part of an REdef.

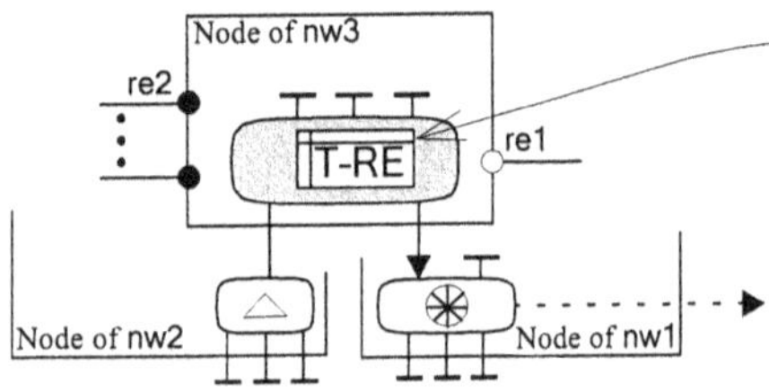

REi	REdef			
	NWi	AddrPar		
		NA	APi	CHi
re1	nw1	na1	ap1	—
re2	nw2	—	—	ch3

In general:
REi→REdef, where
REdef::={hostingNWi OPT, AddrPar OPT}
In this case:
REdef::=CHOICE {NW1,NW2}, where
NW1::={{nw1},NA,API}
NW2::={{nw2},CHi}

SRi (Substructure-relation identifier):

Defines which network levels the relation connects.

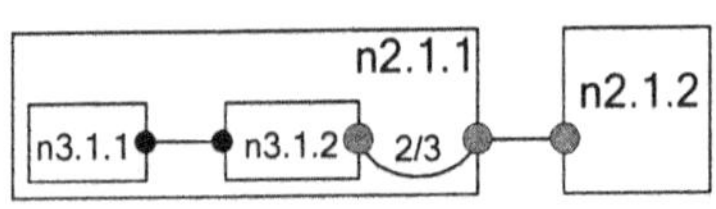

n <network level>.<network identifier>.<node identifier>

REMARKS:

ASi, CEi, and Ci are on-demand parameters. Remaining parameters are permanent or semi-permanent, which means they are set by configuration management.

LIi and PA are used only in model execution.

C2 Tables

AMLn tables are based on 15 basic parameters:
AAli, ALi, APi, As, ASi, CEi, CHi, Ci, GT, Li, Lli, NA, PA, NWi, REi.
Eleven configuration management tables are defined:
T-AL, T-AP, T-LI, T-PA, T-RE, T-NWg, T-NWd, T-ROa, T-ROs, T-GT, T-L.
In addition, there are three dynamic tables for on-demand structures:
T-AS, T-CS, T-CO.

Tables in common agents:

T-AL, actor-layer table:
AAli ↔ ALi
T-AS, association table:
ASi → AAi

Tables for connectivity layers. On each logical network level:

T-AP, access-point table (a table in a hosting node):
APi → hostedNWi
T-RE, route-endpoint table (a table in a hosted node):
REi → REdef, where
REdef::={hostingNWi OPT, AddrPar OPT}, where
AddrPar::={NA, APi, CHOICE{CEi, CHi}} --all optional

Tables in connectivity layers that translate between network levels:

(these are tables that relate horizontally-partitioned logical networks to each other)

T-NWg, layer interface translation table (a table in a hosting node):
NWi (hosted network, level i) → REi (hosting network, level i+1)
T-NWd, layer interface translation table (a table in a hosted node):
NWi (hosting network, level i) → REi (hosting network, level i+1)

Routing tables:

T-ROa, routing table (in a common agent):
AA → REi or ASi → REi
T-ROs, routing table (in a switching element):
NA → REi

Miscellaneous tables:

T-CO, connection table:
  Ci ↔ {hostedNWi,CEi}
T-CS, control store:
  inlet->outlet, using CHi, VCi, and Li for inlet/outlet definitions, depending on the type of switch.
T-GT, global-title table:
  GT → {REi, destinationAPi}
T-L, link table:
  Li ↔ AAli
T-LI, layer-interface table:
  NWi → LIi
T-PA, platform table:
  LIi → {hostingPA, hostedPA}

REMARKS:

T-AS, T-CO and T-CS are set on-demand. Remaining tables are managed configuration tables.

Note that combining a T-GT and a T-RE table delivers GT → REdef translations. This is the method used in e.g. DNS, since the TCP/UDP layers do not handle routes.

T-LI and T-PA are needed only for simulation and model execution.

Appendix D: AMLs and AMLp in Short

D1 Introduction

AMLn is used for modeling the layer structure and the node structure of a network system, as described in this book. AMLn can not, however, be used to specify how the modeled system will behave. To that regard we need another language that can be used for specifying interfaces of the AMLn model, and the behavior of its functional elements. This language may be AMLs,[1] where the "s" stands for "service" (for historical reasons).

The only interfaces that cannot be specified by AMLs alone are network interfaces and their message protocols. In AMLn, network interfaces are structures of message protocols and have no associated behavior, only message protocols have. A message protocol describes the behavior between two OSI layer elements or AMLn agents in terms of the messages they exchange. To that regard we need AMLp. AMLp is a method for specifying message protocols for (primarily) control layers. The "p" stands for "protocol".

As has been discussed in this book, the separation of an OSI layer into one or several layer state machines (LSM) and a layer protocol machine (LPM) is a large improvement with respect to protocol specification since it separates two rather independent issues, which are:

1. *The LPM part*, i.e., the definition of message types, their encoding and any additional function that deals with pure message handling (e.g., error detection and fragmentation).
2. *The LSM part*, i.e., the activities that lie behind sending a message and are an effect of receiving a correct message.

Note, however, that this separation concerns control layers only. Connectivity layers are LPMs only.

Both LPMs and LSMs have a behavior, each with its own states. Both aspects are often described in a single protocol specification based only on messages. The

[1] A number of other languages could serve the purpose as well. For example, both UML and SDL are intended for similar purposes. IDL, the CORBA interface language, can be used for interface specification. However, all these languages are initially developed for software system modeling, and therefore exhibit some assumptions that become unwanted constraints in the context of network system modeling, e.g., software object-orientation and week support for asynchronous communication.

result may be that LSM and LPM states multiply, with a very complex specification as a result.

When the specification of an OSI layer is separated into LSMs and an LPM, it can be divided into the following five parts, (see the figure below). Note that this principle is valid for every control protocol, not only for protocols that rely on a common agent layer (such as TCAP).

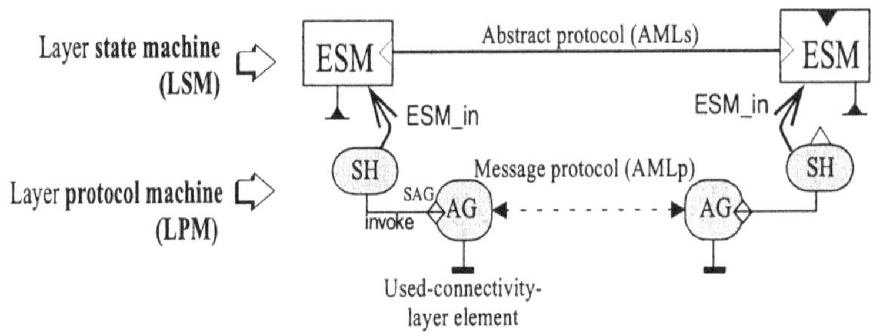

1. A specification of the LSMs (a single LPM can support many LSMs).
2. A specification of meta-primitives for the SAG (i.e., l-invoke and r-invoke), also using AMLs. Meta-primitives must be designed with the actual LSM description language in mind. The generic primitives defined in AMLn are designed to support the AMLs abstract machine system and abstract object structures, including AMLs operation classes. However, the important thing is not which specification language is used but that meta-primitives are defined and they support the chosen LSM language.
3. If the layers support associations, a specification of association control primitives in SAGs.
4. A specification that describes how invokes and association control primitives are mapped on message types.
5. Any other additional specification, describing other agent layer functions, such as error control, fragmentation, etc.

AMLs is a specification method that can be used for the first three specification types. AMLp supports specification of the last two, i.e., of LPMs.

D2 AMLs

AMLs is described in detail in Chap. 7 of Muth (2001). In the present chapter we give a summary of AMLs. It has two basic characteristics:

1. It is intended for specifying behaviour in terms of *state machines*, which is a very suitable method for describing the behavior of controlling functional elements (of which there are many in a network system). Since every system today is supposed to be object oriented, it is important to understand that there is nothing in the concept of a state machine that says that it must be object oriented.

With the exception of some application layers in some network systems, most existing standardized layers are not.

2. AMLs models systems of elements, where the possible distribution of individual elements is transparent to them in communication. Their communication is therefore described as if they were all running on the same platform. Since that might not be the case in reality, the AMLs way of describing communication is an *abstraction* of message communication. The type of invocations between AMLs elements is called **abstract service primitive** (ASP), a commonly used concept in ITU–T standards, for example. An instance of an ASP is referred to as an (abstract) **event**. The notion of abstract communication and abstract elements permeates AMLs.

A consequence of these two characteristics is that the events that trigger transitions in an AMLs state machine are ASP events, not message types (as they are in most other specification languages).

The basic element that is modeled in AMLs is an **abstract machine** (AM). Abstract machines are representations of resident functional elements in a system, i.e., candidates for being implementation elements in a system (e.g., a SW unit or a system of such units). The model below shows an **abstract machine system**.

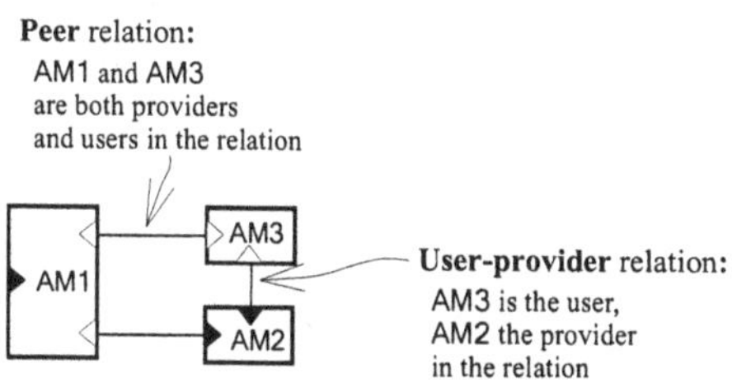

Since every AM is a state machine, the depicted system as a whole is a state machine as well. The model tells which machines can communicate with each other. The relation between two machines is described by an **abstract protocol** in terms of event interactions. Both **user-provider** and **peer** relations may exist (more about that later).

As mentioned, the basic invocation element is the event, which is an instance of an ASP. An ASP describes a type of an asynchronous invocation, which means that the communication between two AMs can be described as a sequence of events, **generated** by one AM and **detected** by the other (since these invocations are not messages, we talk about "generating" and "detecting" instead of "sending" and "receiving").

The basic difference between a message type and an ASP is that the specification of a message type describes just a structured data type, but says nothing about how the receiver will react upon message reception (normally described separately in (often) lengthy natural-language descriptions in protocol specification). The

ASP on the other hand tells what the detector is supposed to do by defining an operation and adding parameter values that are relevant to the referred operation.

Thus, ASP communication is *operation-oriented*. The syntactical description of an ASP in AMLs follows practices in most methods. For example, if one AM (AM1, see the previous figure) wants the other (AM2) to "create a connection" and return a confirmation of that action, AM1 will generate a connectRequest(address), and AM2 will (hopefully) generate a connectConfirm(connId) when it has created the connection

The operation referred to in this example is connect and relevant parameters are address and connId. An ASP does however also include information about if it is a request for an operation (Request) or a confirmation (Confirm) that the operation has been executed successfully. These information elements denote the **primitive type**, and are important for resource state control (between a Request and the Confirm, the requested connection is obviously in an unknown state, as seen from AM1).

The primitive type concept exists, in one form or another, in most ASP-based specification methods (a primitive-type parameter can also be identified in most protocol messages, although called anything but "primitive type"). For example, the OSI RM defines four famous primitive types: request, indication, response and confirm. AMLs (inspired by the ROSE standard of the OSI RM) takes this concept one step further by defining six combinations of primitive types that can be used for an operation. These combinations are called **operation classes** (see the figure below).

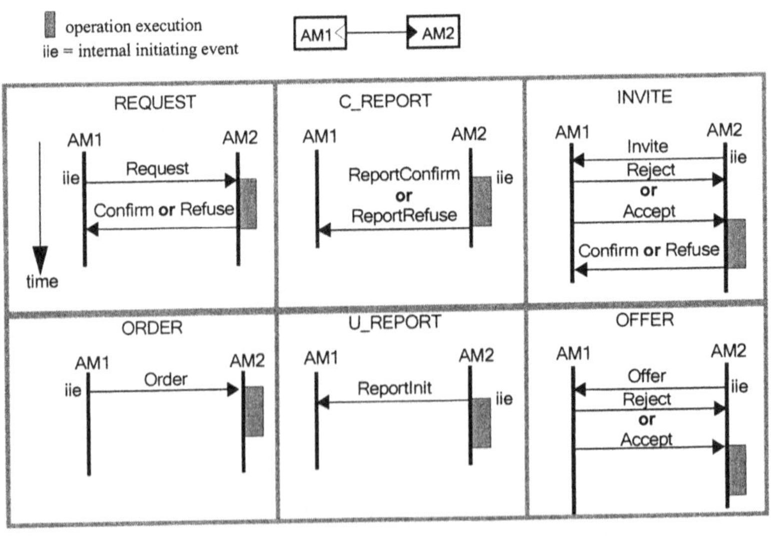

In all these cases, AM1 is the **user** and AM2 the **provider**, since AM2 runs all operations. The operation classification system does, however, allow both users

and providers to be the initiator of an operation execution. It allows for unconfirmed and confirmed operations and, in the latter case, it always defines different primitive types for successful and unsuccessful operation executions.[2]

AMLs also defines the difference between synchronous and asynchronous operations in abstract protocols (**operational synchronism**), as shown by the figure below.

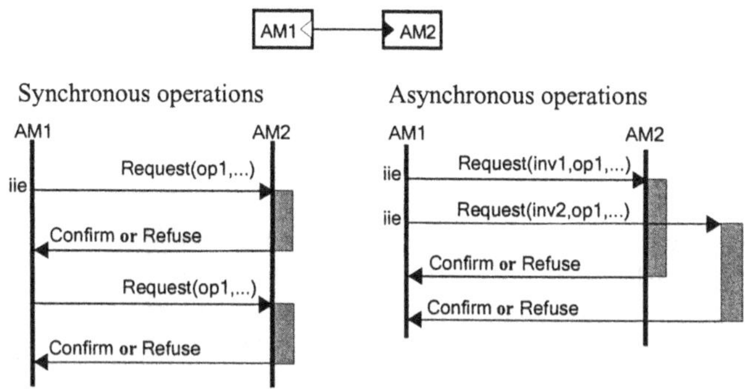

This concept can also be used for particular operations in an interface, i.e., some operations may be synchronous, others asynchronous. Thus operational synchronism may be defined for an abstract protocol as a whole, or on the level of operations (in **operation specifications**).

Abstract machines can also have a **peer** relation, as mentioned before, which implies that they both can execute operations on behalf of the other AM. The next figure gives two examples of that.

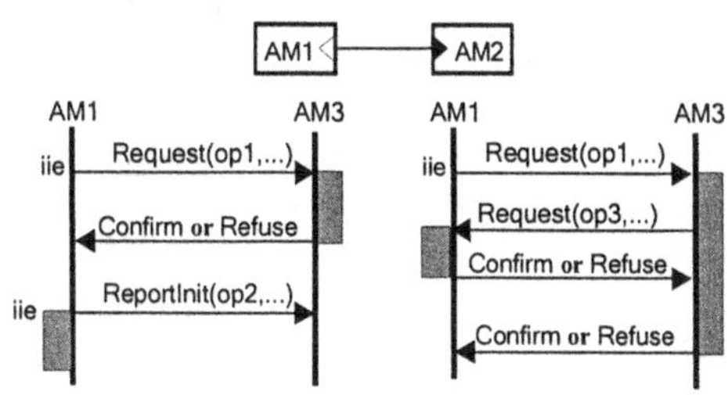

Management layers and layers realized on CORBA platforms are control layers that define a kind of objects. To support object-oriented interfaces between abstract machines (without presuming any particular OO design language), AMLs defines

---

[2] The purpose is similar to the ACK and NACK messages in some older protocols. It is also common for many protocols to add a confirmation of a previous message as a "piggy-back" bit in the next message sent. This demonstrates one aspect off the difference between ASPs and messages (messages being realizations of ASPs).

the **abstract object** (AO). The model below shows such an interface between AM1 and AM2, based on three abstract objects.

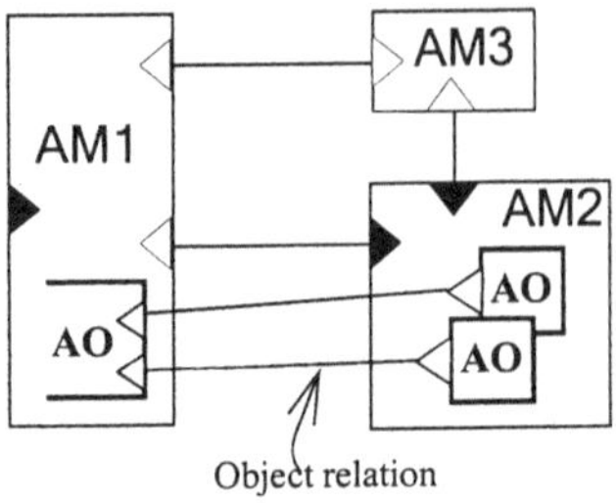

AOs exist conceptually inside an abstract machine in order to not reveal anything about their implementation (an AM that defines no AOs in a user–provider relation is equivalent to a single AO). AOs are only visible to a using AM through the interface that the providing AM offers. As an example, let's assume that AM2 is an abstract machine that can create connections (actually: connection endpoints). A created connection endpoint can be modeled as an AO on which operations such as send, receive and reset could be ordered. These operations are uniquely defined for the AO type "connection endpoint."

The use of AOs affect the way an interface between AMs will look like and the way ASPs will be defined. Objects in general can be volatile, which means that the providing AM may have to offer special operations for object handling (such as createObject and deleteObject). Furthermore, the user of specific object operations must refer the identity of the object. A send event on a connection endpoint object will therefore be defined as connId-sendRequest(data), for example.

In AMLs we say that an AM that knows the identity of an AO has an **object relation** to that AO. For volatile objects the object relation is an **object identifier** (the connId is such an identifier). For permanent AOs, the object identifier is replaced by the name of an object. The object relation concept allows us to create models of systems of communicating AOs, called **(abstract) object structure**,[3] as exemplified in the model below.

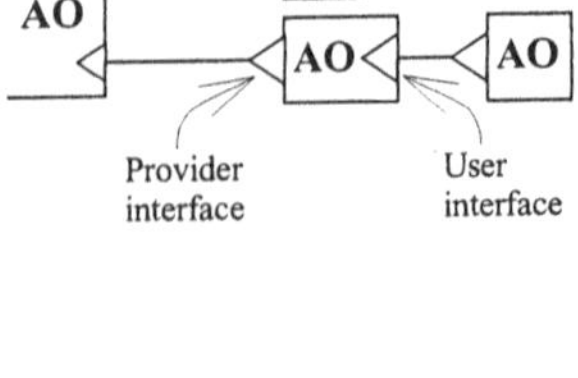

[3] In Muth (2001) the terms "service object," "service relation" and "service network" are used instead of "abstract object," "object relation" and "object structure". We use the latter here in order to be more in line with current terminology.

The notation used for AOs differ from AMs so that an abstract machine system model and an object structure model will not be mixed up. Note that an abstract object is always a provider and may also be a user of other objects. It can never, however, be a peer (only abstract machines can). Note also the difference between a provider interface (defines which operations the object can provide) and a user interface (defines which operations are actually used by the using object). Thus, the relations in an object structure are defined by user interfaces.

From a using AM's point of view, an AO is something that is only visible by an object relation offered by another AM. This allows us to define a whole object structure as a network-wide AO (using the general **substructure** relation), as depicted below.

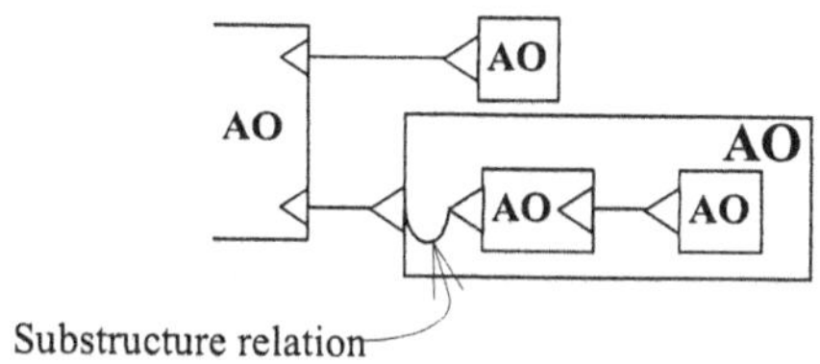

The fact that an AO definition is not necessarily confined to a function in a single node is one of the features of AOs that distinguishes this concept from object-oriented approaches in other modeling and design languages.

Abstract machine systems and object structures are the model types we use for *behavior* descriptions. Both AMs and AOs can be described as state machines, and so can the system and the structure as a whole. An advantage with defining the functionality of AMs in terms of AOs is that the behavior description of an AM becomes simpler. This is because states that relate to operations on AOs are separated from states that relate to operations that create and delete AOs. The behavior description of an AM thereby becomes divided into one description that describes its object handling, and a number of (often rather simple) descriptions of the behavior of the object types that the AM offers.

The behavior of an AO depends on how it reacts on events in all its interfaces. It is therefore necessary, for each AO, to produce an **interface specification** for its provider interface. This specification will comprise all its **operation specifications**, and, in addition, a **state machine description** for that interface (diagram or tabular). The behavior of an AO that depends on other AOs is then constructed by combining the state machine of its provider interface with the state machines of all provider interfaces of the objects on which it depends. The behavior of AMs is constructed in a similar way. The details of this specification technique are given in Chaps. 6 and 7 in Muth (2001).

D3 Applying AMLs in AMLn Models

The behavior of an AMLn model is completely confined to its *layer structure*. The only interface type in such a model that cannot be specified by AMLs are network interfaces and message protocols (see AMLp in the next section). Thus, AMLs can be used for describing operations, interfaces and behaviors of:

- All kinds of layer interfaces (i.e., service points and control points).
- Interfaces between the ESMs of an LSM (control layers only). These interfaces are abstract protocols, since they are mapped on a message protocol by the LSM–LPM separation. Since an ESM is an abstract machine and an LSM is an abstract machine system, the whole range of behavior description techniques of AMLs can be used to describe the behavior of ESMs and LSMs.
- SAGs and SACs (generic primitives for these interfaces are defined as parts of AMLn).

Abstract objects and object structure specifications may be used wherever an LSM, a layer interface, a SAG or a SAC acts in an object-oriented fashion. Typical examples, that have been discussed in this book are management layers. The model below shows the LSM that we defined for TMN and relies on abstract objects called **managed objects** (MO).

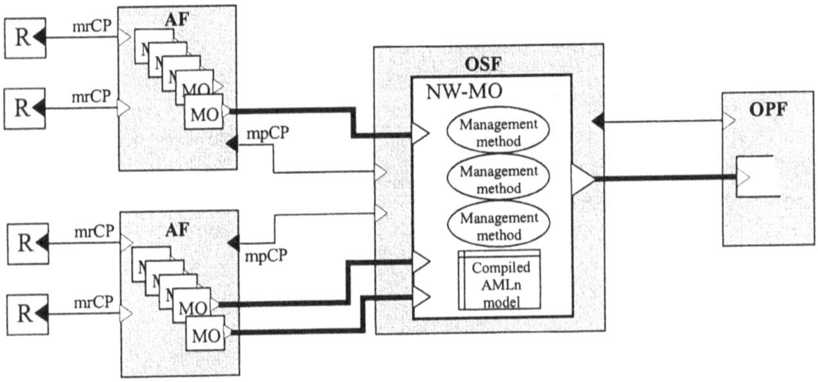

D4 AMLP

Message communication is very different to communication based on ASPs. ASPs are strictly functional-oriented, which implies that confirmation of an operation request must be defined as a separate ASP, for example. The message protocol that realizes this communication does not necessarily define a separate response message type (could be just a piggy-back bit of another message). Messages are optimized for requiring minimum bandwidth, processing capacity, and response time, which implies that there is often no one-to-one relation between an ASP and a message type. Optimizations of message types that are commonly applied in message protocols include:

- The information elements of an ASP (object identifier, operation identifier, primitive type, etc.) are given a fixed position in the sequence of message parameters that carries them. The order of position is not bound to the syntax of the ASP, but optimized for processing efficiency.
- Another typical optimization is: the designer of an ASP does not really care in which order parameters of an ASP are defined in messages, which in the used data typing language may be declared by defining the parameter field as a SET and naming or "tagging" each parameter individually. The corresponding message type may change this and define a fixed order (a SEQUENCE) and remove tags if possible.
- A generator of events may want to generate a whole series of events, addressed to the same detector at the same time. Provided that the underlying connectivity stratum offers unlimited size of data, the message protocol may define a message format that concatenates information elements of several events. TCAP has a function called "component handling" that does that.
- With the exception of some older transport protocols that define separate ACK and NACK messages for the primitive types Confirm and Refuse, respectively, most protocols define a single message type (or parameter) for all types of responses.
- Control protocols that rely on narrow bandwidth (typical for older cellular networks), may use the same message parameter for different ASP elements, depending on the state of the control procedure. This kind of "smart coding" may serve a bit or two in transmission, but is also very confusing, especially if the corresponding ASPs and their mapping on messages is not defined.
- Most protocols that carry ASPs define a single message parameter called "message type" that in fact often corresponds to the combination of an operation identifier and a message type identifier. This is not always obvious by the names selected for message types, which makes the understanding of the protocol unnecessarily difficult. You have to think in terms of ASPs and mapping ASPs on messages to see which primitive type (if any) is represented by the name of the message type.[4] Furthermore, the approach results in a larger a set of message types than ASPs.
- The most common reason for messages and events not corresponding one-to-one is the need for fragmentation, because the used connectivity layer does not

---

[4] A typical example is the use of the message type SETUP in the DSS1 L3 protocol (discussed in Sect. 2.3.2.1): the same message type is used by the network for offering a call to a terminal (in which case it corresponds to a callInvite(...) primitive), and by a terminal to request a service from the network (in which case it corresponds to a serviceRequest(...) primitive). The set of parameters of these two primitives are not equivalent. Since the DSS1 L3 protocol defines no ASPs, the mapping between these imaginary primitives and the SETUP message format is given through many textual comments in the DSS1 L3 protocol specification.

accept unlimited size of data in a single invocation. An event may therefore be represented by a large set of messages, of the same or of different message types. Normally this should not bother event generators and detectors since the agent layer, or (better) the connectivity layer on which it relies, should perform fragmentation. This is often not so, however. As a result, the boundary between the abstract protocol and the message-protocol level becomes blurred. ASPs may have to be "fragmented", resulting in numbered operation names such as connect1, connect2, etc. This may be so even in cases when a separate agent layer is defined. For example, actor layers that rely on TCAP must consider message size limitations since TCAP runs on a non-fragmenting layer (SCCP) and does not fragment messages itself.

AMLp is described in Chap. 5 of Muth (2001). Its application is different for control layers and connectivity layers:

- For **control layers** it is assumed that meta-primitives are defined and AMLs is used for defining operations on the LSM level. An AMLp tool is assumed to be used to generate a single message type for invokes, including all possible parameters and considering only possible length constraints of parameters defined by the invoke specification. The tool also creates suitable message types for association control primitives in SAGs (if any). Code generation comprises only encoding algorithms, control that valid message types are used, message structures do not violate the message type specification, and parameter values are within defined value range. The tool can be designed for message error indications to actors and remote agents.
  A basic assumption in this method is that ASN.1 is used for data typing, both of parameters of LSM–ASPs and of parameters of SAG primitives. A variant of this scheme is therefore that more compact data types are used (including encoding algorithms defined by users), and the tool user can edit message type specifications that are suggested by the tool. Thereby, message formats that require fewer bits to be transmitted, or result in more efficient message processing, can be created. You can read more about this subject in Muth (2001).
- The previous method assumes that the agent layer protocol runs on a connectivity layer that is 100% reliable and offers unlimited length message size. If that is not the case, the agent layer must be regarded as a **connectivity layer** as well. A framework for asynchronous connectivity protocols that includes encoding can be generated by AMLp if the SAC primitives (relay and terminate) between the agent and the relaying actor are specified, and AMLp data types are used.
  However, most of the properties of the layer depend on the difference between the properties of services the layer offers and the services it uses, and of optimization issue. The layer may have to deal with many different issues, such as *error detection*, *error correction*, *fragmentation*, *sequencing*, *flow control*, etc. Automating the design of such a layer is not generally possible (or wise). Modeling techniques based on different types of queue models for service property validation are used in this area. This subject is outside the scope of AMLp.

References

Muth T (2001) Modeling Telecom Networks and Systems. Springer, Berlin Heidelberg New York

Orfali R, Harkey D, Edwards J (1996) The Essential Distributed Objects Survival Guide. Wiley, New York

Rose M (1990) The Open Book, A practical Perspective on OSI. Prentice Hall, Englewood Cliffs, New Jersey

Selic B, Gullekson G, Ward P T (1994) Real-Time Object-Oriented Modeling. Wiley, New York

Shlaer S, Mellor S (1992) Object Lifecycles: Modeling the World in States. Prentice Hall, Englewood Cliffs, New Jersey

Tanenbaum A S (1996) Computer Networks. Prentice Hall, Englewood Cliffs, New Jersey

Index

FSC
www.fsc.org
MIX
Papier | Fördert gute Waldnutzung
FSC® C083411

Kolibri 360 GmbH
Ferdinand-Jühlke-Straße 7
99095 Erfurt, Deutschland
produktsicherheit@kolibri360.de